Topics in Intelligent Engineering and Informatics

Volume 13

Series editors

Imre J. Rudas, Budapest, Hungary
Anikó Szakál, Budapest, Hungary

More information about this series at http://www.springer.com/series/10188

Ryszard Klempous · Jan Nikodem
Péter Zoltán Baranyi
Editors

Cognitive Infocommunications, Theory and Applications

 Springer

Editors
Ryszard Klempous
Faculty of Electronics
Wrocław University of Science and
  Technology
Wrocław, Poland

Péter Zoltán Baranyi
Széchenyi István University
Győr, Hungary

Jan Nikodem
Faculty of Electronics
Wrocław University of Science and
  Technology
Wrocław, Poland

ISSN 2193-9411   ISSN 2193-942X   (electronic)
Topics in Intelligent Engineering and Informatics
ISBN 978-3-030-07124-0   ISBN 978-3-319-95996-2   (eBook)
https://doi.org/10.1007/978-3-319-95996-2

This Springer imprint is published by the registered company Springer Nature Switzerland AG
The registered company address is: Gewerbestrasse 11, 6330 Cham, Switzerland

Foreword

In the series of Springer editions entitled *Topics in Intelligent Engineering and Informatics*, high-quality volumes, which contribute inter alia to the development of innovative technologies, are published.

This book consists of 20 chapters and integrates systematic view on how cognitive processes may evolve with the development of appliances and methods of communication. Traveling through chapters, we discover the cognitive infocommunication as emerging, crucial, and exhilarating field of study, where the capabilities of the human brain could be extended not only through the equipment, regardless of geographical distance, but they can also interact with the capabilities of any artificial cognitive system.

Cognitive infocommunication is connecting and extending the possibilities of cognition and is focused on engineering in which artificial or natural cognitive systems allow more effective cooperation. Cognitive infocommunication based on new fields of cognitive media, cognitive informatics, and cognitive communication is both a quest to understand contemporary society, and is a key necessary for innovation.

The editors because of their experience and also on their knowledge and skills ensure an exciting scientific level. I am deeply convinced that cognitive infocommunication and the content of the chapters guarantee inspiring and interesting reading for anyone who lays claim to be thinking about human future.

Naples, Italy
January 2018

Anna Esposito
Università della Campania Luigi
Vanvitelli and International Institute
for Advance Scientific Studies

Preface

Cognitive infocommunications (CogInfoCom) is an interdisciplinary research field that has emerged as a synergy between infocommunications and the cognitive sciences. One of the key observations behind CogInfoCom is that through a convergence process between these fields, humans and information and communication technologies are becoming entangled at various levels, as a result of which new forms of cognitive capability are appearing. Crucially, these capabilities are neither purely natural (i.e., human), nor purely artificial; therefore, it is suggested that they should be treated in a way that unifies both engineering and human-oriented perspectives, and based on concepts that underline the fact that a new form of co-evolutionary combo is replacing human–machine interactions.

This book is a special edition on CogInfoCom research relevant to a variety of application areas, including among others data visualization, brain–computer interfaces or speech technologies. The goal is to provide an overview of areas that are often viewed as having little or no relationship to one another (as is indeed the case with regard to the examples listed above), but which at the same time can nevertheless be approached under the same umbrella within CogInfoCom by focusing on the kind of cognitive capabilities that are being analyzed and developed. Based on this common ground, it may become possible to see new opportunities for synergy among disciplines that were heretofore viewed as being separate.

The book, then, is comprised of 20 chapters. There are those that focus on modeling human cognitive states and aptitudes in order to better understand what the user of a system is capable of comprehending and doing. The chapters by Wortelen et al., Sperandeo et al., Alam et al., Irastorza et al., Rusko et al., Lewandowska-Tomaszczyk et al., Crivello et al., Ujbanyi et al., and Dobrovsky et al. can be seen as falling into this category. Although this is not to say that the topics of these papers are not very different from each other, the patterns of exploration and the specific tools that they use can certainly be of interest and of great relevance to all of them, and therefore for all researchers who focus on problems of modeling human states and aptitudes.

In a similar vein, several chapters focus on ways to influence cognitive states and aptitudes in order to facilitate learning or more generally improve performance in certain cognitive tasks such as decision-making. The chapters by Török et al., Jaksa et al., Horváth and Bocewicz et al. can be seen as belonging to this category.

If instead of considering cognitive states and aptitudes, one focuses on cognitive capabilities for acting on the world, the notion of the human-AI spectrum becomes important. Some capabilities are purely human, while others are purely artificial, but in general this distinction is rarely clear-cut. Therefore, when discussing new human cognitive capabilities, the technological background which makes them possible cannot be neglected, and indeed often plays a central role in the discussion (as is the case in the chapters by Katona et al. and Papp et al. which both focus on novel human capabilities through new kinds of devices). At the other end of the human-AI spectrum, several papers focus on artificial cognitive capabilities, but as expected, rarely without reference to human behaviors and sometimes even active human contribution. The chapters by Kovács et al., Navarretta, Staš et al., Garai et al., and Ito et al. can all be mentioned as being relevant from this point of view.

Based on the above, as editors of this special edition, our goal was to help highlight the synergy between various fields that are generally seen as being disparate, but nevertheless perfectly fit under the umbrella of CogInfoCom and all somehow contribute to understanding and developing new, human-AI hybrid capabilities. It is our view that the rate at which such kinds of hybrid, merged capabilities are currently appearing and the importance of the role they play in everyday life are unique to the cognitive entity generation that is currently growing up. We hope that this book will help readers better appreciate this view and contribute to the merging of humans and AI.

Our special thanks belong to the reviewers of this book: Attila Borsos, Wojciech Bożejko, Ádám Csapó, Tamás Gábor Csapó, Anna Esposito, Károly Hercegfi, Konrad Kluwak, Maria Koutsombogera, Miroslav Macik, Köles Máté, Carmen Paz Suárez Araujo, Jerzy Rozenblit and Dávid Sztahó.

We would like to thank the Foreword author—Prof. Anna Esposito, Università della Campania *Luigi Vanvitelli* and International Institute for Advance Scientific Studies, Italy for her reliable and substantive work in evaluating the presented chapters and the final product of this eminent selection.

CogInfoCom conference was organized by the team from Széchenyi István University of Győr as well as Wrocław University of Science and Technology (WUST).

We would like to thank the FIEK program (Center for cooperation between higher education and the industries at the Széchenyi István University, GINOP-2.3.4-15-2016-00003) for its support.

Significant essential, financial, and organizational support was obtained from the Rectors of WUST, Prof. Tadeusz Wieckowski (previous) as well as Prof. Cezary Madryas (present).

Wrocław, Poland   Ryszard Klempous
Wrocław, Poland   Jan Nikodem
Győr, Hungary   Péter Zoltán Baranyi
January 2018

Contents

1   **Using Deep Rectifier Neural Nets and Probabilistic Sampling
for Topical Unit Classification**   1
György Kovács, Tamás Grósz and Tamás Váradi

2   **Monte Carlo Methods for Real-Time Driver Workload
Estimation Using a Cognitive Architecture** 25
Bertram Wortelen, Anirudh Unni, Jochem W. Rieger, Andreas Lüdtke
and Jan-Patrick Osterloh

3   **Cognitive Data Visualization—A New Field
with a Long History** 49
Zsolt Győző Török and Ágoston Török

4   **Executive Functions and Personality from a Systemic-Ecological
Perspective** 79
Raffaele Sperandeo, Mauro Maldonato, Enrico Moretto
and Silvia Dell'Orco

5   **Mirroring and Prediction of Gestures from Interlocutor's
Behavior** 91
Costanza Navarretta

6   **Automatic Labeling Affective Scenes in Spoken
Conversations** 109
Firoj Alam, Morena Danieli and Giuseppe Riccardi

7   **Tracking the Expression of Annoyance in Call Centers** 131
Jon Irastorza and María Inés Torres

8 Modeling of Filled Pauses and Prolongations to Improve Slovak Spontaneous Speech Recognition 153
Ján Staš, Daniel Hládek and Jozef Juhár

9 Enhancing Air Traffic Management Security by Means of Conformance Monitoring and Speech Analysis 177
Milan Rusko, Marián Trnka, Sakhia Darjaa, Jakub Rajčáni, Michael Finke and Tim Stelkens-Kobsch

10 Compassion Cluster Expression Features in Affective Robotics from a Cross-Cultural Perspective 201
Barbara Lewandowska-Tomaszczyk and Paul A. Wilson

11 Understanding Human Sleep Behaviour by Machine Learning 227
Antonino Crivello, Filippo Palumbo, Paolo Barsocchi, Davide La Rosa, Franco Scarselli and Monica Bianchini

12 Electroencephalogram-Based Brain-Computer Interface for Internet of Robotic Things 253
Jozsef Katona, Tibor Ujbanyi, Gergely Sziladi and Attila Kovari

13 CogInfoCom-Driven Surgical Skill Training and Assessment 277
László Jaksa, Illés Nigicser, Balázs Szabó, Dénes Ákos Nagy, Péter Galambos and Tamás Haidegger

14 Cognitive Cloud-Based Telemedicine System 305
Ábel Garai, István Péntek and Attila Adamkó

15 Pilot Application of Eye-Tracking to Analyze a Computer Exam Test 329
Tibor Ujbanyi, Gergely Sziladi, Jozsef Katona and Attila Kovari

16 The Edu-coaching Method in the Service of Efficient Teaching of Disruptive Technologies 349
Ildikó Horváth

17 3D Modeling and Printing Interpreted in Terms of Cognitive Infocommunication 365
Ildikó Papp and Marianna Zichar

18 Constraints Programming Driven Decision Support System for Rapid Production Flow Planning 391
Grzegorz Bocewicz, Ryszard Klempous and Zbigniew Banaszak

19 Improving Adaptive Gameplay in Serious Games Through Interactive Deep Reinforcement Learning 411
Aline Dobrovsky, Uwe M. Borghoff and Marko Hofmann

20 A Study on a Protocol for Ad Hoc Network Based on Bluetooth Low Energy 433
Atsushi Ito and Hiroyuki Hatano

Index 459

Chapter 1
Using Deep Rectifier Neural Nets and Probabilistic Sampling for Topical Unit Classification

György Kovács, Tamás Grósz and Tamás Váradi

Abstract In the interaction between humans and computers as well as in the inter-action among humans, topical units (TUs) have an important role. This motivated our investigation of topical unit recognition. To lay foundations for this, we first create a classifier for topical units using Deep Neural Nets with rectifier units (DRNs) and the probabilistic sampling method. Evaluating the resulting models on the HuComTech corpus using the Unweighted Average Recall (UAR) measure, we find that this method produces significantly higher classification scores than those that can be achieved using Support Vector Machines, and what DRNs can produce in the absence of probabilistic sampling. We also examine experimentally the number of topical unit labels to be used. We demonstrate that not having to discriminate between variations of topic change leads to better classification scores. However, there can be applications where this distinction is necessary, for which case we introduce a hierarchical classification method. Results show that this method increases the UAR scores by more than 7%.

G. Kovács (✉) · T. Váradi
Research Institute for Linguistics of the Hungarian Academy of Sciences,
Budapest, Hungary
e-mail: gykovacs@inf.u-szeged.hu

T. Váradi
e-mail: varadi.tamas@nytud.mta.hu

G. Kovács
MTA SzTE Research Group on Artificial Intelligence, Szeged, Hungary

T. Grósz
Institute of Informatics, University of Szeged, Szeged, Hungary
e-mail: groszt@inf.u-szeged.hu

© Springer International Publishing AG, part of Springer Nature 2019
R. Klempous et al. (eds.), *Cognitive Infocommunications, Theory and Applications*,
Topics in Intelligent Engineering and Informatics 13,
https://doi.org/10.1007/978-3-319-95996-2_1

1.1 Introduction

To better understand how participants in a dialog act (when they are not contributing to the conversation in any meaningful way, when they are advancing the current topic, and when they are deviating from it, either by starting an entirely different topic, or by modifying slightly the direction of the conversation), is to better comprehend human-human interaction. This conviction spurred us to investigate topical units in speech communication. Another strong motivation was the possible implications of the findings for human-computer interactions: HCI solutions (e.g. dialog systems) would greatly benefit if the computer were able to tell whether its human interlocutor is contributing to the conversation, and if so, how. For these reasons we will focus on topical unit recognition, where our first goal is to create a topical unit classifier. This means that our task in this study was to create a model that assigns topical unit labels to conversation segments. These labels are the following:

- Topic initiation: the topic is changed by a speaker. Here, the preceding conversation motivates the change, and the new topic is in harmony with it. Verbal cues that might indicate this category include such expressions as "I think...", and "What do you think about...", and utterances in this category are sometimes accompanied by a posture change.
- Topic change: the topic is changed by a speaker. Here—unlike in the previous case—the preceding conversation does not motivate the change, what is heard after it is not in harmony with what is being discussed. A change like this customarily occurs in the form of a question or an imperative, and it is often accompanied by such discourse makers as "otherwise", "you know", and "by the way" [1].
- Topic elaboration: the current topic is elaborated on by a speaker, by adding more details to it, or by specifying it more precisely.
- No-contribution: this label is used when none of the above-described labels apply. This means that while we treat it as a label, no-contribution is more like the absence of labels.

Here, we will approach the problem from a HCI perspective (and of course other perspectives could also be applied, like the perspective of cognitive infocommunications [5]), hence we only attempt to classify the topical units covering one party of the dialog (but we still do this classification for the entire dialog).

Other researchers when working on similar problems mostly focused on using lexical information [16, 17], prosodic information [4, 25], or a mixture of both [18, 24]. Some have attempted to make use of other sources, such as turn-taking and speaker identity [20], but here we go one step further and attempt to exploit visual information as well (such as posture and gaze tracking), following [12].[1]

---

[1]The current study is an extended version of this conference paper.

1.2   Experimental Data

1.2.1   HuComTech Corpus

The corpus we used in our experiments is one result of the HuComTech project [2], carried out at the University of Debrecen. Within the framework of this project a multimodal corpus of approximately 50 h in length was recorded using 111 speakers, both male and female, between the age of 19 and 27, coming from various backgrounds (urban and rural areas). Each speaker participated in the recording of two interviews (one simulating a job interview, and the other involving everyday topics such as friends, jokes and pleasant/unpleasant experiences) that were then processed by annotators. The annotation work of these interviews was carried out based on the audio modality, the video modality or both in 39 tiers. Most tiers recount the behaviour of the interviewee, some recount the behavior of the interviewer, and others recount the behavior of both. While this study contains brief descriptions of all tiers, for a more comprehensive description of the corpus we refer the reader to other publications associated with this project [1, 2, 9, 14], which we strongly recommend for background reading. One aspect of the corpus—the unbalanced distribution of labels describing topical unit categories—should be elaborated on here, because it can cause serious issues in machine learning (see Sect. 1.3).

1.2.2   Data Preparation

A crucial and difficult task in topical unit classification was to convert the annotations in such a way that allows us to use the resulting information for our purposes. To achieve this goal, we first trimmed each conversation based on the begin conversation and end conversation markers (where available), and then based on the beginning and the end of the shortest tier, so as to eliminate the issue of missing labels. After this step, 47 h of annotated data remained. Other key steps in the preparation of data were the establishment of the train set, development set and test set, and feature extraction. During the latter we also adjusted label borders in various tiers to those in the tier annotating topical units, and divided these tiers into time slices of uniform length (where each slice—or frame—covered 320 ms).

1.2.3   Data Splitting

Generally speaking, in machine learning applications we have three important tasks to handle, namely model training, the fine-tuning of the meta-parameters (as we will later see, each machine learning algorithm has its own set of meta-parameters that influence the resulting model), and lastly, model evaluation. Ideally each of

Table 1.1 Ratio of topical unit classes in the different subsets

Label information		Train set (%)	Dev. set (%)	Test set (%)	Full set (%)
Ratio of class occurrences against occurrences of all classes (of speaker contribution) in the same set	Topic change	7.21	7.37	7.23	7.23
	Topic initiation	32.62	32.90	32.65	32.65
	Topic elaboration	60.17	59.73	60.13	60.12
Ratio of frames labelled with a given class against frames labelled with a class (of speaker contribution) in the same set	Topic change	3.89	3.86	3.82	3.87
	Topic initiation	18.61	18.91	18.55	18.63
	Topic elaboration	77.51	77.23	77.63	77.50
Ratio of class occurrences against occurrences of the same classes in the full set	Topic change	74.74	10.21	15.05	100.00
	Topic initiation	74.86	10.09	15.05	100.00
	Topic elaboration	75.00	9.95	15.05	100.00
Ratio of frames labelled with a given class against frames labelled with the same class in the full set	Topic change	75.07	9.83	15.10	100.00
	Topic initiation	74.75	10.01	15.24	100.00
	Topic elaboration	74.84	9.83	15.33	100.00

these phases is performed on a separate subset of the available data so as to avoid overfitting. These subsets are the train set (for model training), the development set (for meta-parameter tuning), and the test set (for evaluation). The splitting of the full set into non-overlapping subsets can be performed using various ratios. Here, we decided that 75% of the available data should be in the train set, 10% should be in the development set, and the remaining 15% should be in the test set. The ratio of these subsets, however is not the only vital factor. It is also important that these subsets should be representative of the full set of data.

Here, we try to ensure this by demanding two requirements. One we expect to see in our partitioning is that when it comes to labels corresponding to meaningful contributions to the topic (i.e. every label that is not "no-contribution"), each subset should have the same distribution of labels (both in terms of the number of occurrences of the label in the annotation and the number of frames associated with the label[2]) as that of the full set. For instance, if the topic elaboration label is around 60% of labels in the full set, we expect that the same label should also make up 60% of the labels in the train set, the development set, and the test set respectively. Similarly, if among the frames associated with the topic change, the topic initiation or the topic elaboration label, the latter should take up 77.5% in the full set, the same should be true for all subsets as well. From Table 1.1 it is clear that the observed proportions for each label in the subsets closely approximate those in the full set, which means that our first requirement is met.

Another requirement we had was that the number of instances one set should contain for each label (both in terms of the number of occurrences of the label

---

[2]The difference between the two is due to the different length in the occurrences of various topical units in the annotation. A 3.2 s long topic elaboration and a 32 s long topic elaboration both count as one occurrence in the annotation, but the number of frames associated with the former will be 10, while the number of frames associated with the latter will be 100.

in the annotation, and the number of frames associated with the label) should be proportional to the size of the given set relative to the full set. Again, using an example: given the proportions of data selected for the various sub-sets, approximately 75% of instances of topic elaboration (and 75% of all frames classified as topic elaboration) should be in the train set, approximately 10% of the instances of topic elaboration should be in the development set, and lastly 15% of the instances of topic elaboration should be in the test set. We expect the same to be not only true for the topic change label and the topic initiation label, but also for the number of frames associated with these labels. Again, looking at Table 1.1 we see that the selected sets approximate this requirement quite well.

1.2.4 Feature Extraction

One phase of data preparation we have not addressed yet is that of feature extraction. In this phase our goal is to take the labels used by the annotators, and transform them into a format that is useful for the machine learning methods to be applied on the task. In our study, this generally meant creating a dummy binary representation for each attribute. The exact details of the transformation, however, may vary from tier to tier. This is why we will briefly explain this process for each tier, so that we can clarify how the final feature vector (with 221 features) came about.

1.2.4.1 Feature Extraction from Audio Information

Here, the annotation work was performed at a phrasal level. This means that even if an attribute applied only to a short segment of the phrase, the whole phrase was still labeled according to the given attribute. As an example, let us consider the case where the interviewee utters a phrase that is 3 s long, and let us assume that for the first half of this phrase the interviewer is also speaking. The annotator in this case will label the whole 3 s long phrase as overlapping speech, despite the fact that in a big portion of the phrase only one person is speaking. This would mean that a simple transformation of the segment-wise labeling into frame-wise labels would introduce many inaccuracies. For this reason, from the annotation of the audio data we only utilized the information on emotions, as in our opinion we can safely assume that the expressed emotion of the interviewee does not change over such a short period. We encoded this information in a 9-bit binary representation (corresponding to the number of categories the annotators were able to select from for labeling the emotion in the audio).

1.2.4.2 Feature Extraction from the Syntax

The annotation of syntax was also performed based on audio information. Here, 20 features were extracted, based on the following sources of information:

- Categories that are missing from the current clause. We created 14 binary features to encode this information, thirteen of which corresponded to the possible missing categories [9], and one binary feature representing the presence or absence of inherently unmarked grammatical relationships.
- The position of the current clause relative to the sentence (e.g. whether it is the first, second, etc. clause). We created 1 feature of an integer type to encode this information.[3]
- The number of clauses coordinated with the current one. We created 1 feature of an integer type to encode this information.
- The presence or absence of embedded clauses. To encode this information we created 1 binary feature that was one in the case where there were clauses embedding the current clause, and zero otherwise.
- The presence or absence of embedding. To encode this information we created 1 binary feature that was one in the case where there were clauses embedded in the current clause, and zero otherwise.
- The number of clauses that the current one is subordinated to. We created 1 feature of an integer type to encode this information.
- The number of clauses that are subordinated to the current one. We created 1 feature of an integer type to encode this information.

1.2.4.3 Feature Extraction from the Prosody

The annotation of prosody was also performed based on the audio data. For this purpose the ProsoTool algorithm was used [21] that had been developed as a Praat script within the framework of the HuComTech project. While the eventual purpose of this tool is to handle the intonation, intensity, stress, and speech rate in recordings, in its current state the algorithm only provides information regarding the fundamental frequency and intensity. Here, 37 features were extracted based on the following information:

- Fundamental frequency values measured at the borders of the given section. We created 2 features of real type to encode this information.
- Fundamental frequency levels measured at the borders of the given section. There are five possible levels of fundamental frequency values (L_2, L_1, M, H_1, H_2), based on speaker-dependent thresholds (T_1, T_2, T_3, T_4). Values of fundamental frequency were placed into levels using the following formula: $L_2 < T_1 \leq L_1 < T_2 \leq M < T_3 \leq H_1 < T_4 \leq H_2$. We created 10 binary features to encode this information

---

[3]It should be mentioned here that before applying our machine learning methods, we normalized all non-binary features so as to have a zero mean and unit variance.

(five binary features for encoding the information coming from either sides of the given segment).

- The change in fundamental frequency values between the beginning and end of the given section, based on the stylization of the smoothed F0 curve. We created 5 binary features to encode this information, based on the five categories available for annotation. The categories are the following: fall, descending, upward, and rise (depending on the direction and rapidness of the change) if the difference between the value of the fundamental frequency at the beginning and at the end of the segment was bigger than a speaker-dependent threshold. And if the difference was below this threshold, the segment was labeled as stagnant.
- The mean of the raw fundamental frequency values measured within the given frame. We created 1 feature of real type to encode this information.
- The ProsoTool algorithm also supplied us with information about whether the current segment (and thus the frames in the current segment) is voiced or unvoiced. We created 2 binary features to encode this information. The first binary feature was one if the current segment was voiced, and zero otherwise. Another binary feature was one if the current segment was unvoiced, and zero otherwise. The reason for having two binary features instead of one was that some segments were labeled as neither voiced nor unvoiced.
- Intensity values measured at the borders of the given section. We created 2 features of real type to encode this information.
- Intensity levels measured at the borders of the given section. We created 10 binary features to encode this information, the same way information on the fundamental frequency levels was encoded.
- The change in intensity values between the beginning and end of the given section, based on the stylization of the smoothed curve of intensity. We created 5 binary features to encode this information, the same way fundamental frequency movement information was encoded.

1.2.4.4   Feature Extraction from Video Information

The annotation work for this was performed in two subsuming tiers, namely a functional tier and a physical tier. An important difference between the two is that annotators who worked on the functional tier (annotating emotional information and emblems) were allowed to use the audio information as well. Here, 111 features were extracted based on the following information:

- The emotion that was reflected by the expression the interviewee had on his or her face. We created 7 binary features to encode this information, as the annotation differentiated between seven types of emotion (happiness, sadness, feeling surprised, recalling, tenseness, natural, and other).
- The direction in which the interviewee was looking. We created 6 binary features to encode this information, five corresponded to the five basic directions, and one

corresponded to the case where the direction cannot be ascertained due to the eyes of the interviewee being closed.

- The way the interviewee moves his or her eyebrows (raising/furrowing them). We created 4 binary features to encode this information, two corresponded to the left eyebrow, and two corresponded to the right eyebrow.
- The way the interviewee moves his or her head. We created 8 binary features to encode this information. Four of these denote whether the interviewee is doing one of the following: head shaking, nodding, raising/lowering his or her head. Another two binary features denote if the interviewee's head is turned or being tilted, while the last two binary features denote whether the head turning/tilting is to the right or to the left.
- The way the interviewee holds his or her hand. We created 15 binary features to encode this information. Three denoting shapes formed using both hands, the others denoting shapes formed using the left or the right hand.
- Auto-manipulation of the interviewee, more specifically the interviewee scratching or touching one of his or her body part (for the purposes of the annotation, glasses, in the case where interviewees were wearing them, were also classified as a body part). We created 30 binary features to encode this information. Fifteen features described actions carried out using the left hand, and another fifteen describing actions carried out using the right hand.
- The interviewee's posture. We created 10 binary features to encode this information, as ten categories were used in annotation (arm crossing, head holding, leaning back, leaning forward, leaning left, leaning right, rotating to the left, rotating to the right, sitting with shoulders up, sitting in an upright position).
- Deictic gestures of the interviewee. We created 10 binary features to encode this information. Five features denoted gestures performed using each hand.
- The emotional state the annotator assigned to the interviewee based on the video and audio information. We created 7 binary features to encode this information. Emotional categories here coincided with those used in describing the facial expression of the interviewee.
- Emblematic gestures performed by the interviewee. We created 14 binary features to encode this information.

1.2.4.5 Feature Extraction from Unimodal Information

Here, unlike in the above, annotation work was based solely on video information. To aid the annotation work, a special program called Qannot was created in the HuComTech project [1]. Here, 15 features were extracted based on the following information:

- The novelty of information in the conversation from the perspective of the interviewee. We created 1 binary feature to describe this information that is one if the interviewee is receiving new information, and zero otherwise.
- The attitude of the interviewee regarding attention (i.e. asking for attention or giving attention). We created 2 binary features to describe this.
- Turn-taking from the perspective of the interviewee. We created 5 binary features to describe this information (whether he or she shows an intention to speak, is interrupting the interviewer, is beginning to speak, has finished speaking, or is currently speaking).
- The agreement levels displayed by the interviewee. We created 7 binary features to encode this information.

1.2.4.6   Feature Extraction from Multimodal Information

Once again, the Qannot program was used to extract audiovisual information from the conversation. In this case each category of features was extracted twice: once, to describe the actions of the interviewee, and again, to describe the same actions of the interviewer. Here, 29 features were extracted based on the following information:

- The novelty of information in the conversation from the perspective of the interviewee/interviewer. Here (unlike in the above) we distinguish between the case where no information is given and when the information given is not new to the person receiving this information. Because of this, here we created 3 binary features to encode this information.
- Acts of communication performed by the interviewee/interviewer. Given that during annotation work seven categories were available here (acknowledging, commissive, constative, directive, indirect, other or none), we created 7 binary features to encode this information.
- Supporting acts of the interviewee/interviewer. We created 4 binary features to encode this information (given that four such labels were always used during the annotation work, which is backchanneling, expressing politeness, repair or other supporting act).
- The topical unit labels assigned to the interviewee/interviewer. Here, feature extraction is only performed for the interviewer, as the corresponding information for the interviewee will be used as the targets. Here, we created 3 binary features to encode this information. All binary features were set to zero when the interviewer was not contributing to the conversation in a meaningful way.

1.3   Methods

As we noted earlier in Table 1.1, different topic change labels only make up approximately 40% of the meaningful contributions. At the frame level this imbalance is even worse: regarding all the labels, topic change and initiation constitutes just 12% of all frames. As this can lead to certain difficulties, the first two methods introduced below will chiefly be employed to handle the imbalance in class distribution.

1.3.1 Unweighted Average Recall (UAR)

This imbalance can affect the evaluation because the rarer classes making up 12% of the cases means that a classifier could achieve a seemingly reasonably high accuracy of around 88% without correctly identifying a single instance of those classes. Thus accuracy in such cases is perhaps not the most suitable measure. The UAR might be a more appropriate metric here [15]. To compute its value, we first have to create a confusion matrix A. A is a square matrix whose size is the square of the number of class labels (N^2). Here, the value in each position A_{ij} is given by the number of instances our model predicted to be from the i-th class that are in reality in the j-th class. Given the A matrix, we can calculate the UAR as follows:

$$UAR = \frac{1}{N} \sum_{j=1}^{N} \frac{A_{jj}}{\sum_{i=1}^{N} A_{ij}} \quad (1 \le i, j \le N). \tag{1.1}$$

1.3.2 Probabilistic Sampling

Another potential effect of an imbalanced class distribution is a bias in the trained models against the more rare classes, which leads to lower UAR scores on those classes [13]. One way to overcome this problem is by changing the number of instances used from each class during training. Omitting examples would mean losing important data, while adding extra samples is usually not feasible. Another approach is that of probabilistic sampling, where a random selection of class is followed by drawing a sample from the selected class. If for each class a $P(c_i)$ probability is given:

$$P(c_i) = \frac{\lambda}{N} + (1 - \lambda) \cdot Prior(c_i), \tag{1.2}$$

we can treat the selection of a class as sampling from a multinomial distribution [8]. Here $Prior(c_i)$ is c_i class's prior probability, while the parameter λ controls the uniformity of the distribution: $\lambda = 0$ results in the original distribution, while $\lambda = 1$ (also known as "uniform class sampling" [23]) results in a uniform distribution.

1.3.3 Support Vector Machine (SVM)

In our study we applied SVMs as a baseline of comparison. As the SVM classifier uses hyperplanes to separate data points into two categories (in other words, two classes), it lends itself more readily to two-class problems. SVMs, however, can also be used for classification problems with multiple class targets, such as topical unit classification. The LibSVM implementation [6] we apply here utilizes the one-against-one approach for this, which means having a support vector trained for each pair of classes. The meta-parameters of the algorithm were selected from a pool of a hundred and ten candidates based on the UAR scores we obtained on the development set using the different candidate settings.

1.3.4 Deep Rectifier Neural Net (DRN)

The method we propose for the classification task is DRNs trained using the probabilistic sampling approach. The DRN architecture, where the traditionally applied sigmoid activation function is replaced with the rectifier activation function ($rectifier(x) = max(0, x)$—shown in Fig. 1.1), has become more prominent in recent years, as we can see in, for example, speech technology applications [22]. One good point about this architecture is that owing to the function returning zero for every negative input, the resulting model tends to be more sparse, which is advantageous computationally. Another benefit of this architecture is the mitigation or elimination of the vanishing gradient issue, due to the fact that the rectifier function does not saturate. Because of this we can effectively stack multiple hidden layers on top of one another. In our experiments we used three such layers, each containing one thousand rectifier linear units (ReLUs). On top of this, a softmax output layer was placed that

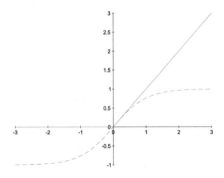

Fig. 1.1 Plot of the rectifier activation function (solid line), and a sigmoid activation function (dashed line)

consisted of as many neurons as there were target labels in the given task. We trained
our neural networks on the train set using early stopping and a learn rate scheduler
that in all but one case was based on the UAR scores got on the development set.

1.4 Results and Discussion

1.4.1 Experiments with no Temporal Context

Let us first consider the simple case where each machine learning algorithm has
to make its decision on a frame based solely on the frame itself, which means that
the only temporal context available for it is the 320 ms covered by the frame. We
conducted these experiments using SVMs, DRNs using accuracy for early stopping,
DRNs using UAR for the same purpose (DRN+UAR), and lastly DRNs that were
trained using the probabilistic sampling approach (DRN+UAR+PS). The UAR scores
we got in these experiments are summarized in Table 1.2. When we compare the
scores obtained with different DRNs, we notice that by using the UAR metric in the
training phase, better results can be attained. Given that the evaluation of models is
also based on this metric, this comes as no real surprise.

The results also tell us that by adopting the probabilistic sampling method we can
achieve better UAR scores than those obtained using the baseline DRNs, regardless
of the exact value of the λ parameter. It is also true, however, that a carefully selected
parameter value (which in our case—based on the development set—is $\lambda = 1$) leads
to even higher UAR scores (a 10.8% relative improvement over the performance
of the DRN+UAR model). Looking at the results produced by the various machine
learning methods used as baselines for comparison, we notice that the best performance is achieved with the DRN+UAR setup. This setup slightly outperformed the
SVM method, which in turn outperformed the DRN setup by a larger margin. The
1% relative improvement got using the DRN+UAR setup compared to the results
obtained with the SVM may not seem that big. But the fact that the SVM was not
even the best performing baseline combined with the grid search on its parameter
values requiring more than 16 days of CPU time (when executed on several servers
in parallel) meant that SVM training was excluded from all subsequent experiments.

In order to gain a deeper understanding of the classification performance, we
examined the confusion matrix produced by the DRN+UAR and the DRN+UAR+PS
setup using the test set (see Fig. 1.2). We may recall from Sect. 1.3.1 how these
matrices are created: the value in the i-th row and j-th column tells us in how many
cases our model classified a data sample into the i-th class when in fact it was a
member of the j-th class (naturally if $i = j$ we get the number of correctly classified
data samples from the given class). This means that the 4990 displayed in the second

Table 1.2 Classification results (UAR) on the HuComTech corpus; no context

Method	Meta-parameters	Dev. set (%)	Test set (%)
DRN	–	48.1	48.0
SVM	$C = 2^3, \gamma = 2^{-5}$	51.4	50.3
DRN+UAR	–	51.3	50.8
DRN+UAR+PS	$\lambda = 0.1$	51.0	51.2
	$\lambda = 0.2$	50.9	51.6
	$\lambda = 0.3$	52.6	51.5
	$\lambda = 0.4$	53.0	53.5
	$\lambda = 0.5$	53.7	54.1
	$\lambda = 0.6$	54.4	54.7
	$\lambda = 0.7$	55.4	54.7
	$\lambda = 0.8$	55.8	**56.7**
	$\lambda = 0.9$	**56.8**	**56.9**
	$\lambda = 1.0$	**57.1**	56.1

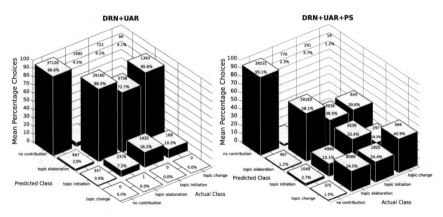

Fig. 1.2 Confusion matrix derived from the results of the DRN+UAR and DRN+UAR+PS obtained on the test set of the HuComTech corpus

column and third row of the confusion matrix on the right hand side of Fig. 1.2 signifies that in 4990 cases the DRN+UAR+PS model labeled instances from the topic elaboration class as members of the topic initiation class. Here the confusion matrix is not only supplemented with the percentage values of the instances from each class being classified in each class (so looking at the same row and column as before, we not only see that 4990 instances are classified into the topic initiation category instead of the topic elaboration category, but we can also tell that it is roughly 15.1% of all instances of the topic elaboration class), but these percentages are also indicated by the height of each cell in the matrix.

In Fig. 1.2, we notice that the DRN+UAR setup practically disregarded the category of topic change, as only one instance was classified in that category, and even this was an erroneous classification. This means that this particular model produced a recall rate of 0% on the topic change class, which can offer an insight into the relatively poor performance of the classifier. This seems to confirm our concerns mentioned in Sect. 1.3.1, as here we have a model that distinguishes the topic elaboration and no-contribution classes relatively well, and achieves an accuracy score of over 83%, but it rarely, if ever, recognizes instances from the rarer classes. We, however, observe a very different behavior in the confusion matrix of the DRN+UAR+PS setup. Here, the recall rates of the topic initiation and especially the topic change class have dramatically improved, but alas this is at the expense of decreased recall rates of the topic elaboration class.

1.4.2 Experiments with a Wider Temporal Context

During the experiments conducted earlier, the context available for DRNs was constrained by the length of frames (i.e. 320 ms). It is reasonable to expect, however, that better performance scores could be obtained using a wider context. A set of experiments was designed to investigate this assumption. Here, the number of frames available for the model was increased, one at a time, up to twenty one (ten on each side of the current frame). For each size we evaluated the resulting DRN and DRN+UAR model on the development set (see Fig. 1.3). We observe above that the scores got using the DRN+UAR setup steadily improved throughout the experiments, while the scores got using the DRN setup reached their maximum at a context size of seventeen frames. But in order to avoid differences in the performance arising from the difference in the size of the available context, more experiments on all DRN setups were performed using a context of seventeen frames (the current frame, and eight of its neighbors from both sides).

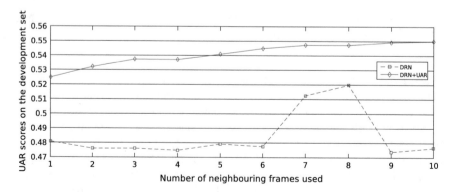

Fig. 1.3 Classification results (UAR) attained on the development set of the HuComTech corpus as a function of the context size [12]

Table 1.3 Classification results (UAR) on the HuComTech corpus with a context of 17 frames

Method	Meta-parameters	Dev. set (%)	Test set (%)
DRN	–	52.0	50.9
DRN+UAR	–	54.7	54.0
DRN+UAR+PS	$\lambda = 0.7$	58.3	57.5
	$\lambda = 0.8$	**59.3**	**60.0**
	$\lambda = 0.9$	**59.3**	58.9
	$\lambda = 1.0$	**59.0**	59.3

The UAR scores got from these experiments are listed in Table 1.3. The results justify our opinion on the benefit of having a wider context available for classifiers. For each case, the introduction of a wider context improved our results by at least 5% relative to those we got earlier. This improvement was most noticeable for probabilistic sampling, and it led to to a relative improvement of about 11.5% compared to that of the baseline.

Just like before, we created confusion matrices from our results on the test data (see Fig. 1.4). When examining the confusion matrix of the DRN+UAR model, it becomes apparent that the wider context markedly increased the recall rate of the topic initiation class. At the same time, however, the recall rates of the no-contribution and topic elaboration classes deteriorated slightly. And although here we got a slightly better recall rate for the class of topic change, the majority of topic change detections were false positives. When examining the confusion matrix produced based on the results of the DRN+UAR+PS method, we see that expanding the context had practically no effect on the recall rates of the no-contribution class, but the recall rate of the topic elaboration and topic initiation classes dramatically improved. The only setback occurred in the recall rate of the topic change class.

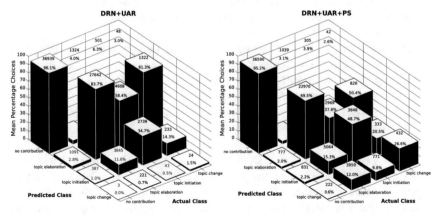

Fig. 1.4 Confusion matrix derived from the results of the DRN+UAR and DRN+UAR+PS obtained on the test set of the HuComTech corpus with the use of a seventeen frame-wide context

However, even for this class, the decline in false positives far outweighs the decline in true positives. But even with this decline in false positives, the precision for the class of topic change was below 10%. We suspect that the low precision might be caused by one of the following phenomena.

The topic change is so rare that any increase in the number of its instances used during training that is sufficiently big (i.e. big enough so that it leads to a model that has a recall rate of more than 0% for the topic change class) inevitably leads to a spike in false positives. The difference between motivated and non-motivated topic change is not significant enough for reliable models to be trained on it, either because the two cases are too similar, or because the annotation was inconsistent. This low precision, regardless of its cause, motivated us to find out whether it would be beneficial to abandon the use of four target labels, and combine the topic initiation and topic change categories (in other words, stop trying to distinguish between different topic change types).

1.4.3   Experiments with Three Target Labels

Owing to the reasons outlined above, here we investigated the case where the model is not required to distinguish between different types of topic change. This means that in the following experiments only three categories of topical unit (topic change, topic elaboration, and lack of the former two) will be used. To make this possible, we relabeled all frames classified as topic initiation with the topic change label. This step was not only motivated by a relatively poor classification performance on the four-class problem, but also by the fact that the task of distinguishing between the topic change and topic initiation labels proved difficult even for the human annotators. Here we did not repeat the experiments where we sought to optimize the size of the context used, but instead we applied the same seventeen frame context (the current frame complemented with eight of its neighbors from both sides), as in our previous experiments.

Table 1.4 summarizes the UAR scores got using three target labels. One thing we notice is the gap between results obtained by baseline methods rose markedly: from a relative improvement of approximately 6% to a relative improvement of 20%. This suggests that in the three-class problem the use of UAR in the learn rate scheduler by itself yields a much bigger improvement over the simple DRN that uses accuracy scores for the same purposes. At the same time the improvement in performance we got using the method of probabilistic sampling relative to the results of the DRN+UAR baseline did not change substantially. This means that even in the case of three-class labeling, the probabilistic sampling method offers a marked improvement. Moreover, the results in Table 1.4 agree with our findings about the three-class and four-class problems, as for each setup the UAR scores we got for the three-class case are much better than those we got for the four-class case (the relative improvement being between 25 and 36%). Although part of this improvement could easily be due to the UAR metric itself and the fact that by combining the two classes

Table 1.4 Classification results (UAR) on the HuComTech corpus using three target labels with a context of 21 frames

Method	Meta-parameters	Dev. set (%)	Test set (%)
DRN	–	63.1	62.8
DRN+UAR	–	72.0	70.6
DRN+UAR+PS	$\lambda = 0.7$	75.0	73.1
	$\lambda = 0.8$	**75.8**	74.1
	$\lambda = 0.9$	**75.8**	**74.8**
	$\lambda = 1.0$	**76.0**	74.0

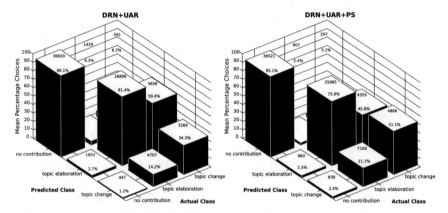

Fig. 1.5 Confusion matrix derived from the results of the DRN+UAR and DRN+UAR+PS obtained on the test set of the HuComTech corpus on the three-class task, using a seventeen frame-wide context

with the lowest recall rates, the results of the classes with higher recall rates are only counterweighted by one under-performing class. What makes us think that the three-class approach may be way to handle the task of topical unit classification, however, is the degree of improvement we observe in the UAR scores.

This assessment of ours is seemingly strengthened by examining the confusion matrices obtained in the three-class scenario (see Fig. 1.5). We see that in the DRN+UAR+PS setup the no-contribution class has approximately the same recall rates as it had before, but the recall rates of the topic elaboration class are notably higher. Moreover, the recall rates of the class we got by combining the two types of topic change are markedly higher than the recall rates we got earlier for the topic change and topic elaboration classes separately. What is more, this is the first instance where the recall rate of all classes is above 50%. The confusion matrix for the DRN+UAR case (on the left hand side of Fig. 1.5), however, is quite different. Here the recall rate of the topic elaboration class declined markedly, and the recall rate of the combined topic change class is slightly lower that that of the topic initiation class in the four-class case.

Furthermore, even in the case of probabilistic sampling, it is not clear whether the source of the improvement is solely due to the merging of the two topic change classes before the evaluation step, or whether it is also beneficial to execute the training step with three classes instead of four. We can get a better insight into this if we create a confusion matrix from the results of the four-class experiments using probabilistic sampling (see the right hand side of Fig. 1.4) after first merging the two topic change classes, hence not taking into consideration the within-group confusions for that class. This is a technique that, among other things, is commonly used in speech recognition for recognizing phones [7, 11]. Using this evaluation method, the recall rate for the merged class is about 56.5%, which combined with the 95.2% recall rate for the no-contribution, and the 69.5% recall rate for the topic elaboration task, results in a better UAR score of the model for the three-class problem, namely 73.8%. This score is just slightly lower than the 74.0% listed in Table 1.4, corresponding to the case which gave the best performance on the development set.

1.4.4 Training and Classification Using Different Targets

Above we concluded that it is probably more beneficial to classify topical unit labels into three categories. However, as we also found, this does not necessarily mean that we should merge the topic change and topic initiation labels before training. To test this proposition, we examined two evaluation methods for the three-class problem, suitable for models trained using four labels. One, as we saw previously, worked by changing each topic change and topic initiation label predicted to a merged topic change label. The other worked by summing the predicted probability values of the topic change and topic initiation labels, and assigning the summed probability values to the merged topic change class.

With these changes we decided to repeat our previous experiments on probabilistic sampling. Here, however, we did not restrict ourselves to the number of neighboring frames that proved to be the most beneficial earlier. Instead we repeated the experiments where we sought to determine the optimal context to be used. Also, to offset any numerical instability that might arise from the random initialization of weights at the beginning of neural net training, we trained three independent models for each setting. To visualize the results of these experiments on the development set, we plotted the average of the results obtained with the best λ score for each label- and neighbor number. The resulting curves are depicted in Fig. 1.6. Here we can see that as the number of frames included grows, the gap in the performance between the models trained using three classes and four classes also grows. We also notice that the opposite is true for the different evaluation methods used on the four-class models: with a small context the gap here is bigger, while with a bigger context the difference may vanish or even be reversed. The graphs also tell us that the optimal context size to be used differs for the different settings. To ensure that we compared the best results for both settings, we decided to use a different number of neighbors when evaluating different settings on the test set.

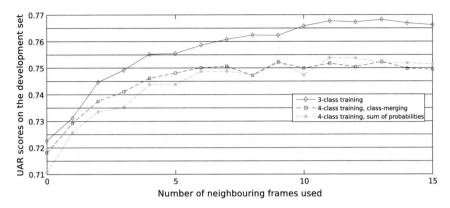

Fig. 1.6 Results of the three-class training, and the four-class training evaluated with different methods on the development set

Table 1.5 Classification results (UAR) on the three-class problem, using models trained with three class labels, and four class labels (the average of three independently trained neural nets)

No. of classes	Evaluation method	No. of frames altogether	λ	Dev. set (%)	Test set (%)
3	Traditional	27	1.0	**76.8**	**75.8**
4	Class-merging	27	0.9	75.2	74.5
4	Sum of probability values	23	0.8	75.4	74.5

Table 1.5 summarizes results got in these latter experiments. This table tells us that there is no significant difference in the results obtained with the models trained with four labels regardless of the evaluation method. We also see that the models trained with three labels here perform markedly better, yielding a relative improvement of about 5% on average.

To better understand what led to the difference between the performance of the models trained using three classes, and that of the models trained on four classes, let us look at the confusion matrix of a model corresponding to the settings in the first row of Table 1.5, and the confusion matrix of a model corresponding to the settings in the second row. These confusion matrices are illustrated in Fig. 1.7. As we now see, the difference between the recall rates of the no-contribution class is minimal: both models are able to distinguish this class from the others relatively well. There is a big difference, however, in the recall rates of the two other classes, and more importantly in how the rates of the two classes compare with one another. The model trained on three classes yields a balanced recall performance on these two classes, while its counterpart trained on four classes gives a much better performance on the topic change class, at the expense of a markedly lower performance in the topic elaboration task.

1.4.5   Towards a Hierarchical Classification

As we saw in the confusion matrices of Fig. 1.7, the models trained using three labels
and those models trained using four labels have different strengths and weaknesses, as
the two worked well on different classes. This may suggest that a combination of the
two methods could bring about a gain, especially in the four-class task, where there
is more room for improvement. Given the hierarchical nature of our taxonomy (see
Fig. 1.8), and the popularity of hierarchical classification methods [3, 10] especially
in imbalanced cases, one other way for combining the models trained using different
number of labels might be a partial-hierarchical classification. This could be carried
out in many different ways [19]. Here, we decided to apply a very simple method,
namely the predicted probability values of the three-class model that performed the
best on the development set will be used as features, and they will be concatenated
with the feature vector used by the models trained on four class labels.

Now, we are going to evaluate this combination method on the four-class problem.
One reason for it, as we already remarked, is the poor performance of our original
model on this task. Also hierarchical classification lends itself more readily to this
version of the problem. And even though we can achieve a much better UAR score
on the three-class problem, we might still need a four-class labeling of the data. For
these reasons, we decided to modify the feature vectors of the four-class problem by
concatenating them with the probability estimates got from the model that performed
best on the three-class problem—based on the development set. We then repeated
our experiments on parameter optimization using these new feature vectors. There,
we attempted to find the best context length and λ parameter value for the given task.

Once again, for each context size, one UAR score is reported, which is the average
performance of the nets corresponding to the λ parameter value that produced the
best results. A comparison of the results got with this new combination method with

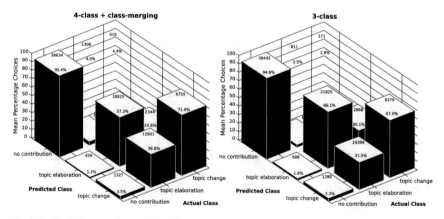

Fig. 1.7 Confusion matrix created using results obtained on the test set, comparing a model trained
on the four-class problem, and evaluated on three classes, after merging two classes, and a model
trained and evaluated with three classes

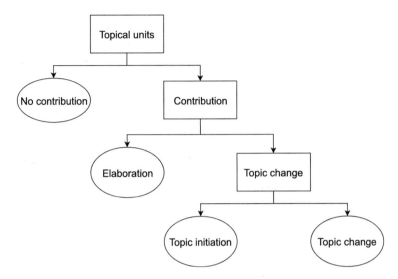

Fig. 1.8 Hierarchy of the class labels ordered in a tree structure, with the classes of the four-class task used as the leaves of the tree

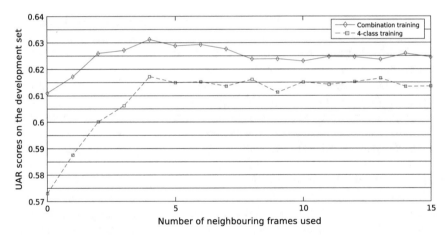

Fig. 1.9 Results of the combinational training, and four-class training using the development set

those of a simple four-class model on the development set can be seen in Fig. 1.9. As the reader will notice, the difference between the two methods is the largest when they use no additional context. This was expected, as the predicted probability values in the features of the combination model already use a context of 8.6 s. What is surprising however is that with both settings the best result is obtained using nine frames, meaning that with this combination method a bigger context (11.1 s, altogether) can be successfully applied.

As the reader will notice in Table 1.6, the combination method performs much better on the test set as well. Using the proposed hierarchical combination method, we

Table 1.6 Classification results (UAR) got on the four-class task, using combinational training and four-class training (reported scores are the averages of three independently trained neural nets)

Training	No. of frames	λ	Dev. set (%)	Test set (%)
4-class	4	1.0	61.7	61.2
4-class (best)	14	1.0	61.6	62.6
Combination	4	1.0	**63.1**	**64.1**

achieved a relative gain of 7.5% over the simple four-class classification approach. One could argue that this is in large part due to the bigger context. However, we should point out that for both cases, the optimal context was decided based on the results attained on the development set. Furthermore, even if we compare the results got with the combinational method with the best result obtained by applying our original method on the test set, it would still produce a relative gain of 4%.

1.5 Conclusions and Future Work

In our studies with topical units, the long-term goal is to create a TU recognizer. Here, in the first stage towards this goal, we focused on the task of topical unit classification. For this, we constructed a set of features based on the labels of the multimodal corpus created in the HuComTech project. We also split this data into three separate sets (namely a train set, a development set, and a test set) to assist the machine learning methods we applied. Despite the fact that in our early experiments the optimal performance on the development set and the optimal performance on the test set in some cases were attained using different meta-parameter values, choosing the meta-parameter values based on the results on the development set proved to be a good strategy as it led to an acceptable performance. Thus it appears that we were able to construct a dependable partitioning of the full set.

What is more, regarding the classification results, we demonstrated that the proposed method of combining deep learning with probabilistic sampling performed better than deep learning by itself or support vector machines. This was true independently of the actual value chosen for the parameter λ, but with a carefully selected parameter value we were able to improve our results still further. In spite of the improvements achieved with this method, the classification results we obtained using four target labels were still far from optimal. We were however able to noticeably improve these results by fusing the two categories of topic change. We also demonstrated that the gain cannot be attributed solely to the characteristics of the UAR metric, but using three target labels for training was also beneficial. Moreover, the new three-class task not only led to better UAR scores in each case, but models trained on this task also contributed to a better classification of topical units in the four-class model as well. This surely means that a new approach should be considered if we are to work towards our long-term goal of creating a recogniser for topical units.

During the experiments described in this study we attempted to exploit all the information that the annotators marked, regardless of the utility of that information in TU classification. It may be beneficial however, to investigate the contribution of different modalities and an-notational tiers for it. This might help us to select a set a small set of features that are the best for our main goal. Furthermore, given the contribution of the context to correct classification, it would also be advisable to experiment with neural net structures that are better suited to time series, such are Recurrent Neural Nets (RNNs), Long-Short Term Memory networks (LSTMs), and Gated Recurrent Units (GRUs). Lastly, to achieve our goal of TU recognition, we hope to integrate our TU classifiers into a hybrid HMM/ANN (or HMM/RNN) model and then apply it.

Acknowledgements The research reported in the paper was conducted with the support of the Hungarian Scientific Research Fund (OTKA) grant # K116938. Tamás Grósz was supported by the ÚNKP-16-3 new national excellence programme of the Ministry of Human Capacities.

References

1. Abuczki A (2011) A multimodal analysis of the sequential organization of verbal and nonverbal interaction. Argumentum 7:261–279
2. Abuczki A, Baiat GE (2013) An overview of multimodal corpora, annotation tools and schemes. Argumentum 9:86–98
3. Babbar R, Partalas I, Gaussier E, Amini MR (2013) On flat versus hierarchical classification in large-scale taxonomies. In: Advances in neural information processing systems, vol 26. Curran Associates, Inc., pp 1824–1832
4. Baiat GE, Szekrényes I (2012) Topic change detection based on prosodic cues in unimodal setting. In: Proceedings of the CogInfoCom, pp 527–530
5. Baranyi P, Csapó A, Gyula S (2015) Cognitive infocommunications (CogInfoCom). Springer International, Cham, Switzerland
6. Chang CC, Lin CJ (2011) LIBSVM: a library for support vector machines. ACM Trans Intell Syst Technol 2:27:1–27:27
7. Demichelis P, Rinotti A, Martin JCD (2005) Performance analysis of distributed speech recognition over 802.11 wireless networks on the Timit database. In: Proceedings of the VTC, pp 2751–2754
8. Grósz T, Nagy I (2014) Document classification with deep rectifier neural networks and probabilistic sampling. In: Proceedings of the TSD, pp 108–115
9. Hunyadi L, Szekrényes I, Borbély A, Kiss H (2012) Annotation of spoken syntax in relation to prosody and multimodal pragmatics. In: Proceedings of the CogInfoCom, pp 537–541
10. Jr CS, Freitas A (2009) Novel top-down approaches for hierarchical classification and their application to automatic music genre classification. In: Proceedings of the IEEE SMC, pp 182–196
11. Kai-Fu Lee HWH (1989) Speaker-independent phone recognition using Hidden Markov Models. IEEE Trans Acoust Speech Signal Process 37(37):1641–1648
12. Kovács G, Grósz T, Váradi T (2016) Topical unit classification using deep neural nets and probabilistic sampling. In: Proceedings of the CogInfoCom, pp 199–204
13. Lawrence S, Burns I, Back A, Tsoi AC, Giles CL (1998) Neural network classification and prior class probabilities. In: Orr GB, Müller KR (eds) Neural networks: tricks of the trade. Springer, Heidelberg, Berlin, pp 299–313

14. Pápay K, Szeghalmy S, Szekrényes I (2011) Hucomtech multimodal corpus annotation. Argumentum 7:330–347
15. Rosenberg A (2012) Classifying skewed data: importance weighting to optimize average recall. In: Proceedings of the Interspeech, pp 2242–2245
16. Sapru A, Bourlard H (2014) Detecting speaker roles and topic changes in multiparty conversations using latent topic models. In: Proceedings of the Interspeech, pp 2882–2886
17. Schmidt AP, Stone TKM (2013) Detection of topic change in IRC chat logs. http://www. trevorstone.org/school/ircsegmentation.pdf
18. Shriberg E, Stolcke A, Hakkani-Tür D, Tür G (2000) Prosody-based automatic segmentation of speech into sentences and topics. Speech Commun 32(1–2):127–154
19. Silla CN Jr, Freitas AA (2011) A survey of hierarchical classification across different application domains. Data Min Knowl Discov 22(1–2):31–72
20. Su J (2011) An analysis of content-free dialogue representation, supervised classification methods and evaluation metrics for meeting topic segmentation. PhD thesis, Trinity College
21. Szekrényes I (2015) Prosotool, a method for automatic annotation of fundamental frequency. In: Proceedings of the CogInfoCom, pp 291–296
22. Tóth L (2013) Phone recognition with deep sparse rectifier neural networks. In: Proceedings of the ICASSP, pp 6985–6989
23. Tóth L, Kocsor A (2005) Training HMM/ANN hybrid speech recognizers by probabilistic sampling. In: Proceedings of the ICANN, pp 597–603
24. Tür G, Hakkani-Tür DZ, Stolcke A, Shriberg E (2001) Integrating prosodic and lexical cues for automatic topic segmentation. CoRR 31–57
25. Zellers M, Post B (2009) Fundamental frequency and other prosodic cues to topic structure. In: Workshop on the discourse-prosody interface, pp 377–386

Chapter 2
Monte Carlo Methods for Real-Time Driver Workload Estimation Using a Cognitive Architecture

Bertram Wortelen, Anirudh Unni, Jochem W. Rieger, Andreas Lüdtke and Jan-Patrick Osterloh

Abstract Human-machine interaction gets more and more cooperative in the sense that machines execute many automated tasks and cooperate with the human operator, who also performs tasks. Often some tasks can be executed by both, like a car that can autonomously keep the lane or is steered actively by the driver. This enables the human machine system to dynamically adapt the task sharing between machine and operator in order to optimally balance the workload for the human operator. A prerequisite for this is the ability to assess the workload of the operator in real-time in an unobtrusive way. We present two Monte Carlo methods for estimating workload of a driver in real-time, based on a driver model developed in a cognitive architecture. The first method that we present is a simple Monte Carlo simulation that gets as input the information that the driver can perceive, but does not take the actions of the driver into account. We evaluate it based on a driving simulator study and compare the workload estimates with functional near-infrared spectroscopy (fNIRS) data recorded during the study. Afterwards the shortcomings of the simple approach are discussed and an improved version based on a particle filter is described that takes the driver's action into account.

B. Wortelen (✉)
Cognitive Psychology Lab, C.v.O. University, Oldenburg, Germany
e-mail: bertram.wortelen1@uol.de; bertram.wortelen1@uni-oldenburg.de

A. Unni · J. W. Rieger
Applied Neurocognitive Psychology Lab, C.v.O. University, Oldenburg, Germany
e-mail: anirudh.unni@uol.de

J. W. Rieger
e-mail: jochem.rieger@uol.de

A. Lüdtke · J.-P. Osterloh
OFFIS - Institute for Information Technology, Oldenburg, Germany
e-mail: luedtke@offis.de

J.-P. Osterloh
e-mail: osterloh@offis.de

© Springer International Publishing AG, part of Springer Nature 2019
R. Klempous et al. (eds.), *Cognitive Infocommunications, Theory and Applications*,
Topics in Intelligent Engineering and Informatics 13,
https://doi.org/10.1007/978-3-319-95996-2_2

2.1   Introduction

People are interacting with complex technology on a daily basis especially at work: driving trucks, navigating ships, controlling power plants, performing surgeries. The listed examples are all from safety critical domains. In these domains it is essential that the human operator performs well and makes no fatal error. High workload can decrease the performance of the human operator [47]. A typical design goal for technical systems is therefore to avoid high workload conditions. In the automotive domain this goal is not easy to achieve. Driving can be a very complex task, especially when the driver has to deal with many other traffic participants and when it is difficult to assess the situation. At other times, driving is nearly effortless. Driving on highways with low traffic places a low demand on the drivers that they are often engaged in other tasks like listening to the radio, letting their mind wander, interactions with other passengers, or even worse, texting, or phoning. The increasing level of automation in road vehicles will free even more resources for the driver to engage in secondary tasks. This is a beneficial development for drivers as long as sufficient resources of the driver are spent on the driving task. It is known that reducing the task load up to the point that the driver is little involved in the task, can reduce driver's workload especially in recovery situations [7]. Intelligent assistance systems that adapt to the current workload level of the driver could help to achieve this balance.

This paper presents a model-based method to estimate the workload of a driver in real-time by simulating driver models in a cognitive architecture. It is an extension of work originally presented at the CogInfoCom conference 2016 [53], where we introduced a very simple model-based workload assessment method. This paper first illustrates the initial simple simulation method shown in [53] (©2017 IEEE. Reprinted, with permission from [53]) and then discusses shortcomings and introduces an improved method based on a particle filter algorithm. The remainder of the paper is structured as follows. Existing workload assessment techniques are listed and categorized in the next section. Afterwards the cognitive architecture CASCaS is briefly described in Sect. 2.3. We tested our approach in a highway driving situation by simulating a driver model with CASCaS and compare the simulated data with data from a previous driving simulator study [43], which is outlined in Sect. 2.4. Subsequently, we show a very basic driver model. Using this driver model we demonstrate a simple variant of our model-based workload assessment method [53] and compare the predicted workload levels with data gained form brain activity measures. Finally, we discuss shortcomings of the simple variant and introduce an improved simulation concept.

2.2   Measuring and Simulating Cognitive Workload

One objective of our work is to enable the measurement of human workload in interaction with technical systems. A large number of approaches to measure workload can be found in literature. In this section we highlight some of these and categorize different approaches. Finally we describe where we advance the state of the art in workload assessment.

O'Donnell and Eggemeier [28] group workload assessment techniques into three categories: (1) subjective measures, (2) performance-based measures and (3) physiological measures. Task demand is also often mentioned as additional measure for mental workload [4, 5, 50]. However, de Waard [5] notes that task demand is not the same as workload. Task demand is a major external factor that induces workload. The effect of task demand on workload differs based on individual capabilities [6] and the result can be observed using a measure from one of the aforementioned categories. For a detailed discussion see [5].

One problem with measuring workload is that there are several definitions of human workload in literature that often differ slightly. Most definitions have in common that the workload is described as a degree of utilization of a limited resource in a limited time frame, e.g., [5, 15, 18, 54]. In many studies workload is broken down into subcomponents. McCracken and Aldrich [25] identified four subcomponents that refer to four different processing resources: visual, cognitive, auditory and psychomotor (VCAP). This differentiation gives credit to the fact that humans are able to execute two tasks in parallel more easily, if the two tasks do not both rely heavily on the same resource [36], e.g., both rely heavily on cognitive processing or on visual processing. Another differentiation is made by the NASA Task Load Index (NASA-TLX) [13]—a multidimensional, subjective questionnaire for workload assessment. It identifies 6 subcomponents: mental demand, physical demand, temporal demand, performance, effort, frustration. Wu and Liu [54] linked these components also to the utilization of different cognitive processing components.

Wickens and Hollands [47] view workload from a very generic perspective as the amount of time spent working on a task in relation to the total time available. This definition easily allows to observe workload simply by looking at the time spent on a task as a *performance-based* measure. In the automotive domain workload is often viewed in a very similar way. However, typically it is not the time spent on the primary task of driving that is of interest. When designing new in-vehicle infotainment or assistance systems, the interaction with these systems is viewed as a secondary task compared to the primary task of driving. In this context, workload is defined by the competition between primary and secondary task [2, 39]. One design goal is that the workload induced by the interaction with the in-vehicle systems should degrade the driving performance as little as possible. This again can be measured. So not only the time-on-task for the secondary task, but also the effect on the primary task can be used as workload measure. Several tests have been developed to specifically test driver workload induced by secondary tasks, like the lane change task defined in ISO 26022 [16] or the peripheral detection task [44]. These performance-based tests are conducted during the design-time of a new in-vehicle system in experimental settings.

Measuring the effect of workload on drivers' behaviour is not sufficient for designing better in-vehicle systems. A more detailed understanding is needed about which aspect of the secondary task affects which component of workload and how it interferes with the driving task. A common approach for a detailed analysis of workload is to make a task analysis and identify all actions and operations that a human has to do in a specific situation. The basic idea is then to infer workload based on a workload model that associates human actions (elements of the task description) with the components of workload. In this way workload is measured by an analysis of the *task demand*. McCracken and Aldrich [25] developed such a workload model for the aeronautics domain that offers a scale, which assigns workload levels for the four VCAP-subcomponents to specific kinds of human actions like *recall, memorize* or *visually inspect*. McCracken and Aldrich used descriptive models of the operator's task in their studies. They simply analysed the sequence of actions that are typically performed for a task and inferred the associated workload. The simplicity of their approach makes it easy to apply it to very different application domains. Indeed, several others adopted it. Weeda and Zeilstra [46] relied on it as basis for their Objective Workload Assessment Technique (OWAT), which they used to estimate workload of suppliers of railroad travel information.

Using descriptive models of the operator's task enables only the analysis of a prototypical task execution. It does not take into account how the workload is affected by the dynamics of the situation. To account for the dynamic aspects, two recent approaches used the McCracken and Aldrich model to annotate behaviour primitives in cognitive models created with the cognitive architectures MIDAS (man-machine integration design and analysis system) [11] and CASCaS (cognitive architecture for safety critical task simulation) [8] with associated workload levels. A similar approach was used with the QN-MHP (queueing network—model human processor) cognitive architecture [54], but based on the workload components defined in NASA-TLX. Cognitive architectures try to simulate the information processing within the human cognitive system and are able to execute a task description in a psychological plausible way. Most architectures also simulate motoric and perceptual processes. With this ability, cognitive architectures can interact with the real or a simulated environment in a human-like way. Thus, a detailed closed-loop simulation is achieved that does take the dynamics of a situation into account. In the automotive domain, there are several driver models available for different cognitive architectures, like ACT-R [35], CASCaS [52], Soar [1] or QN-MHP [41, 54]. In the current work, we build upon the driver model developed for CASCaS [52].

The approaches to analyse the workload of a human operator presented so far are offline-methods that cannot measure workload in real-time. They are used during system design time in experimental settings or simulations. Other approaches are needed to measure workload while the operator is working. *Subjective* measures are questionnaires and self reports, which are easy to apply in most work environments. Commonly used examples are the NASA-TLX [13] questionnaire or the Subjective Workload Assessment Technique (SWAT) [34], which have been widely applied and validated in very different application domains. However, subjective measures are not suitable for real-time workload assessment, because they are applied after the human operator finished his or her work.

For *performance-based* measures the behaviour of the operator is observed. These measures are applicable for real-time and offline workload assessment. The lane change task, mentioned above, is such a measures for offline workload assessment that uses the steering behaviour and response times to infer the workload of the driver. Nakayama et al. [26] developed a similar concept based on steering entropy that can in principal be used in real-time in cars. Because behavioural measures of the driver can often be recorded in a very unobtrusive way, they are well suited for real-time workload assessment.

For the same reason, *physiological* measures are quite popular as they can continuously record the operators response without actually intruding into the operators task. The most commonly used physiological measures for workload assessment are electrocardiogram (ECG) and electrodermal activity (EDA). Previous researches have consistently demonstrated that increased workload levels lead to increased heart rate (HR) and decreased heart rate variability (HRV) [20]. Solovey et al. [38] recorded ECG and EDA while driving and were able to discriminate three driving situations with increasing control demand. However, it is not clear how specific these somatophysiological recordings are for different cognitive demands. Even driver's speech can be an indicator for workload. Wood et al. [50] found evidence, that changes in the variance of the pitch and durations of pauses in driver's speech indicated changes in driver's cognitive workload.

Cognitive workload levels can also be assessed using neurophysiological measurements. Kojima et al. [19] measured brain activity using functional near-infrared spectroscopy (fNIRS) with several optodes placed over the forehead while train drivers controlled a train simulator in manual and automatic conditions. These authors reported higher oxygenated-hemoglobin (HbO) concentrations in the manual condition. In the driving domain, fNIRS was used by Unni et al. [43] to predict cognitive workload and by Tomioka et al. [40] to compare brain activity in pre-frontal areas between elderly people with and without Alzheimer disease in a collision avoidance task in a driving simulator. Here, the Alzheimers disease group showed a less prominent increase of HbO concentration in the recorded prefrontal cortices. Lei and Roetting [21] estimated driver workload from EEG spectrum modulation while participants performed a combination of lane-change and n-back tasks in a driving simulator. They showed that 8–12 Hz (alpha) oscillatory power decreased and 4–8 Hz (theta) power increased with increasing workload.

In the remainder of the paper, we describe, how we use driver simulations in a dynamic driving environment with the cognitive architecture CASCaS to assess workload based on the task-demand. We improve a previous approach [8], that relied on a direct annotation of the VCAP workload components on the task simulated by CASCaS. Instead, we measure directly the utilization of components in CASCaS during the task simulation. We compare the simulated workload with an independent physiological measure. For that, we measured brain activation using continuous and spatially resolved fNIRS measurements, which we extended to frontal, parietal and temporo-occipital brain areas. Finally we describe an improved simulation concept based on particle filtering, which also simulates the driver model based on the current state of the environment as input, but improves the workload estimation by also taking the observed actions of the driver into account. In this way, we combine performance-based measures with measures based on task-demand.

2.3 Integration of Workload Measures in a Cognitive Architecture

The cognitive architecture for safety critical task simulation (CASCaS) was originally developed as a simulation-based means to detect human errors in safety critical scenarios [23]. It is structured like many other cognitive architectures. Figure 2.1 shows the overall structure. It contains a memory component, which simulates information storage and retrieval. A dedicated component processes the information on three different layers according to Fitts and Posners [9] three levels of skill learning. The implementation of these layers differ based on Rasmussens [32] model of performance levels. For the autonomous layer, an API is provided for developing simple stimulus-response models, for example using a PID controller [52]. Skilled behaviour on the autonomous layer can be processed in parallel. In contrast, behaviour simulated on the associative layer is sequential in nature and is implemented by a rule processing engine. Finally, the cognitive layer simulates problem solving and planning processes [24]. Behaviour simulated in CASCaS is always goal-directed. A dedicated component organizes the current task goal and handles multi-tasking and task switches [51]. Like many other general cognitive architectures, CASCaS does not only contain components describing purely cognitive processes, but also components describing motor and perceptual processes. A simulation of CASCaS is not deterministic, because most of these processes contain probabilistic decision processes (e.g., for the selection of the currently active goal) or noise components (e.g., for the duration of eye movements) [30].

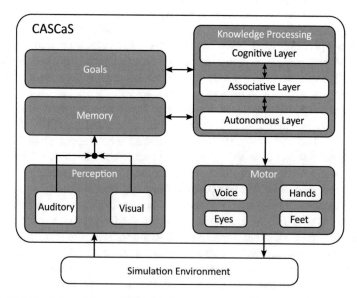

Fig. 2.1 High-level view of the cognitive architecture for safety-critical task simulation (CASCaS)

Table 2.1   Measures of workload in CASCaS

Workload measure	Measured by
Perceptual load (visual/auditory)	Amount of perceived information (rate of information elements written to memory by perception component)
Working memory load	(a) Rate of information elements written to memory by associative and cognitive layer
	(b) Rate of rules fired by associative layer
Visual attention load	Rate and size of gaze movements
Motor load	Rate and size of hand movements
Multitasking load	(a) Number of active tasks
	(b) Rate of task switches

Feuerstack et al. [8] already used CASCaS to predict operator workload by annotating behaviour primitives of a model developed using CASCaS with the workload levels of McCracken and Aldrich [25]. The objective of the current work is to estimate the workload level directly by analysing the data processing in the different components of CASCaS during the simulation of an operator model. Because workload is commonly defined as the utilization of a limited resource, we identified internal variables of the components that describe the degree of utilization of the respective component. These measures are listed in Table 2.1. It includes measures for the VCAP components of McCracken and Aldrich [25].

We expect that the amount of information perceived visually or auditory by the perception component is an indicator for the visual, respectively auditory load. Another measure linked to the visual load is the rate and size of gaze movements. Similarly, the rate and size of hand movements gives an estimate for the psychomotor load. We expect that the working memory load is strongly linked to the amount of information that needs to be processed in working memory for the active tasks.

In order to measure the workload online, we use CASCaS in a kind of open-loop simulation. We simulate a virtual driver with CASCaS and provide it with all the information that is available to the driver. However, the actions of the driver model do not change the environment. Only the actions of the real driver affect the environment. Thus, the feedback which is typical for analytical offline models is not available.

2.4   Method

In order to test our approach, we used data from a driving simulator study which was conducted under different workload conditions. The study is published in [42, 43]. In this section a brief description of the experiment design and data recording is reproduced from [42]. We used the driving data as input for the simulation of a driver model in CASCaS. Furthermore, drivers' brain activity was recorded during the experiment. This data is used to compare the driver model output with physiological measures.

2.4.1 Participants

Three male participants aged 26–31 years participated in this study. In this age group performance in working memory tasks with high workload should be good [49], while at the same time participants had some years of practice in driving. All participants had a valid German driving license. The participants were informed prior to the experiment and gave written consent according to the guidelines of the German Aerospace Research Centre and the Carl von Ossietzky University Ethics Committee, which approved the experimental protocol.

2.4.2 Experiment Design

The n-back task has been used as a benchmark in the field of psychology and neuro-science to manipulate cognitive workload. In our experimental paradigm, we introduce a novel approach to combine an n-back speed regulation task as secondary task while the subject controls a driving simulator. A detailed explanation of the paradigm can be found in [42].

The driving scenario consists of a highway with moderate traffic. We introduced five different levels of workload (i.e. 0-back to 4-back). Each n-back task lasted around 3 min and consisted of 10 trials. The subject encountered nine different speed signs randomly varying from 60 to 140 km/h in steps of 10 km/h. A new speed sign was encountered approximately 20 s on average. We had two repetitions for each n-back task. The n-back tasks were randomly distributed to avoid sequencing effects. The whole driving experiment lasted for about 30 min. The n-back task can be better understood from Fig. 2.2.

For a 0-back task, as soon as participant encounters the 80 km/h speed sign, he or she has to adjust and maintain the speed close to 80 km/h. For a 1-back task, the participant had to drive at the speed indicated by the previous speed sign (hence the name 1-back), here 140 km/h, and remember 80 km/h which he or she would drive at the next speed sign. For a 4-back task, the participant had to drive at the speed that occurred four speed signs previously, here 60 km/h, and remember the four element speed sequence, e.g., 100, 120, 140 and 80 km/h. This way, the participants had to update, memorize and recall a sequence of n speed signs and adjust their speed accordingly. Participants were prompted by a visual message on the screen to stay in the correct speed range when they drove more than 5 km/h above or below the target speed. They were allowed to take 3 s before and after passing a new speed sign to adjust their speed. Because the task was performed with concurrent traffic,

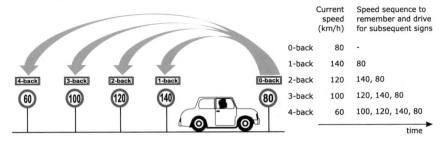

Fig. 2.2 Example of n-back experimental paradigm to manipulate cognitive workload. Consider a scenario where the participant is about to pass the 80 km/h speed sign and the previous four speed signs were as shown on the left. For the corresponding n-back task, participants had to memorize the last n speed signs and drive at the n-th speed sign which occurred previously

participants had to consider the current traffic situation throughout the task which included lane changes.

For any n-back task, the participant needs to first pass through n successive speed signs before regulating the speed. For this reason, when a new n-back task begins, the participant was instructed to drive at the speed shown by the first speed sign.

2.4.3   FNIRS Data Recording

Participants brain activity was measured using fNIRS. FNIRS uses low-energy optical radiation in the near infrared range to measure absorption changes in the subsurface tissues and obtain local concentration changes of oxy-hemoglobin (HbO) and deoxy-hemoglobin (HbR) as correlates of functional brain activity using modified Beer-Lambert law [37, 45]. We used two NIRScout (NIRX Medical Technologies) systems in tandem to acquire fNIRS signals. The system uses two wavelengths of 760 and 850 nm and outputs relative concentration changes of HbO and HbR. 32 optical emitters and detectors each were used to obtain full coverage of the frontal, parietal and temporo-occipital cortices. In order to avoid crosstalk between the two systems, we arranged the optodes such that one system covered the frontal areas and the other system covered parietal and temporo-occipital areas. We had 78 channels (combinations of emitters and receivers) in total for measuring HbO and HbR over nearly the whole head. The average distance between an emitter and detector was approximately 3.5 cm. The sampling frequency of the tandem system was nearly 2 Hz.

2.5 Evaluation

To evaluate our simulation approach, we created a driver model for CASCaS, which can be simulated with the driving data recorded in the driving simulator experiment. In the following we describe the driver model and the simulation procedure, followed by an interpretation of the simulation results and a description of the procedure for validating the results with the fNIRS data.

2.5.1 Driver Model

We created a CASCaS model that simulates the tasks of the participants. To illustrate the high-level structure of the model, we use Concur Task Trees (CTTs), which are used to model tasks in interactive applications [29]. Such a task tree for the driver model is shown in Fig. 2.3 as a Concur Task Tree using the graphical notation introduced by Nóbrega et al. [27]. As can be seen in the figure, there are sub-tasks related to driving and some related to the n-back task. On a closer look, it is shown that the driver model simulates four parallel sub-tasks. The two driving related sub-tasks i.e. lateral control (steering) and longitudinal control (keeping speed) are typical tracking and control tasks. The driver has to look at the road or the speedometer and has to make adjustments in the lane position while adjusting to the current speed. The definitions of these tasks were taken from an existing driver model [52]. The n-back task is modelled by two parallel sub-tasks. The participants had to (1) look out for speed limit signs by looking to the right lane edge and updating their internal list of the last n speed limit signs, if a new speed limit sign occurred and (2) rehearse their internal list of the last n speed limit signs. Juvina and Taatgen [17] showed that participants in their experiment adopted two different strategies for the n-back task: (1) a high-control strategy based on rehearsal and (2) a low-control strategy based on time estimation. Because the inter-stimulus interval in this study [17] was far smaller than in our experiment, we assume that participants are unlikely to adopt a low-control strategy. Therefore, our model implements the high-control strategy based on rehearsal.

The four sub-tasks related to driving and the n-back task (see Fig. 2.3) should be executed in parallel. However, the associative layer only allows sequential process-ing of tasks. After it executed one task, it immediately selects the next tasks to be executed. CASCaS uses an implementation of the Salience-Effort-Expectancy-Value (SEEV) model for the selection, as described in [51]. Tasks are selected based on two of the SEEV model parameters per task, which describe the expected rate of information processed by the task and the importance of the task. We used the lowest ordinal heuristic proposed by Wickens et al. [48] to estimate these parameters.

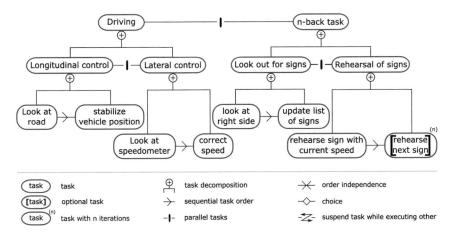

Fig. 2.3 CTT of the four parallel tasks of the CASCaS model. Graphical notation based on [27]

2.5.2   Simulation

In the driver model shown in Fig. 2.3, the execution of each of the four parallel tasks *Longitudinal control*, *Lateral Control*, *Rehearsal of signs* and *Lookout for signs* is nearly independent of each other. Furthermore, the task structure contains no major sequential elements. Take the longitudinal control task as an example. The driver model only corrects its speed after it looked at the speedometer. Thus, the longitudinal control task can only be in one of three states: *<inactive>*, *look at speedometer* and *correcting speed*. This is similar for all tasks. So what concerns the task structure is that the driver model can only be in one of a very limited number of states. In each of these states, every data sample of the recorded data from the driving simulator study is plausible for the driver model and can be processed. Therefore, we did not try to estimate the actual state of the driver. This simple open-loop Monte Carlo approach is shown as pseudocode in Listing 2.1. In the listing the state of the i-th simulation at time step k is denoted by the state vector x_k^i. At the beginning, N independent simulations are started in a Monte-carlo fashion by drawing initial states x_0^i for all simulations from a distribution of initial states p_{x_0}. In this case the system is the driver. While the driver is driving, in each simulation step k the observations y_k^e of the environment are collected and used as input to simulate the driver model. The workload is then estimated as the mean of the workload calculations of all N simulations. In our study we used the rate of information elements written to memory by associative and cognitive layer as workload measure.

Listing 2.1 Workload estimation using simple open-loop Monte Carlo simulation

int k = 0

//draw initial states
$x_0^i \sim p_{x_0}$, $i = 1, \dots, N$

```
while (true) do
begin   //do next simulation step
    k = k + 1

    // simulate next step using CASCaS
    for (i = 1 to N) x_{k+1}^i = simulate(x_k^i, y_k^e)

    // calculate workload estimate
    workload = 1/N · Σ_{i=1}^N getWorkload(i)
end
```

This is a very simplistic approach that is sufficient for the driving scenario used in our study. In Sect. 2.7 we will discuss some problems of this approach and propose an improved method.

2.5.3   Simulation Results

In a first step, we simulated the CASCaS model in an ideal environment. In this environment, the vehicle always drove at the correct speed and in the centre of the lane. We use an open-loop simulation in which the steering and acceleration actions of the model have no effect on the environment. Thus, the demands for the lateral and longitudinal control tasks were low because there was no need to correct speed or stabilize the vehicles lane position. For 1000 s, we simulated a sequence of speed limit signs identical to the sequence used in the experiment but we used a fixed n-back level. We even used n-back levels up to n = 10, which exceeded the experiment conditions (n ≤ 4). CASCaS contains several probabilistic elements. Therefore, we computed Monte Carlo simulations of the above described model. Figure 2.4 shows the rate of information elements written to memory by associative and cognitive layer (f_M) as a measure of working memory load. For each simulation f_M was calculated over the entire simulation duration. Figure 2.4 shows that the working memory load increases with the n-back level. However, the cognitive processing resource is limited and other tasks compete against the n-back task. Thus, this rate does not increase linearly but asymptotically approaches a maximum.

The shape of the curve can be explained by the model. Consider a model with k parallel tasks (here: $k = 4$, see Fig. 2.3). Each task i has its own rate of information elements written to memory $f_{M,i}$, which it would achieve if it is the only task that is executed. In a multitasking setting, f_M is the sum of each $f_{M,i}$ times the probability e_i that the respective task is currently executed:

$$f_M = \sum_{i=1}^{k} (e_i \cdot f_{M,i}) .$$ (2.1)

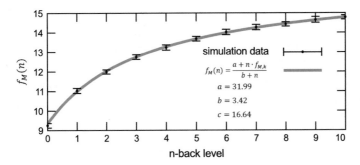

Fig. 2.4 Rate of information elements written to memory by the cognitive and associative layer in relation to the n-back level

The execution probability e_i of task i is equal to the product of the probability s_i that task i was selected and the typical duration of it (d_i), compared to the sum of this product over all k tasks:

$$e_i = \frac{s_i \cdot d_i}{\sum_{j=1}^{k} (s_j \cdot d_j)} . \tag{2.2}$$

As mentioned before, the s_i parameters are calculated by the SEEV model [51]. We are interested in one of the tasks: the n-back task. Let task k be the n-back task. Changing the n-back level (n) does not affect the selection probability of the n-back task but it changes the duration (d_k) of it. For simplicity, we assume that it is affected by a constant factor: $d_k = n \cdot c$. Combining (2.1) and (2.2) gives:

$$\begin{aligned}
f_M &= \sum_{i=1}^{k-1} \left(\frac{s_i \cdot d_i}{\sum_{j=1}^{k} (s_j \cdot d_j)} \cdot f_{M,i} \right) + \frac{s_k \cdot d_k}{\sum_{j=1}^{k} (s_j \cdot d_j)} \cdot f_{M,k} \\
&= \frac{\sum_{i=1}^{k-1} (s_i \cdot d_i \cdot f_{M,i}) + s_k \cdot d_k \cdot f_{M,k}}{\sum_{j=1}^{k} (s_j \cdot d_j)} \qquad | d_k = n \cdot c \\
&= \frac{\sum_{i=1}^{k-1} (s_i \cdot d_i \cdot f_{M,i}) + s_k \cdot n \cdot c \cdot f_{M,k}}{\sum_{j=1}^{k-1} (s_j \cdot d_j) + s_k \cdot n \cdot c} \qquad | \cdot \frac{(c \cdot s_k)^{-1}}{(c \cdot s_k)^{-1}} \\
&= \frac{\sum_{i=1}^{k-1} (s_i \cdot d_i \cdot f_{M,i}) \cdot (c \cdot s_k)^{-1} + n \cdot f_{M,k}}{\sum_{j=1}^{k-1} (s_j \cdot d_j) \cdot (c \cdot s_k)^{-1} + n}
\end{aligned} \tag{2.3}$$

Equation (2.3) shows that f_M is a degree 1 rational function of the n-back level n. Substituting the constant terms with parameters a and b, we get

$$f_M = \frac{a + n \cdot f_{M,k}}{b + n}. \tag{2.4}$$

We used the method of least-squares to fit the function to the data, which works very well in this ideal environment ($a = 31.99$, $b = 3.42$, $f_{M,k} = 16.64$). The resulting function is plotted as a solid grey line in Fig. 2.4. However, this function does not take into account several interactions between the tasks, which affect the $f_{M,i}$

Fig. 2.5 Error bars show
mean and standard deviation
of the rate of information
elements written to memory
by associative and cognitive
layer (fM) averaged over all
three participants. The solid
line shows the error
probabilities for each n-back
level. Note that the model
and the subject data are
plotted against different
ordinates. Therefore the
offset has no interpretation

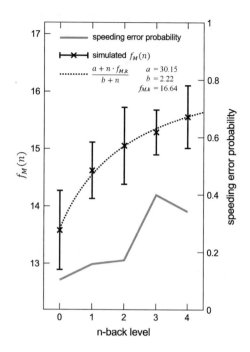

and d_i parameters. These interactions are more relevant in more complex scenarios. For example, increasing the n-back difficulty can reduce the longitudinal control quality, which increases the probability that speed corrections are required, which again increases the mean duration d of the longitudinal control task.

To test the model in more complex scenarios, we used the data of the three participants from the experiment to generate a simulation environment, in which the model perceives the speed that the participants were driving. Thus, the model also had to correct speed errors of the participants. The probabilities that a participant drove with a speed deviation of more than 5 km/h for more than 5 s are shown in Fig. 2.5 as a grey solid line. The actions of the model had no effect on the models environment (i.e. the simulated speed of the vehicle and position in the lane). We had to exclude this feedback for the model because here our objective is to develop an open-loop simulation model for online workload assessment.

The model now also had to change the n-back level during simulation. This happened exactly at the same time and with the same sequence of changes, as it did for the participants. We simulated the model with each of the data sets of the three participants 80 times. We used the simulation data and the method of Byrne [3] to determine the minimum number of simulations required to achieve a 99 % confidence interval for f_M that is half the size of the smallest expected effect. The smallest expected effect is between the two highest n-back levels (3-back and 4-back) for which a minimum number of 23 simulations was calculated. Thus, the 80 simulations were already sufficient.

The model was simulated with input data from the three participants. The error bars in Fig. 2.5 show for each n-back condition the mean and standard deviation of f_M for the input data of all three participants over all 240 (3×80) simulation runs. Each participant drove an n-back level twice and so did the model. We fitted Eq. (2.4) to the data of the more complex scenario using the Levenberg-Marquardt algorithm but we kept $f_{M,k} = 16.64$ like in the ideal environment, because we do not expect it to change. The fitted function is shown in Fig. 2.5 as dotted line. It shows a very good fit to the simulation data (reduced $\chi^2 = 1.01$). Thus, $f_M(n)$ seems to be a good indicator for workload. Even though the data shows stronger individual deviations due to the more realistic input data, the good fit lets us assume that the task interactions in our setup are not very significant. In future studies, we want to use the model in more complex situations and also in closed-loop simulations in order to investigate how it affects the workload measurement.

2.5.4 Multivariate Cross-Validated Prediction of Working Memory Load Level from fNIRS Data

As described in [42] we used multivariate lasso regression [14] implemented in the Glmnet toolbox[1] to find channel-wise weights for instantaneous fNIRS HbR data to predict the working memory load level. The λ parameter which determines the overall intensity of regularization was optimized internally by Glmnet in the training phase of the cross-validation of the lasso regression model.

We used a standard nested cross validation procedure [14] to train the model and test generalization performance. Each loop implemented a 10-fold cross-validation. The outer cross-validation loop tested the generalization of the regression model with the optimized hyperparameters (number of PCs and λ). The hyperparameters were optimized in an inner cross-validation loop which was implemented in the training phase. Cross-validation avoids overfitting of the data to the model and provides an estimate of how well a decoding approach would predict new data in an online analysis [33].

2.6 Comparison to Neurophysiological Measure with fNIRS

Figure 2.6 depicts the time course of working memory load induced by the n-back task (thick gray curve) together with the working memory load predicted by the fNIRS multivariate regression model (thin black curve) for participant 3 as shown in [42]. The Pearsons correlation between the two curves is almost 0.8. It can be seen from Fig. 2.6 that the predicted working memory load nicely follows the induced working memory load and could be used to predict variations in cognitive working memory

---

[1] http://web.stanford.edu/hastie/glmnet_matlab/.

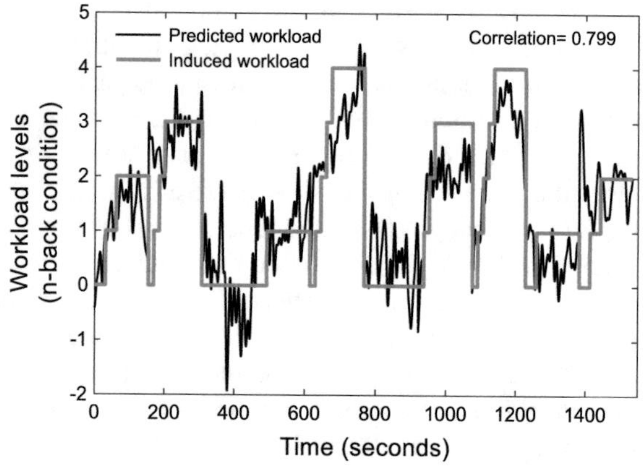

Fig. 2.6 Ten-fold cross-validated prediction of workload from HbR fNIRS measurements using multivariate regression analysis for an example participant [42]

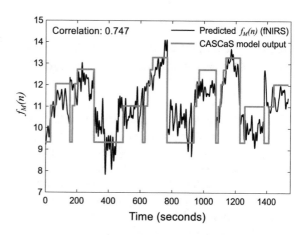

Fig. 2.7 Ten-fold cross-validated prediction of CASCaS workload indicator $f_M(n)$ from HbR fNIRS measurements using multivariate regression analysis for an example participant

load levels. There are intervals when the model seems to over- or under-estimate the working memory load level. This may be due to the incomplete model which currently neglects the workload imposed by the concurrent driving task in the changing traffic situations. In spite of this, we are able to achieve satisfactory predictions. The mean Pearsons correlation and the standard deviation across the three participants were 0.70 and 0.1 respectively which were all statistically significant ($p < 0.05$). All multivariate correlations were determined in a 10-fold cross-validation to evaluate generalization to new data the model had not 'seen' before to approximate an online analysis. A detailed description can be found in [42].

Figure 2.7 shows the results obtained after we repeated the analysis procedure to compare the rate of information elements written to memory by the associative and cognitive layer $f_M(n)$ which represents the workload indicator of CASCaS (thick gray curve) with that of the predicted $f_M(n)$ from the HbR fNIRS data (thin black curve) for the same participant. The correlation between the CASCaS model output and the fNIRS data for a ten-fold cross-validation is almost 0.75 which is very similar to the correlation result achieved in Fig. 2.6 between the n-back task and HbR fNIRS data. The correlation results for the above two cases were comparable in the remaining two participants and were statistically significant for a 95 % confidence interval.

2.7   An Improved Simulation Concept for Real-Time Analysis

The presented results were achieved by performing Monte-Carlo simulations using data recorded during the driving simulator study as input for the driver model. This means the results were achieved offline. Our vision is to enhance this method for real-time estimation of workload. The basic idea is to perform the Monte-Carlo simulations in real-time while the driver is driving and provide a real-time estimation of workload to enable driver assistance systems that adapt to the current workload of the driver. In this study, we identified several problems and shortcomings that render this approach—as it is now—impossible for practical applications. In this section, we describe the issues we identified and describe a conceptual framework to overcome these.

For the study shown above, we used a basic driving scenario and a very simple driver model. Only few variables of the scenario were used as input for the driver model and the driver model is only able to handle driving situations where it has to keep the lane and a target speed (free driving). More complex models will impose some problems. This can be illustrated on an example. Figure 2.8 shows parts of a Concur Task Tree for a driver model that is able to handle car following and overtaking situations besides free driving situations. In this case the temporal order of the task execution gets more complicated.

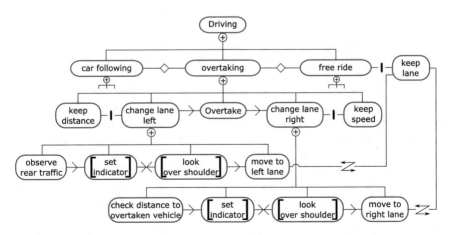

Fig. 2.8 CTT of a driver model that includes overtaking ability

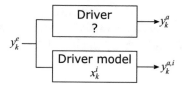

Fig. 2.9 The driver's behaviour (y_k^a) is a reaction to the state of the environment (y_k^e), which is also input to the driver model. The internal state of the driver is not observable, in contrast to the state of the driver's model (including workload)

The task structure is different for each of the driving conditions: *car following*, *overtaking* and *free ride* (only *overtaking* is shown in Fig. 2.8). These tasks contradict each other. Thus, only one is active at a time (choice relation $-\diamondsuit-$). Looking at the task structure for *overtaking*, it can be seen, that overtaking consist of many subtasks, which are mainly organised sequentially. However, the order of the subtasks is not always fixed (order independence $\rightarrowtail$). This is the case for *set indicator* and *look over shoulder*, which can even be omitted (optional task [ ]). Drivers should not omit it, but this is not a nominal model. Thus, the temporal order of sub-tasks is far more complex than it is for the simple model (see Fig. 2.3) used until now.

With the complex model shown in Fig. 2.8, this approach is not viable. Some input data is not plausible in certain states. For example, input data indicating that the driver is driving on the right lane is not plausible in the state *overtake*, or setting the indicator in state *free ride*. Thus, for estimating the workload using Monte-Carlo simulations in real-time while the driver is driving, it needs to be ensured that the driver model at time k in each of the simulation runs is in a state x_k^i that is plausible with regard to the observations y_k. The observations consists of measurements of the environment y_k^e (e.g., speed of the vehicle, current lane or road signs) and the actions of the driver y_k^a (e.g., steering, accelerating or use of indicators). The plausibility with regard to y_k^e is achieved by the nature of the simulation itself, because it is the input that drives all N simulations (see Fig. 2.9). But the driver's actions y_k^a depend on the internal state of the driver, which cannot be observed directly. Only the state of the driver model can be observed. Therefore the likelihood of being in a certain state needs to be estimated.

Estimating the current state of a system (here the driver) given its input and output is a typical filtering problem. Because a CASCaS simulation model can become quite complex and involves non-linear components, common filtering approaches for linear systems, like Kalman filters, are not applicable. Instead we propose to use Sequential Monte Carlo methods (particle filter) [10, 12].

But before describing the proposed filtering algorithm, we give an idea about what the state x_k^i represents. A CASCaS simulation is not deterministic. A model simulated with CASCaS contains many probabilistic elements and therefore the same input to the driver model can lead to very different states of the driver model. However, [30, 31] argued that the magnitude of the effect that these different probabilistic elements of CASCaS have on the simulated behaviour is not equally strong. They also simulated driver models in CASCaS and reduced the state space by only considering probabilistic state transitions in CASCaS that most likely have a great impact on the model's behaviour. In their study they only focused on the goal selection and rule selection processes. CASCaS associates each task with a task goal (see [51] for details). The goal selection process defines which task (see Fig. 2.8) is currently processed. To execute a task goal CASCaS uses a rule based system [51]. The rule selection process defines which rule is applied, to achieve the current task goal. A rule might specify actions, that reflect visible actions of the driver (y_k^a), like steering, setting the indicator or head movements.

Using the same argumentation as [30], we represent the state of the driver model by the current set of active task goals and the last selected rule. As the visible actions of the driver model mainly depend on the current goal and the selected rule, we can calculate based on the definition of the driver model the likelihood $p(y_k^a|x_k^i)$ that the current observations of driver's actions would occur, if we assume, that the driver's internal state is x_k^i. This is required for the particle filter.

Listing 2.2 Workload estimation using particle filtering (based on [12])

```
int k = 0
// draw initial states
x₁ⁱ ~ pₓ₀, i = 1, ..., N

// set initial weights
w₀ⁱ = 1/N
while (true) do
begin
    // do next simulation step

    // calculate workload estimate as weighted average
    workload = ∑ᴺᵢ₌₁ wₖⁱ getWorkload(i)

    // simulate next step using CASCaS (state propagation)
    for (i = 1 to N) xₖ₊₁ⁱ = simulate(xₖⁱ, yₖ₊₁)

    // update weight based on measurement probability
    //and compensate for proposal distribution
    for (i = 1 to N) wₖ₊₁ⁱ = wₖⁱ · p(yₖ|xₖⁱ) · p(xₖ₊₁ⁱ|xₖⁱ)/q(xₖ₊₁ⁱ|xₖⁱ,yₖ₊₁)

    //calculate normalization weight
    cₖ = ∑ᴺᵢ₌₁ wₖⁱ

    // normalize weights
    for (i = 1 to N) wₖⁱ = wₖⁱ/cₖ

    // resample
    if (depleted)
      for (i = 1 to N) resampled[i] = 0
      for (i = 1 to N)
      begin
          j ⟵ draw random integer j in 1...N with p(j) = wₖ₊₁ʲ
          resampled[j]++
```

```
    end
    for  (i = 1  to N)
    begin
        if  (resampled[i] == 0) stopSimulation(i)
        else for  (j = 2 to  resampled[i]) clone(i)
    end
end
//simulation step finished
k = k + 1
end
```

Listing 2.2 shows the particle filtering algorithms. The notation of variables is based on the description of particle filters given in [12]. The particle filter algorithm draws from a distribution p_{x_0} of initial system states N initial states x_0^i, $i = 1, \ldots, N$ and independently simulates these using discrete time simulation. In this case the system is the driver, who is simulated using the CASCaS driver model. For particle filters a weight w_k^i is needed for each simulation run (particle) that reflects the relative probability of the particle compared to the relative probability of all other particles. Weights are initialized equally for all particles and are continually updated. In each simulation step the weights are used to estimate the workload as a weighted sum over the workload estimations of all simulation runs. Then the current observations are used to continue the simulations. Afterwards the weights are updated based on the measurement probability $p(y_k^a | x_k^i)$ of observing the driver actions y_k^a given that the driver is currently in state x_k^i. This can be derived from the driver model as explained above. Furthermore a compensating factor $p(x_{k+1}^i | x_k^i)/q(x_{k+1}^i | x_k^i, y_{k+1})$ is multiplied to the weight that will be explained later. The weights are then normalized to reflect the relative probabilities of the simulation runs. Each loop iteration ends with an optional resampling step. Different strategies for resampling are possible. We propose to resample, when sample depletion occurs. This happens, when most of the weights are nearly zero [12] and can be measured by the effective number of samples [22]. Resampling is then done by drawing particles with replacement from the set of all particles (see [12]). The probability of selecting a particle is equal to its weight. The simulations of particles that are not selected are stopped, while simulations that are selected multiple times are cloned.

2.7.1 Proposal Distribution

Using just the observations of the environment y_k^e to simulate the next step with CASCaS as it is done in the simple approach (see Listing 2.1) might lead to states x_{k+1}^i that are not plausible with regard to the recent driver's actions y_{k+1}^a. Preventing the simulation to reach states that are not plausible makes the simulation more effective. This can be done by changing the probabilities of the internal state transitions. Instead of using the nominal state transition probabilities $p(x_{k+1}^i | x_k^i)$, we use a proposal distribution $q(x_{k+1}^i | x_k^i, y_{k+1})$ that optimizes the probabilities with regard to the observed actions of the driver (see [12] for details on the general approach).

However, this results in biased simulation runs and therefore biased workload esti-mates. To compensate for the biased a compensating factor is multiplied to the weights. A unique feature of CASCaS is that it already has a programming interface that allows to observe the state transition probabilities during runtime of a simulation and also to change it to guide the simulation. It was originally developed to observe simulation probabilities and guide the simulation onto rare but critical path [30].

2.8   Discussion

In the current experiment, we investigated the cognitive workload levels of drivers in a simulated driving environment. We manipulated the cognitive workload level by introducing an n-back task that is linked to the longitudinal control task. Using fNIRS as a physiological measurement of workload, we could show that the fNIRS data correlates well with the difficulty of the secondary workload task. As a comple-mentary measure of cognitive workload, we used the cognitive architecture CASCaS to develop a model of the driving task and the n-back task. As input data for the model, we used the data that was also available to the participants (current speed, etc.). By analysing the activity within the cognitive and associative layer of CASCaS infor-mation processing component, we were also able to achieve a correlation between the CASCaS model output and the fNIRS brain activation measurements similar to the correlation between the n-back task and HbR fNIRS data (Figs. 2.6 and 2.7).

However, the approach did not take into account the actions of the driver, which greatly limits the applicability of the approach. Therefore we described a concept for improving the simulation process, by using the driver model with a particle filter technique.

The presented work is only a first step. It needs to be validated in different driving scenarios with other measures of workload. We will do this in subsequent studies using brain activity measures based on fNIRS data. There are further issues that needs to be addressed in subsequent work.

We only looked at cognitive workload, but mentioned that other aspects like visual or motor workload might be linked to activities within other internal components of CASCaS. It is of great interest to analyse if such activities can be correlated with actual brain activity data in different brain areas. In fact, we aim to investigate whether brain activity measures like fNIRS and the presented model-based approach can complement each other.

Acknowledgements The authors thank their project partners from the German Aerospace Research Centre, Braunschweig for their technical assistance while performing the experiment. This work was supported by the funding initiative Niedersächsisches Vorab of the Volkswagen Foundation and the Ministry of Science and Culture of Lower Saxony as a part of the Interdisciplinary Research Centre on Critical Systems Engineering for Socio-Technical Systems.

References

1. Aasmann J (1995) Modelling driver behaviour in soar. Ph.D thesis, Rijksuniversiteit Groningen
2. Auflick JL (2015) Resurrecting driver workload metrics: a multivariate approach. Procedia Manuf 3:3160–3167
3. Byrne MD (2012) How many times should a stochastic model be run? An approach based on confidence intervals. In: Proceedings of 12th international conference on cognitive modeling (ICCM t'12), pp 445–450
4. da Silva FP (2014) Mental workload, task demand and driving performance: What relation? Procedia Soc Behav Sci 162:310–319
5. de Waard D (1996) The measurement of drivers' mental workload. PhD thesis, Rijksuniversiteit Groningen, Groningen, NL
6. Dudek B, Koniarek J (1995) The subjective rating scales for measurement of mental workload—Thurstonian scaling. Int J Occup Saf Ergon 1(2):118–129
7. Endsley MR, Kaber DB (1999) Level of automation effects on performance, situation awareness and workload in a dynamic control task. Ergonomics 42(3):462–492
8. Feuerstack S, Lüdtke A, Osterloh JP (2015) A tool for easing the cognitive analysis of design prototypes of aircraft cockpit instruments: the human efficiency evaluator. In: Proceedings of 33rd annual conference of the European Association of Cognitive Ergonomics (ECCE). ACM, Warsaw, Poland
9. Fitts PM, Posner MI (1973) Human performance. Prentice-Hall, Basic Concepts in Psychology
10. Gordon NJ, Salmond DJ, Smith AF (1993) A novel approach to nonlinear/non-Gaussian Bayesian state estimation. IEE Proc Radar Signal Process 140:107–113
11. Gore BF, Hooey BL, Wickens CD, Socash C, Gosakan M, Gacy M, Brehon M, Foyle DC (2011) Workload as a performance shaping factor in MIDAS v5. In: 2011 Conference presentation held at the behavioral representation in modeling and simulation (BRIMS)
12. Gustafsson F (2010) Particle filter theory and practice with positioning applications. IEEE Aerosp Electron Syst Mag 25(7):53–81
13. Hart SG, Staveland LE (1988) Development of NASA-TLX (task load index): Results of empirical and theoretical research. In: Hancock PA, Meshkati N (eds) Human Mental Workload, Advances in Psychology, vol 52. Elsevier Science. Amsterdam, North-Holland, pp 139–183
14. Hastie T, Friedman JH, Tibshirani R (2009) The elements of statistical learning. Data mining, inference, and prediction, 2nd edn. Springer
15. Hockey GRJ (1997) Compensatory control in the regulation of human performance under stress and high workload: a cognitive-energetical framework. Biol Psychol 45
16. ISO (2010) Road vehicles – ergonomic aspects of transport information and control systems – simulated lane change test to assess in-vehicle secondary task demand. Standard ISO 26022:2010, International Organization for Standardization, Geneva, CH
17. Juvina I, Taatgen N (2007) Modeling control strategies in the n-back task. In: Proceedings of eighth international conference on cognitive modeling (ICCM t'07), Taylor & Francis/Psychology Press, Oxford, UK, pp 73–78
18. Kahneman D (1973) Attention and Effort. Prentice-Hall, Englewood Cliffs, New Jersey
19. Kojima T, Tsunashima H, Shiozawa T, andTakuji Sakai HT, (2005) Measurement of train driver's brain activity by fNIRS. Opt Quantum Electron 37(13–15):1319–1338
20. Kramer AF (1991) Physiological metrics of mental workload: a review of recent progress. In: Damos DL (ed) Multiple Task Performance, Taylor & Francis, chap 11, pp 279–328
21. Lei S, Roetting M (2011) Influence of task combination on EEG spectrum modulation for driver workload estimation. Human Factors 53(2):168–179
22. Liu JS, Chen R (1998) Sequential monte carlo methods for dynamic systems. J Am Stat Assoc 93(443):1032–1044
23. Lüdtke A, Osterloh JP, Weber L, Wortelen B (2009) Modeling pilot and driver behavior for human error simulation. In: Duffy VG (ed) Digital human modeling. Lecture Notes in computer science, vol 5620/2009. Springer, Berlin, pp 403–412

24. Lüdtke A, Osterloh JP, Mioch T, Rister F, Looije R (2009) Cognitive modelling of pilot errors and error recovery in flight management tasks. In: Jean Vanderdonckt MW, Palanque P (eds) Proceedings of the 7th working conference on human error, safety and systems development systems development (HESSD), vol 1, LNCS 5962. Springer, pp 54–67
25. McCracken JH, Aldrich TB (1984) Analyses of selected LHX mission fucntions: Implications for operator workload and system automation goals. Technical Report, Technical Note ASI479-024-84, U.S. Army Research Institute Aviation Research and Development Activity, Fort Rucker, AL, USA
26. Nakayama O, Futami T, Nakamura T, Boer ER (1999) Development of a steering entropy method for evaluating driver workload. SAE Technical Paper Series: #1999-01-0892: Presented at the SAE International Congress and Exposition, Detroit, MI, USA
27. Nóbrega L, Nunes NJ, Coelho H (2006) Mapping concurTaskTrees into UML 2.0. In: Gilroy SW, Harrison MD (eds) Proceedings of the 12th international workshop on interactive systems design specification and verification, pp 237–248
28. O'Donnell RD, Eggemeier FT (1986) Workload assessment methodology. In: Boff KR, Kaufman L, Thomas JP (eds) Handbook of perception and human performance, vol II, Cognitive processes and performance. Wiley, New York, pp 41/1–42/49
29. Paternó F (2003) ConcurTaskTrees: an engineering notation for task models. In: Diaper D, Stanton NA (eds) The Handbook of task analysis for human-computer interaction, chap 24. Lawrence Erlbaum Associates, Mahwah, New Jersey, pp 483–503
30. Puch S, Wortelen B, Fränzle M, Peikenkamp T (2012) Using guided simulation to improve a model-based design process of complex human machine systems. In: Klumpp M (ed) Proceedings of the 2012 European simulation and modelling conference, EUROSIS-ETI, pp 159–164
31. Puch S, Wortelen B, Fränzle M, Peikenkamp T (2013) Evaluation of drivers interaction with assistant systems using criticality driven guided simulation. In: Duffy VG (ed) Digital human modeling and applications in health, safety, ergonomics and risk management. LNCS, vol 8025. Springer, pp 108–117
32. Rasmussen J (1983) Skills, rules, and knowledge; signals, signs, and symbols, and other distinctions in human performance models. IEEE Trans Syst Man Cybern 13(3):257–266
33. Reichert C, Fendrich R, Bernarding J, Tempelmann C, Hinrichs H, Rieger JW (2014) Online tracking of the contents of conscious perception using real-time fMRI. Front Neurosci 8(116)
34. Reid GB, Nygren TE (1988) The subjective workload assessment technique: a scaling procedure for measuring mental workload. In: Hancock PA, Meshkati N (eds) Human mental workload, advances in psychology, vol 52. Elsevier, pp 185–218
35. Salvucci DD (2006) Modeling driver behavior in a cognitive architecture. Hum Fact 48:362–380
36. Salvucci DD, Taatgen NA (2008) Threaded cognition: an integrated theory of concurrent multitasking. Psychol Rev 115(1):101–130
37. Sassaroli A, Fantini S (2004) Comment on the modified Beer-Lambert law for scattering media. Phys Med Biol 49:N255–N257
38. Solovey ET, Zec M, Perez EAG, Reimer B, Mehler B (2014) Classifying driver workload using physiological and driving performance data: two field studies. In: Proceedings of the SIGCHI conference on human factors in computing systems (CHIt'14). ACM, New York, NY, USA, pp 4057–4066
39. Tijerina L (1995) Final report–program executive summary: Heavy vehicle driver workload assessment. Tech. Rep. DOT HS 808 467, National Highway Traffic Safety Administration–Office of Crash Avoidance Research
40. Tomioka H, Yamagata B, Takahashi T, Yano M, Isomura AJ, Kobayashi H, Mimura M (2009) Detection of hypofrontality in drivers with Alzheimer's disease by fNIRS. Neurosci Lett 451(3):252–256
41. Tsimhoni O, Liu Y (2003) Modeling steering using the queueing network—model human processor (QN-MHP). In: Proceedings of the human factors and ergonomics society annual meeting, vol 47, no 16, pp 1875–1879

42. Unni A, Ihme K, Jipp M, Rieger JW (2017) Assessing the driver's current level of working memory load with high density functional near-infrared spectroscopy: a realistic driving simulator study. Front Hum Neurosci 11(167) (distributed under Creative Commons Attribution License (CC BY))

43. Unni A, Ihme K, Surm H, Weber L, Lüdtke A, Nicklas D, Jipp M, Rieger JW (2015) Brain activity measured with fNIRS for the prediction of cognitive workload. In: Proceedings of 6th IEEE conference on cognitive infocommunications (CogInfocom t'15). IEEE Press, pp 349–354

44. van Winsum W, Martens M, Herland L (1999) Effects of speech versus tactile driver support messages on workload, driver behaviour and user acceptance. Technical report TM-99-C043, TNO Human Factors Research Institute, Soesterberg, NL

45. Villringer A, Planck J, Hock C, Schleinkofer L, Dirnagl U (1993) Near infrared spectroscopy: a new tool to study haemodynamic changes during activation of brain function in human adults. Neurosci Lett 154(1–2):101–104

46. Weeda C, Zeilstra M (2013) Prediction of mental workload of monitoring tasks. In: Dadashi N, Scott A, Wilson JR, Mills A (eds) Rail human factors—supporting reliability, safety and cost reduction, chap 67. Taylor & Francis, pp 633–640

47. Wickens CD, Hollands JG (2000) Engineering psychology and human performance, 3rd edn. Prentice Hall

48. Wickens CD, Goh J, Helleberg J, Horrey WJ, Talleur DA (2003) Attentional models of multitask pilot performance using advanced display technology. Hum Fact 45(3):360–380

49. Wild-Wall N, Falkenstein M, Gajewski PD (2011) Age-related differences in working memory performance in a 2-back task. Front Psychol 2(186). https://doi.org/10.3389/fpsyg.2011.00186

50. Wood C, Torkkola K, Kundalkar S (2004) Using driver's speech to detect cognitive workload. In: Paper presented at 9th international conference on speech and computer (SPECOM 2004)

51. Wortelen B, Baumann M, Lüdtke A (2013a) Dynamic simulation and prediction of drivers' attention distribution. Transp Res Part F: Traffic Psychol Behav 21:278–294. https://doi.org/10.1016/j.trf.2013.09.019

52. Wortelen B, Lüdtke A, Baumann M (2013b) Integrated simulation of attention distribution and driving behavior. In: Kennedy WG, Amant RS, Reitter D (eds) Proceedings of the 22nd annual conference on behavior representation in modeling and simulation. BRIMS Society, Ottawa, Canada, pp 69–76

53. Wortelen B, Unni A, Rieger JW, Lüdtke A (2016) Towards the integration and evaluation of online workload measures in a cognitive architecture. In: Baranyi P (ed) Proceedings of 7th IEEE international conference on cognitive infocommunications (CogInfoCom t'16). IEEE, Wroclaw, Poland, pp 11–16

54. Wu C, Liu Y (2007) Queuing network modeling of driver workload and performance. IEEE Trans Intell Transp Syst 8(3):528–537

Chapter 3
Cognitive Data Visualization—A New Field with a Long History

Zsolt Győző Török⊙ and Ágoston Török⊙

Abstract Cognitive data visualization is a novel approach to data visualization which utilizes the knowledge of cartography, statistical data representation, neuroscience and ergonomic research to help the design of visualizations for the human cognitive system. In the current chapter, we revisit some benchmark results of research in cartography in the last half a millennium that shaped the ways how we think of and design visualizations today. This endeavor is unique since typical earlier reviews only assessed research in the past century. The advantage of our broader historical approach is that it not only puts cognitive data visualization in wider cultural context, but, at the same time, it calls attention to the importance of reconsidering the proceedings of earlier scholars as a crucial step in directing exploratory research today. In this chapter, we first review how conventions in data visualization evolved in time, then we discuss some current and pressing challenges in modern, cognitive data visualization.

3.1 Introduction

The methods of effective visual communication has been in the focus of research within various fields, from cartography to statistics in the past decades [1–3]. Numerous benchmark books and several papers endeavored to describe the principles of visual representation of information and authors have recommended practical hints

Z. G. Török (✉)
Department of Cartography and Geoinformatics, Faculty of Informatics,
Eötvös Loránd University, Budapest Pázmány Péter sétány 1/A, Budapest 1111, Hungary
e-mail: zoltorok@map.elte.hu

Á. Török
Systems and Control Laboratory, Institute for Computer Science and Control,
Hungarian Academy of Sciences, Budapest 1111, Hungary
e-mail: torok.agoston@sztaki.mta.hu

Á. Török
Brain Imaging Centre, Research Centre for Natural Sciences,
Hungarian Academy of Sciences, Budapest 1111, Hungary

© Springer International Publishing AG, part of Springer Nature 2019
R. Klempous et al. (eds.), *Cognitive Infocommunications, Theory and Applications*,
Topics in Intelligent Engineering and Informatics 13,
https://doi.org/10.1007/978-3-319-95996-2_3

how to assemble, analyze, and present collections of data. These principles, however, were rarely based on experimental knowledge about the human cognitive system. Although recently the accumulated knowledge about human memory and attention (how much information we can process), emotion (Kahnemann and Tversky) and perception (affordance, Gestalt principles) has begun to affect our views of the ideal way of visual infocommunication, there is still a lot to do. While data visualization has improved significantly, we need to reconsider traditional design principles and find new and effective methods to adapt to the new challenges of cognitive infocommunication [4, 5], namely virtual and augmented reality [6].

Nowadays, emerging technologies facilitate the need of revisiting and extending data visualization guidelines. As data dashboards, interactive visualizations, and mixed reality data displays are becoming widespread the guidelines developed for static, low dimension visualizations has to be adapted to help researchers and practitioners from various fields in creating effective visuals with new technologies. This endeavor is especially important since the price of generating massive amount of data is rapidly increasing, and visualization tools struggle hard to keep up with it [7].

Most books on visualization are dealing with the challenge if how to present the results of statistical or spatial analysis, however visualization actually serves various purposes:

- explore patterns, structures, relations in mass data
- present results of some analysis visually
- support human decision making with graphical interface.

Visualization is therefore a multifaceted tool. As such, forms of visualization does not only include graphs and charts, but also maps, dashboards and other interactive visualization types as well. However it is more than a simple graphical display of raw data. It always includes some kind of abstraction, either in the form of interpretation or simplification. From visualizations users can derive information and knowledge (Table 3.1).

In the following chapter, we first overview the history of graphical methods of visualization and revisit the historical material and the empirical results of practice that helped modern scholars in establishing the principles of data visualization. Thereafter we discuss the challenges raised by the emerging new technologies, and third we formulate an updated set of guidelines that can be applied to data visualization.

Table 3.1 Difference between data and visualization

Data	Visualization
Nonstructured	Structured
No communication purpose	Communication purpose
High resolution	Low resolution
Not interpreted, meaningless	Interpreted, meaningful
As is	Designed

3.2   History of Visualization as a Cognitive Tool

The ability to represent objects or concepts in the external world goes back to prehistory [8]. Rock carvings or cave paintings from different parts of the world demonstrate the important development steps in the human cognitive system. The fundamental issue here is the appearance of the ability to represent something with meaning, reflecting the intention of the human subject. In other words, the external representation and the internal representation (i.e. thinking) had to be linked somehow in one system, material culture. Although our knowledge about the beginning of the use of external representation as a cognitive tool is rather limited, the earliest examples of identifiable, 'meaningful' images were presumably created by *Homo sapiens* more than 12,000 years ago. According to the theory of cultural evolution by Donald [9], this was a decisive invention in human history and meant the dawn of material culture and the end of the mimetic and the episodic eras.

3.2.1   Visualization as a Form of Externalized Memory

From the beginning external representations were projections of the human mind. In particular, they served as an expansion of biological memory: their appearance made it possible for early societies to accumulate and transfer knowledge. It is important to note that, once materialized, the constituents of any graphic bear also spatial attributes. The graphic space of a representation becomes part of the system which is inherently spatial. This is why *mapping* or, in its more developed form, map making is considered here as archetype of *any visualization*. However, one should keep in mind that not only have been maps simply used as visualization tools by highly diverse human cultures for thousands of years. Working with diagrammatic space had tremendous effects on human minds in the course of history leading to modern societies, where map use is common in spatial problem solving. The effective instruments of visualization have made humans able to explore, beyond their geographical environment, large, inaccessible or complex sets of objects, phenomena or even abstract concepts.

Drawing on the definition proposed by Harley and Woodward [10], maps can be interpreted as graphic designs that facilitate a spatial understanding of things, concepts, conditions, processes and events in the human world. This functional approach is in contrast with the professional definitions focusing on the form, structure and content of the modern map. Although these may describe the most important, contemporary types of maps, it is historically misleading to apply modern criteria to all kind of mapping. Indeed, as a social practice mapping was never a monolithic enterprise—how it is suggested by traditional stories about developments of cartography [11]. On the other hand, mapping was always based on the relations between the external representation space and its graphic objects on one side, and also on the internal processes of the human mind on the other.

3.2.2   The First Map

Although the pictorial form of some modern maps may suggest so, historical evidence shows that early maps were highly abstract graphic representations. The first maps made by humans were presumably ephemeral and has not survived. The Mesopotamian clay tablet from c. 2300 BC is generally considered to be the earliest uncontested. It demonstrates how advanced cartographic principles had been applied by the unknown maker of that more than four thousands years old instrument. As the cuneiform script reveals, its direct function was to depict the location of a land property in the geographical context of ancient Mesopotamia. The graphic language, the symbols for water flow, mountain range or settlement are, although abstract in form, easily recognizable for a modern reader who can understand geographical concepts from this very early artifact. The same clay tablet exemplify another remarkable characteristics of cartographic visualization: although a property map the two-dimensional representation is oriented according to the cardinal directions, which were originally marked on the sides. In other words, the first map survived is strong evidence that a universal, geographical reference frame has been used by human civilizations for more than four thousands years.

3.3   History of Graphic Methods

Although map making looks back to a long history, as a regular and systematic social practice it is closely related to the early modern period. In the past five hundred years graphic representation became a common tool to visualize highly complex systems and solve various problems using maps, charts, plans and diagrams. Medieval maps demonstrate the transfer of geographic information and knowledge about methods of data visualization [12]. However, the most common type of world maps, the circular maps also evidences for the importance of social-cultural aspect of visualization.

3.3.1   Early Forms of Infographics

The usually small size and diagrammatic depictions of the division of land and water according to the Bible clearly display the geographical arrangement of the known continents and their spatial relations. Though, the simplicity of the graphic design may be misleading for the modern reader, who is not familiar in Christian symbolism. But one must note that the waters dividing the land form the letter 'T' (*terra*), here also a symbol for the Crucifixion. It is placed in the middle of the circle of the ocean, making a letter 'O' standing for the Latin word '*orbis*', world. Simply drawing these two letters, arranging them this way in the graphical space, medieval scribes would not only draw an initial with map of the world: they also told the reader the whole story

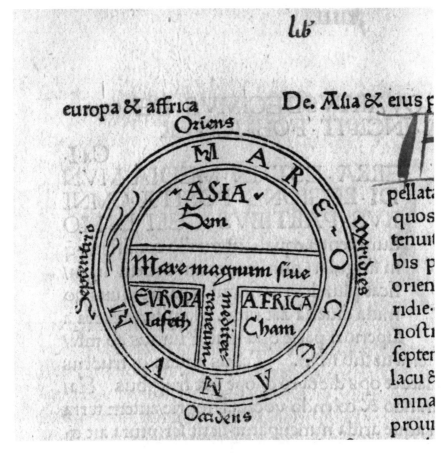

Fig. 3.1 The first printed map: a diagrammatic representation of the Christian Universe (Augsburg, 1472)

of the creation of a Christian Universe. The abstract letters interpreted as a diagram, as a representation in a graphical space, convey much information about the spatial structure, the geography of the world. If it was made today, it was considered an infographics rather than a real map (Fig. 3.1).

3.3.2 Visualization Before Conventional Signs

Signs and symbols on the map, in the graphic space, are located in a spatial reference system and this is the fundamental advantage of graphic representation over verbal or written description. On early modern maps one can find great variety of pictorial and abstract signs, which makes interpretation sometimes difficult. Although the

Fig. 3.2 Wilhelm Crome's economic map of Europe with symbols

signs standing for the same object are similar, each map maker could use its own version. The lack of convention in the sign systems used is a striking characteristics of early modern maps. 16–17th century maps often represent information which is not directly visible in the field and stands for the quality of objects. Signs could be referred to points, but could be distributed to represent the spread of the same type of qualitative information, usually a category (e.g. forest). The distribution of languages on each continents on Gottfried Henschel (1741) or the pioneer 'geognostic' maps by Jean Etienne Guettard demonstrated how early visualization could make invisible, inaccessible objects or phenomena visible and easy to comprehend [13]. By the 18th century scientific research or statistical surveys resulted in large collections of data, including both qualitative and quantitative information.

In 1782 the German economist and statistician Wilhelm Crome published a map of Europe (Fig. 3.2), showing the major products of the countries by signs and letters. A few years later Crome produced a series of comparative diagrams [14] showing the size and population of states in Europe in graphic form. In 1818 he published his

Fig. 3.3 The legend on Korabinszky's pioneer economic map of the Kingdom of Hungary (1791)

first economic map in a new edition with additional pie charts, another contemporary graphic invention.

The first thematic map of a country, Johannes Korabinszky's 1791 map of Hungary, depicted national economy by using 92 different signs and symbols. The information was taken by the author from his own geographical-economic lexicon (1786), an early collection of economical and statistical data (Fig. 3.3). The visualization of the some 15,000 entries, represented by the miniature signs on the map, although neither spectacular not very effective, was highly appreciated by the rational minds of contemporary scholars. Korabinszky's pioneer thematic map was not only used by the traveler and naturalist Robert Townson in 1797, but he added the mineralogical information he collected in the field in 1793 to the map in a new layer. This is a remarkably early example of multiple and interactive visualization.

3.3.3 Coordinates, Charts, Diagrams

Similarly to the geographic coordinate system, which has been in use already in the Antiquity, the graphic visualization of phenomena in a planar coordinate system was already known in the Middle Ages. The 14th century Italian mathematician, Nicole

Oresme explained concepts like time, velocity, distance in a graphical way, using simple *graphs* . The method of modern analytical geometry was introduced by René Descartes in 1637 [15].

In 1765 Joseph Priestly published 'A Chart of Biography' [16], which represented temporal data: the dates of birth and death of important persons were connected to create a *stick chart*. The representation of statistical data by *diagrams* was the novelty of William Playfair's 'The Commercial and Political Atlas' [17] (Fig. 3.4). Although it was called an atlas, the economic data was represented not in maps but by the method called 'linear arithmetic', which meant graphs. Playfair's books were published in different editions and these publications made the methods of statistical data visualization, graphs and diagrams, available for scholars of the 19th century.

3.3.4   The Emergence of Isolines

Edmond Halley is generally considered by historians of cartography as the first thematic cartographer, especially because of his highly influential charts showing the variations of the compass [18]. The novelty of these graphic representations was not only the representations of magnetic declination, but also helped finding the geographical position of ships in the oceans. Halley, who collected magnetic data during his journey in the Atlantic, selected to show the variation of the magnetic compass by 'curve lines', running across points with equal declination (Fig. 3.5). As the magnetic field is a *continuum*, the graphic invention of the *isoline*, a line connecting points of equal value, made map makers able to represent all kind of continua. This technique is now widely used in other fields too, for example in the form of contour plots.

Modern thematic maps similarly represent the spatial distribution of objects or phenomena in the geographical or abstract spaces. For spatial reference a background map is needed and the map theme is layered above that. These types of maps are in contrast both conceptually and graphically with the general map, which shows the spatial location of a set of geographical objects (settlements, rivers etc.) and serves orientation and navigation. One of the most important followers of Halley was Alexander Humboldt, who published his treatise in 1817 with a diagrammatic chart showing the global distribution of temperatures by using lines of equal value, *isotherms* [19] (Fig. 3.6).

3.3.5   Flow Lines

Halley's first thematic map, his 1688 wind chart, also brought a remarkable novelty. The *dynamic* phenomena was symbolized by small strokes. As the author explained, these represented the ship sailing with the wind behind, the direction of the wind was shown indirectly by the narrowing end of the lines. This was a remarkable

THE

COMMERCIAL AND POLITICAL

A T L A S,

Reprefenting, by Means of ·

STAINED COPPER-PLATE CHARTS,

THE

PROGRESS OF THE COMMERCE, REVENUES, EXPENDITURE,

AND DEBTS OF ENGLAND,

DURING THE WHOLE OF THE

EIGHTEENTH CENTURY.

THE THIRD EDITION,

Correĉed and brought down to the End of laft Year.

By WILLIAM PLAYFAIR.

Printed by T. Burton, Little Queen-ftreet, Lincoln's-Inn Fields,

FOR J. WALLIS, NO. 46, PATERNOSTER-ROW; CARPENTER AND CO. BOND-
STREET; EGERTON, WHITEHALL; VERNOR AND HOOD, POULTRY;
BLACK AND PARRY, LEADENHALL-STREET.

1801.

Fig. 3.4   The title page of Playfair's statistical atlas (1801)

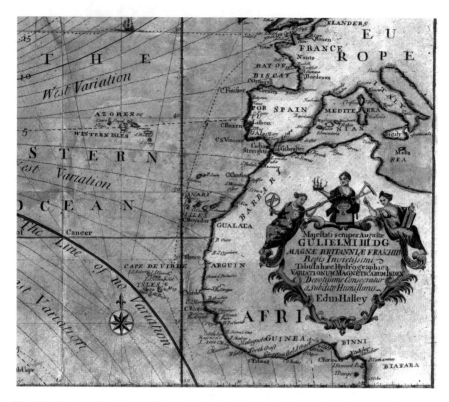

Fig. 3.5 Detail of Edmond Halley's isogonic chart of the Atlantic Ocean: note the 'line of no variation'

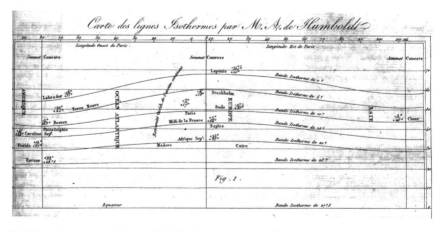

Fig. 3.6 Alexander von Humboldt 1817 chart showing the distribution of average annual temperature

invention to show movement on a static map by *flow lines*. Navigational charts with the direction of the winds appeared regularly in the 18th century, usually applying arrows, of which a few appeared on Halley's chart as well. Another early example designed by the Jesuit scholar Athanasisus Kircher (1665), or the German novelist, Eberhard Happel's 1685 chart displayed the water circulation of the world oceans with streamlines, but, curiously, without direction arrows.

In the 19th century the graphic representations of social and economic activities could use detailed statistical data bases in France, where Charles Joseph Minard produced a series of highly inventive flow charts. He was interested in international economic relations and the movements of a wide range of subjects from people to products. Of his maps, '*carte figuratives*' Minard published some ten thousand copies from the 1850s and his graphic methods became known by a wider public [20].

3.3.6 From the Depth of the Sea to Population Density

However, one must note that isolines appeared much earlier, already in the 16th century to show the depth of water in a river in the Netherlands (Bruins 1584). After this pioneer visualization it took almost two centuries to expand the scope of the method from rivers and seas to continental areas. After important publications with isobaths (e.g. Marsigli 1725, Buache 1752 referenced in [2, 21] in 1782 Bonifas Marcellin du Carla proposed the general use of isolines to show the physical surface of the Earth. To represent the *contour lines* of equal depth or height, however, cartographers needed mass qualitative data about relief. Unfortunately, before remote sensing to measure altitude accurately and economically was cumbersome and costly. In 1791 the French geographer Dupain-Triel published the first map of a country with a few contour lines based on barometric measurements and calculations. A few years later, in 1798–99, to enhance the graphic he added different shades to layers between his contours, following the principle 'the higher the darker', and produced the first layer-tinted map. The first map with hypsometric coloring represented a region in northern Hungary (today Slovakia), and was constructed by the Swedish botanist and explorer, Wahlenberg in 1813. These traditions triggered the use of several color-map schemes in contemporary visualization.

A milestone in the history of data visualization, more specifically thematic cartography, was published in the form of a systematic collection of thematic maps by Heinrich Berghaus in 1838–48 [22]. The sheets of the '*Physikalischer Atlas*' were based on the concept suggested earlier by Humboldt, and they were intended to illustrate his ambitious physical description of the world (Fig. 3.7) five volumes of his '*Kosmos*' (1845–62). To portray meteorological, geophysical phenomena, but also plant geography or anthropogeography a wide variety of data visualization techniques were used by the designers of the maps, including isolines, diagrams and graphs. *To demonstrate quantitative data by lines of equal value, based on the interpolation*

Fig. 3.7 Choropleths and isolines combined in a complex thematic map in Berghaus' atlas (1840)

between localized observations or measurements points became a common place by the mid-19th century. To depict more abstract, e.g. statistical phenomena was the next step. This was proposed by Lalanne in France [23], who extended the ideas of du Carla and Humboldt to statistical data, which was considered as a third dimension superimposed on the general map. In 1857 the naval officer Ravn published such a map [24], showing the density of the population in Denmark by *pseudo-isolines* and using colour tints for his 500 person per square mile intervals. As the result of the evolution of the graphic methods of data visualization a highly effective new method was created, the *isopleth*. Although graphically similar, the *choropleth* represent data related to pre-defined regions or areas (e.g. nodes belonging to the same cluster) whereas *isopleth* is based on quantitative data located to points (e.g. degree of a nodes).

3.3.7 Visualization for the Public: Geographical and Abstract Spaces

By the mid-19th century both the physical surface of the Earth and abstract, statistical surfaces could be represented cartographically, using the same graphic methods, isolines and layer tinting. Despite their simple visual appearance these representations were based on rather sophisticated concepts. For example hypsometric relief representation was used only at small scales because it required a precise measure of altitudes, which before air photogrammetry was cumbersome. Thematic cartography also depended heavily on reproduction methods. The introduction of a new printing method, lithography, made graphical reproduction faster and cheaper in the early 19th century. Another advantage, chromo-lithography offered color printing, and this technical invention had great impact on the distribution of data visualizations in public media.

From the mid-19th century international conferences advanced the emerging discipline of statistics [25]. The graphical methods of data representation were explored and discussed in Wien in 1857, where the maps of Josef Bermann or Carl Czoernig were also displayed. The connection between topographical and thematic map making is best exemplified by the participation of Franz Hauslab, a military cartographer, who proposed the principle 'the higher is the darker' for hypsometric representation

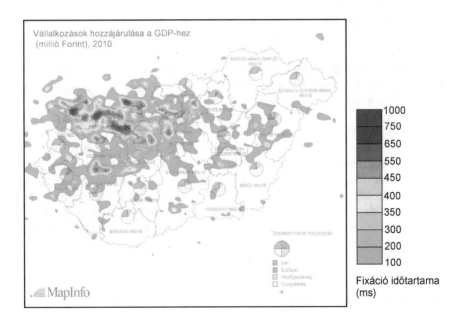

Fig. 3.8 Eyetrack data visualized with hypsometric color scheme (so called 'natural colors')

of the relief (Fig. 3.8). Later special committees regularly reported to the plenary session about graphical methodology issues, first of all categorization and the use of colors.

3.3.8   Cool Heatmaps and Cognitive Issues

In 1885 Émile Levasseur [26] proposed to demonstrate deviation of mean values: red for categories above and blue for categories below the average, giving birth to the jet colormap. This approach had its roots in the antique traditions of cartography, warmer colors were used for land surface and cooler, usually blue and green, colors for waters. This tradition and later convention demonstrates also how real world inspiration drove color choices in visualization. As a reflection to scientific advancements thematic maps were included in general atlases, showing e.g. the global distribution of temperature. Already Humboldt suggested to show summer and winter average temperature (isotheres and isochimenes).

From the meteorological data a statistical surface was created, and the intervals between the *isotherms* were colored according to a legend (i.e. colorbar). These were the real heatmaps, and for the associative use of color, anybody would understand which parts of the world were the warmest or the coldest. Similar maps appeared in school atlases everywhere and the graphical presentation methods of scientists became part of the pictorial language of everybody. The heatmap in statistical data visualization represents the structure of a data matrix and goes back to the 19th century [27].

Considering the great popularity of 'heatmap' we should clearly better understand not only the historical roots of the concept. Although it is well known that modern 'heatmaps' are actually density maps, it is questionable how much about the complexity of the visualization is understood by non-professionals. What is actually represented in the heatmaps used by visualizations of eye tracking data? How would people understand aggregated or dynamically displayed fixations? How different color schemes influence the communicated message? Some of these questions had been addressed in cartography (e.g. the eye tracking study of graphical potential of different GIS softwares by [28]) or data visualization, but many of the graphical principles used today are still based on tradition and have never been seriously evaluated.

This is why a systematic research on the usability of graphical visualization methods is an important and immediate task for the field of cognitive data visualization.

3.4   New Challenges for Data Visualization

For a long time, data visualization has been constrained to two dimensions and was static in format. Both traditional, analogue media (e.g. paper) and the computer screen are primarily suitable for visualizing data in two spatial dimensions. This

Reversed hot heatmap   Jet heatmap

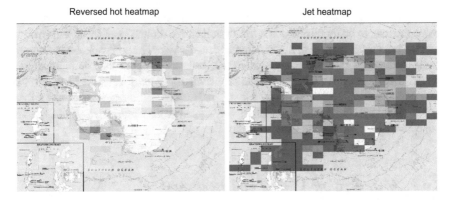

Fig. 3.9 Visualizing eyetracking data using a reversed hot colormap and the frequently used jet heatmap

does not necessarily mean that more dimensions cannot be represented here, it is adequate to say, however, that visualization in higher dimensions require steps of abstraction, both from the author's and from the reader's side. For example, as was described earlier, heatmaps are used in several domains today. In the simplest case, the heatmap uses colors to represent a third dimension. While this is a very useful feature, it works best if the color dimension denotes a qualitatively different measure. For example, if we want to visualize the average annual temperature in a country, the x and y axes should denote location in space, and an added color space should be used to demonstrate temperature. While this is a rather straightforward example, often we visualize data where all dimensions are different, and in this case the author has to make a decision as to which dimension to select for color or attribute coding (Fig. 3.9).

Another challenge with heatmaps is the proper selection of color. Some colors have conventional associations. However, associations such as blue is 'cold' and red is 'hot' are not innate, but develop through cultural influence [26, 29]. This also means that while it is easy to think that all color associations are universal, they are possibly not. For example, associations for red as sign to stop and green to go are particularly strong in western culture, but they are not in eastern cultures, e.g. in China [30]. One should also note that colormaps were often created to depict some graphical resemblance to their signaled quality. Blue for lower values was motivated by the color of water, green as the middle was motivated by grass, whereas yellow by the sun in the jet colormap. This also means for other types of visualization the colormap can and sometimes should be adapted also to the denoted quality.

There are ways to visualize three dimensional structures in two dimensions through projections. Mesh plots, contour plots and surface plots are the most frequent of these, but their use is usually not preferred because due to the nature of projection some parts of the image are not visible unless rotated, which option is rarely available for traditional data visualizations. This is not an issue when representing curved surfaces (e.g. the Earth), how we make maps, but issues may emerge when the data to be represented is relevant in its three dimensional form. This means that the reader has limited options to investigate the visualization which is communicated by the author. The static nature of visualizations on paper and computer screen makes them ineffective when it comes to visualizing high dimensional data. In the following section, we propose how data visualization can be extended to more than two dimensions. In the end we explain why dynamic and interactive visualizations are essential for human intelligence today.

3.4.1 Visualization Above Two Dimensions

The perception of the world around us is essentially three dimensional. The human visual system developed to render the three dimensional information of the environment in the mind. This not only means that we perceive depth information despite the two dimensional nature of the optical image on the retina, but, more importantly, our perceptual system has adapted to the challenges of the physical world. So to speak, the framework of embodied cognition [31] claims that the cognitive system is inseparable from the body [32] and the environment [33]. This evolutionary developed fit between our cognition and the environment makes us able to cope with the vast amount of information reaching our senses at any given moment [31] and quickly react to new information in the environment [34].

3.4.2 Ultra-Rapid Visual Categorization

Embodied cognition is the reason why processing information presented in forms that are not present in nature takes more time and are not straightforward to interpret. Oddly, this suggests that carefully designed two dimensional graphs may take more time to process than a more natural three dimensional scene. This notion is supported by the results of several studies investigating ultra-rapid visual categorization [35–39]. These studies consistently find that complex natural scenes displayed for milliseconds can be categorized under 150 ms as it is revealed by both EEG evidence [38] and saccadic reaction times [39]. Further studies showed that people can process even multiple scenes in parallel with this speed [37], which means that no directed attention is required. Furthermore, ultra-rapid categorization of complex natural scenes is not only highly automatic but is not affected by the familiarity of the exact pictures [36]. Thus, this phenomenon clearly indicates that the visual system is adapted to the complexity of the visual world; consequently, the natural-unnatural dimension is far more important in perception than the simple-complex one.

3.4.3   Multisensory Effects on Visual Perception

Another corollary of the embodied nature of the human cognitive system is that we perceive through all of our senses, and sensory modalities can facilitate each other. Such multisensory enhancements causes decreased reaction times and better performance for multisensory stimuli [40, 41].

The most prominent multisensory phenomena are the visual capture of sounds in the spatial domain, known as ventriloquism [42]; and the auditory capture of visual stimuli in the temporal domain, known as the illusory-flash effect [43] . The ventriloquism illusion is so strong that this is actually the reason why the cinema experience is so natural: our eyes easily makes us believe that the sounds are coming from the mouth of the actor and not from the speakers [44]. The illusory-flash effect is a little less trivial. In the typical experimental situation one flash is presented with two short beep sounds, of which one is concurrent with the flash. The resulting percept is two beeps *with two flashes*. These results show sensory stimulation in multiple modalities interact and shape the final percept. Nevertheless, in both of these cases we were aware that auditory and visual stimulation was also present.

In data visualization these factors may not take a significant role since we usually design visuals and not synchronized stimuli in another modality. However, this is only partly true. Curiously enough, multisensory effects are present also in situations where one would not expect them. There is actually one organ of sense people usually forget about—despite being the most fundamental percept in life. This is the vestibular sense and its contribution to the perception of up and down directions. Our primary senses, eyes, ears, nose, skin, and tongue are all easily observed and have been studied since ancient times. The vestibular sense, however, is located in the inner ear and was discovered only in the beginning of the twentieth century by von Bárány [45]. This is responsible for our sense of balance [46] and contributes to bodily awareness [47]. Studies investigating the neural underpinnings of vestibular sensation found that although there are areas dedicated to vestibular processing, vestibular afferents reach several areas throughout the cortex [48]. Therefore, despite being often subconscious, the vestibular sensation modulates the perceptual processes in other sensory modalities.

One striking example of this is the interaction between visual and vestibular sensation in visual distance perception [34, 49–51]. These studies show that the same visual distance is perceived differently depending on the position of the body [49, 50], the head [34] and the eyes [51]. Things above the horizon seem afar while things below that seem closer. There are reasons to believe that the direction of the effect is in connection with perceived effort [52], but is present also when no effort is included in the task [34]. The effect is also nonlinear: experiments dealing with unnavigable angles (90°) found that because of fear of falling the effect reverses for these extreme angles. From the scope of the current review, the relevance of these results is that the size and layout of a visualization may easily distort the perceived differences between two figures. Since visual distance is inferred from the perceived size and known real size of the object [53] one can easily deduce, that any change in the perceived distance of the same object means change also in the perceived size-since the known size cannot change.

The vestibular perception of gravity affects visual information also on another level. Difference in the speed of motion of an object is differentiated more accurately when the motion is consistent with gravity [54]. Also, even memory for gravity consistent motion is biased [55]. This is most easily seen when in an experiment the participant is required to show the location where an object has disappeared. They consistently find that people show below the location where the object actually disappeared when the motion was consistent with gravity.

The relevance of these effects to data visualization is emphasized for map-like dynamic visualizations. As North is traditionally associated with up and South is associated with down in cartography, this cultural convention shapes our perception of the world. Although we may think it was always so, before the early modern age different orientations were used in cartography. This may have been related to human values maps always presented. Not only size differs on the vertical axis, other studies showed that "up" is associated with good, profit, and higher altitude, whereas "down" is associated with bad, prices, and lower altitude [56–59]. The down-up visual axis is also associated with hierarchy and development. Furthermore, our memory of the world map is biased in the location of the home continent, which is usually remembered larger than actually. Also, Europe is remembered as being larger while Africa as being smaller than its actual size [60]. The strength of the verticality effect can be easily seen if we look at a map where South is associated with up (see Fig. 3.10). Little known is the fact that, although orientation to the North goes back to the mathematical astronomical tradition of geography represented by Ptolemy in the 2nd century AD, until the early modern age maps were oriented to various other directions. Medieval Christian cosmographic diagrams had 'Oriens' at the top (and hence the word 'orientation'), while Islamic cartography adopted South as the primary direction. Even in early modern Europe, after the rediscovery and adoption of the Ptolemaic method appeared maps with other orientations. A famous example is the series of south-oriented, anthrophomorphic maps from the 16th century, representing Europe as a Queen (see Fig. 3.10)

3.4.4   Visualizing in Three Dimensions

There are two areas where three dimensional visualization is especially helpful and already in common use. These fields are architectural design and medical imaging. In architectural design computer generated renderings are used for presentation, marketing, and design purposes. Here three dimensional, more realistic visualization greatly supersedes the use of two dimensional plans. These virtual copies are often used for simulating different light and environmental conditions, and panoramic and renovation effects [61]. As abstract and symbolic representations of the real world architectural renderings are often considered outside of the traditional scope of data visualization [3], but they should be included as data visualizations. Moreover, the

Fig. 3.10   Queen Europe as a map oriented to the West (Sebastian Münster, 1588)

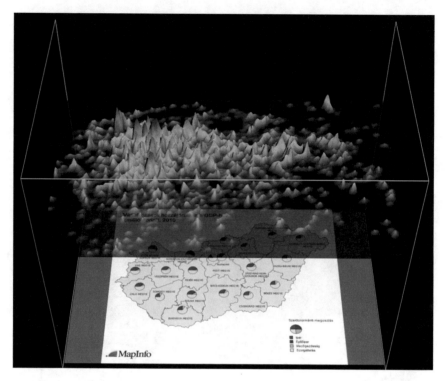

Fig. 3.11 Eyetracking data visualized in three dimensions by Török and Bérces [28]

practical experience gained in the planning of three dimensional rendering softwares is already valuable for other areas, where three dimensional visualization has just begun to emerge (Fig. 3.11).

The other field where three dimensional visualization is already widespread is medical imaging. Medical imaging techniques such as computer tomography (CT), structural and functional magnetic resonance imaging (sMRI and fMRI), and magnetoencephalography (MEG), along with other methods, are available in medical practice for decades. The ultimate aim of these tools to help diagnosis by providing spatial information of lesions or other alterations of tissue. These images are not only used to describe the medical situation but also to prescribe surgery. For example, pharmacologically intractable epilepsy patients undergo surgery based on the MRI + ECoG localization of epileptic foci [62]. Since in these situations millimeters of mislocalization means potential harm to well-functioning brain tissue, the visualization of MR images is of great importance. Medical doctors have been using softwares like SPM [63] to analyze and visualize magnetic resonance imaging data. While these tools are excellent in correcting artifacts and reconstructing image from the original frequency domain information (see more in [64]), they are generally not outstanding when it comes to visualization. Luckily, in recent years more and more tools became available for medical staff to utilize virtual reality to visualize medical images [65]. Visualizing tissues and organs in three dimensions not only lets the viewer to freely observe the surroundings of the given locus, but these views can be easily collaboratively shared with other practitioners. Nevertheless, the spread of mixed reality is still limited by the some factors. First, while head mounted displays are available on the consumer market medical doctors may utilize more augmented interfaces, which can be more easily used for collaboration [66]. Of course, virtual reality and realistic models are not only used for diagnosis but also for teaching and practicing purposes [67–69].

3.4.5   Using Non-Euclidean Spaces

The reason why three dimensional visualization easily spreads in engineering design and medical imaging is that in both cases the actual data of which we are gaining insight is three dimensional. Therefore, using three dimensions for the representation means no significant spatial information loss. Unfortunately, this is not true for higher dimensional problems. Take, for example, a graph created from the co-occurrence matrix of a paper (see Fig. 3.12). The dimensionality of a graph is the least n such that there exists a representation of the graph in the Euclidean space of n dimensions where the vertices are not overlapping and edges are of unit length [71]. This number can easily go very high as the number of edges increase, actually the upper limit of n for graph G is twice its maximal degree plus one [72]. Because of this (and also because of the computational complexity of identifying dimensionality) common graph representations often use not unit length edges. For example, a widespread graph representation—the spring layout—uses physical simulations by assigning forces to edges and their endpoint nodes. This way the resulting layout shows more interconnected regions being closer and less connected vertices are pulled to the extremities of the available space (see also on Fig. 18.2). In these kinds of representations—since the actual physical position of a vertex is not meaningful without the connected vertices—two dimensional embedding of the layout is usually preferred since adding a third dimension would only add another item to the arbitrary position vector.

However, information that may not easily be represented in Euclidean space can still easily be processed by the human brain. The simplest example for this is our social network of friends. If we need to visualize the relationship even just a cluster of our friends we will be in trouble: the information does not fit easily to the two dimensional paper or a three dimensional virtual space. Nevertheless, we can easily 'navigate' between these people because our cognitive map does not need to conform necessarily the norms of the Euclidean space. This means we can conveniently utilize walls/borders, routes, shortcuts, and even subspaces. Shortcuts are probably the most interesting of these since they not even need to be physically possible shortcuts. Studies show that people easily learn to navigate in space with teleportation wormholes [70, 73, 74]. The fact that these can be processed means also that we can design environments where we purposefully place such things. That is we can

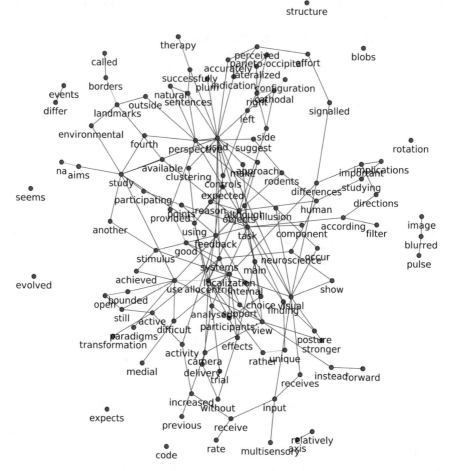

Fig. 3.12 A graph created from the bigrams mined in one previous work [70] of one author of the current review. For presentation purposes only a randomly chosen fraction of the nodes and edges are displayed

visualize graphs like impossible yet interpret figures by defining such shortcuts [6]. Similar, artificial memory spaces have been used by ancient Greeks and other cultures to store large amount of information in memory, also known as the 'method of loci' [75].

Interestingly, these graphic tools differ from those ones we consider conventionally as visualizations in one key factor, which is perspective. While traditional visualizations are viewed from an external perspective, the above mentioned mental visualizations are viewed from an embedded perspective. The difference between these two perspectives is even more pronounced in the brain. Embedded perspective is associated with egocentric reference frame use, while external perspective is associated with allocentric reference frame use [76]. Furthermore, there are two core geometric systems in the brain. One is responsible for analyzing two dimensional forms from an external perspective (e.g. studying a map) and the other is responsible for navigating three dimensional environments from an internal viewpoint (e.g. actual navigation). Studies have found that neither of these two core geometric systems is able to represent correctly all of the fundamental properties of Euclidean geometry, which are distance, angle and directional relationships [77]. Studies showed that from the external perspective length/distance and angle information are correctly identified but shapes are easily mistaken for their mirrored versions. In turn, during navigation length information and direction are parsed easily, but angles are not well remembered. Therefore, changing the perspective in visualization is not only a matter of aesthetics, but requires a cognitive reframing of information.

There are also drawbacks of the non-Euclidean properties of the cognitive map for the representation of three dimensional information. Unlike teleportation, three dimensional rotational movements are proven to be difficult for humans [78]. This is not surprising since spatial perception is essentially a multisensory process where the vertical axis remains the most basic spatial knowledge for humans [34], even if views can be visually similar in any direction. In fact, representation of three dimensional space has only been verified in bats [79, 80]. Bats are flying animals, and they use echolocation as their primary distal sensory system. Importantly, the activity of the hippocampal formation in bats does not exhibit oscillatory activity in the theta band, which, in turn, is an essential functional correlate in both rodents and humans [80]. Therefore, the spatial representation in bats is different from that in rats [81] and presumably from that in humans, too. Thus, although some nervous systems have developed to deal with three dimensional navigation, the human brain has not.

3.4.6 The Niche for Interactive Visualizations

Most researchers would agree that, although dynamic and interactive visualizations may look impressive, they are often not more than useless 'eye-candy'. Many open source (plot.ly, shiny) and proprietary projects (Microsoft PowerBI, Tableau) offer solutions for more interactive visualization, so it may become even more widespread in the near future. In the current section we introduce some examples where dynamic visualizations are favored over static ones, and they can facilitate the better communication of insights.

With dynamic and interactive visualizations authors have to communicate information in a generally interesting way to call users' attention. For example, when someone wants to tell how house renting and buying expenses are related, he or she may need to use several separate graphs to display the factors contributing to costs and proceeds. This kind of visualization is easily skipped by most viewers since the information conveyed—despite being relevant—is too complex. However, if the authors can tailor the message for the actual viewer it will reach its goal easily. This was what Bostock et al. [82] did in their interactive visualizations published in the on-line edition of The New York Times. Here the reader is invited to adjust sliders on the specific factors to reach a conclusion at the end if renting or buying pays off for his/her *specific* case. Therefore, interactive visualization is sometimes useful: when understanding the structure in high dimensional data would require large effort from the reader's side. It can help increasing the incentive value of the visualization and motivate readers to engage in the understanding of the image. Nevertheless, this also means that not the exact same message will be delivered to each reader, thus the variance in the message has to be considered when designing the visualization and interpretations.

In sum, mixed reality brought visualization new challenges. The ability to visualize data in three physical dimensions is sometimes useful (e.g. medical diagnostics), but oftentimes does not contribute to better understanding. However, mixed reality is not only capable of visualizing data in three dimensions but makes us able to place or project a visualization anywhere, not only on computer screens. Embedded and situated visualizations could be easier to understand since the surrounding environment could provide us fundamental context for the interpretation. These visualizations will quickly become widespread—as soon as affordable augmented reality headsets are entering the consumer market [83].

Nevertheless, especially with embedded visualizations one has always to consider that, although visual modality plays a pivotal role in human perception, the process is still affected by other sensory modalities as well, e.g. the vestibular system. Finally, mixed reality can help us to visualize structures that are hard to understand in pictures, but these relations are readily processed once the perspective is not out of the visualization but is internal. Good examples are graphs and other high dimensional structures that are visualized in non-Euclidean ways.

3.5 Summary

Cognitive data visualization is a novel approach to data visualization focusing on the strengths and weaknesses of the human mind in knowledge acquisition. Especially in cases beyond the capacity of human senses our working memory we rely upon external memory tools as projections of the human mind. The graphic representation, mapping in its most general sense, creates spaces of data and information which are open to visual and mental exploration and navigation. As a process analogue to similar activities in real world, physical or geographical spaces, visualization is inherently a visuospatial process resulting in the recognition of relations, patterns or structures in images.

Data visualization has a long history starting with the first spatial representations in ancient times. After the pioneer thematic maps in the early modern age systematic data collection increased in the Enlightenment period, and resulted in new forms of visual knowledge. Graphic data representation methods developed rapidly in the 19th century, when the traditional graphical methods were practically all invented and tested in a great variety, in masses of statistical graphs, diagrams and thematic maps produced and distributed in all societies around the world. Visualization tools were reproduced by lithographic and offset printing and became common not only scientific research but also in popular culture. A good example of this development is the appearance of isothermal charts in school atlases which laid the foundations of the recent popularity of heatmaps. By the 1980 s, when visualization became computer graphics, the traditional methods were so deeply integrated in modern culture that their effectivity was rarely questioned. Only in the new millennium, when new visualization methods in new environments (e.g. virtual and augmented reality, network spaces and big data etc.) became more and more important in human computer interaction, became cognitive issues of data visualization seriously considered.

As it is apparent from recent research issues *visualization have vital importance in future human-computer interaction (HCI)*, where the rapid development of artificial intelligence urgently requires more effective interfaces than the obsolete existing ones. Here plays the human visual mind a key role: with new visualizations developed on empirical research on human cognitive processes the interaction with information and spaces, interactively generated by AI, can be more effective. Based on neuropsychological research findings *cognitive design* can already effectively influence pre-attentive visual processes. However, as we emphasize here, human vision is a product of both *biological and cultural* evolution. Modern researchers can not only learn from the empirical knowledge cumulated by traditional methods, but it is necessary to better know the cultural traditions and history of visualization.

Acknowledgements This multidisciplinary project was supported by a grant from ELTE Tehetséggondozási Tanácsa, Talent Supporting Council, Eötvös Loránd University, Budapest. We would like to thank the comments of Gergely Dobos and Eszter Somos on the first draft of the chapter.

References

1. Bertin J (1983) Semiology of graphics: diagrams, networks, maps. Translated by Berg WJ. University of Wisconsin Press (in French 1967)
2. Robinson AH (1952) The look of maps. University of Wisconsin Press, Madison
3. Tufte E (1991) Envisioning information. Optom Vis Sci 68(4):322–324. https://doi.org/10.1097/00006324-199104000-00013
4. Baranyi P, Csapó Á (2012) Definition and synergies of cognitive infocommunications. Acta Polytech Hungarica 9(1):67–83
5. Baranyi P, Csapó Á, Sallai G (2015) Cognitive infocommunications (CogInfoCom), 1st edn. Springer International Publishing, Cham
6. Török Á (2016) Spatial perception and cognition, insights from virtual reality experiments. Budapest, Doktori disszertäciö
7. Fox P, Hendler J (2011) Changing the equation on scientific data visualization. Science 331(6018):705–708. https://doi.org/10.1126/science.1197654
8. Liebenberg E, Collier P, Török, ZG (Eds.). (2014) History of cartography: lecture notes in geoinformation and cartography. Springer, Berlin, Heidelberg. https://doi.org/10.1007/978-3-642-33317-0
9. Donald M (2001) A mind so rare: the evolution of human consciousness. WW Norton & Company, New York

10. Harley J., Woodward D. (1987). Cartography in prehistoric, ancient, and medieval Europe and the Mediterranean. The History of Cartography(1). Chicago & London. The University of Chicago Press

11. Török ZG (1993) Social context: the selected main theoretical issues facing cartography an ICA report. Cartographica 30(4):9–11

12. Török, ZG (2007) Renaissance cartography in East-Central Europe. The History of Cartography(3) 14501650. Chicago & London. The University of Chicago Press. http://www.press. uchicago.edu/books/HOC/HOC_V3_Pt2/HOC_VOLUME3_Part2_chapter61.pdf

13. Török Z (2007) Die Geschichte der thematischen Kartographie im Karpatenbecken unter besonderer Berücksichtung der ungarischen geowissenschaftlichen Karten. Nova Acta Leopoldina 94(349):25–48

14. Crome AFW (1785) On the greatness and population of all the European states: an agreement to the understanding of the conditions of the states, and the explanation of the new map of Europe; with an illuminated map. Weygand's book

15. Descartes R (1960) Discours de la méthode. Oeuvres de Descartes, 6, Paris. Hachette

16. Priestley J (1803) A chart of biography. M. Carey

17. Playfair W (1786) Commercial and political atlas: representing, by copper-plate charts, the progress of the commerce, revenues, expenditure, and debts of England, during the whole of the eighteenth century. London, Corry

18. Thrower NJ (1969) Edmond Halley as a thematic geographer. Ann Assoc Am Geograph 59(4):652–676

19. von Humboldt A (1817) Des lignes isothermes et de la distribution de la chaleur sur le globe. Perronneau

20. Minard CJ (1861) Des tableaux graphiques et des cartes figuratives. Thunot et Cie, Paris

21. Török Z (2006) Luigi Ferdinando Marsigli (1658–1730) and early thematic mapping in the history of cartography. Térképtudományi tanulmányok = Studia cartologica 13:403–412. http:// lazarus.elte.hu/hun/digkonyv/sc/sc13/52zsolt_torok.pdf

22. Berghaus H (1848) Physical atlas. William Blackwood & Sons, London

23. Lalanne L (1843) Un Million de Faits

24. Ravn NF (1857) Populations Kaart over det Danske Monarki 1845 og 1855. Statistiske Tabelværk, Ny Række, Bind, p 12

25. Houvenaghel G (1990) The first International Conference on Oceanography (Brussels, 1853). German J Hydrograph 22:330–336

26. Levasseur É (1885) La statistique graphique. J Statistic Soc London 218–250

27. Loua T (1873) Atlas statistique de la population de Paris. Paris. J, Dejey

28. Török Zs.Gy., Bérces Á. (2014). 10 Bucks eye tracking experiments: the Hungarian MapReader. In: CartoCon: conference proceedings. Olomouc: Palacky University Press. https://cogvis. icaci.org/pdf/icc2013/Torok.pdf

29. Morgan GA, Goodson FE, Jones T (1975) Age differences in the associations between felt temperatures and color choices. Am J Psychol 125–130. https://doi.org/10.2307/1421671

30. Courtney AJ (1986) Chinese population stereotypes: color associations. Human Fact 28(1):97-99. https://doi.org/10.1177/001872088602800111

31. Haselager P, van Dijk J, van Rooij I (2008) A lazy brain? Embodied embedded cognition and cognitive neuroscience. Handbook Cogn Sci Embodied Appr 5:273–287

32. Proffitt DR (2006) Embodied perception and the economy of action. Perspect Psychol Sci 1(2):110122. https://doi.org/10.1111/j.1745-6916.2006.00008.x

33. Gibson EJ, Walk RD (1960) The visual cliff. WH Freeman Company & Co, San Francisco, CA, US

34. Török Á, Ferrè E, Kokkinara E, Csépe V, Swapp D, Haggard P (2017) Up, down, near, far: an online vestibular contribution to distance judgement. PLoS ONE 12(1):e0169990. https://doi. org/10.1371/journal.pone.0169990

35. Besson G, Barragan-Jason G, Thorpe SJ, Fabre-Thorpe M, Puma S, Ceccaldi M, Barbeau EJ (2017) From face processing to face recognition: comparing three different processing levels. Cognition 158:3343. https://doi.org/10.1016/j.cognition.2016.10.004

36. Fabre-Thorpe M, Delorme A, Marlot C, Thorpe S (2001) A limit to the speed of processing in ultra-rapid visual categorization of novel natural scenes. J Cogn Neurosci 13(2):171180. https://doi.org/10.1162/089892901564234
37. Rousselet GA, Fabre-Thorpe M, Thorpe SJ (2002) Parallel processing in high-level categorization of natural images. Nature Neurosci 5:629630. https://doi.org/10.1038/nn866
38. Thorpe S, Fize D, Marlot C (1996) Speed of processing in the human visual system. Nature 381(6582):520522. https://doi.org/10.1038/381520a0
39. Wu C-T, Crouzet SM, Thorpe SJ, Fabre-Thorpe M (2014) At 120 ms you can spot the animal but you dont yet know its a dog. J Cogn Neurosci 27(1)
40. Alais D, Burr D (2004) The ventriloquist effect results from near-optimal bimodal integration. Curr Biol 14(3):25762. https://doi.org/10.1016/j.cub.2004.01.029
41. Senkowski D, Saint-Amour D, Höfle M, Foxe J (2011) Multisensory interactions in early evoked brain activity follow the principle of inverse effectiveness. Neuroimage 56(4):2200–2208. https://doi.org/10.1016/j.neuroimage.2011.03.075
42. Howard IP, Templeton WB (1966) Human spatial orientation, New York, Wiley. http://www.amazon.com/Human-Spatial-Orientation-Ian-Howard/dp/0471416622
43. Shams L, Kamitani Y, Shimojo S (2000) Illusions: what you see is what you hear. Nature 408(6814):788. https://doi.org/10.1038/35048669
44. Török Á, Mestre D, Honbolygó F, Mallet P, Pergandi JM, Csépe V (2015) It sounds real when you see it. Realistic sound source simulation in multimodal virtual environments. J Multimodal User Interf 9(4):323–331. https://doi.org/10.1007/s12193-015-0185-4
45. Barany R (1906) Untersuchungen über den vom Vestibularapparat des Ohres reflektorisch ausgelösten rhythmischen Nystagmus und seine Begleiterscheinungen. Oscar Coblentz, Berlin
46. Cullen K (2012) The vestibular system: multimodal integration and encoding of self-motion for motor control. Trends Neurosci 35(3):185–196. https://doi.org/10.1016/j.tins.2011.12.001
47. Ferré ER, Vagnoni E, Haggard P (2013) Vestibular contributions to bodily awareness. Neuropsychologia 51(8):1445–1452. https://doi.org/10.1016/j.neuropsychologia.2013.04.006
48. Guldin W, Grüsser O (1998) Is there a vestibular cortex? Trends Neurosci 21(6):254–259. https://doi.org/10.1016/s0166-2236(97)01211-3
49. Di Cesare S, Sarlegna C, Bourdin F, Mestre CD, Bringoux L (2014) Combined influence of visual scene and body tilt on arm pointing movements: gravity matters! PLoS ONE 9(6):e99866. https://doi.org/10.1371/journal.pone.0099866
50. Harris L, Mander C (2014) Perceived distance depends on the orientation of both the body and the visual environment. J Vision 14(12):17–17. https://doi.org/10.1167/14.12.17
51. Ooi T, Wu B, He Z (2001) Distance determined by the angular declination below the horizon. Nature 414(6860):197–200. https://doi.org/10.1038/35102562
52. Bhalla M, Proffitt D (1999) Visual-motor recalibration in geographical slant perception. J Experiment Psychol Human Percept Perform 25(4):1076–1096. https://doi.org/10.1037//0096-1523.25.4.1076
53. Nakamizo S, Imamura M (2004) Verification of Emmert's law in actual and virtual environments. J Physiol Anthropol Appl Human Sci 23(6):325–329. https://doi.org/10.2114/jpa.23.325
54. Moscatelli A, Lacquaniti F (2011) The weight of time: Gravitational force enhances discrimination of visual motion duration. Journal Of Vision 11(4):5–5. https://doi.org/10.1167/11.4.5
55. De Sá Teixeira N (2016) The visual representations of motion and of gravity are functionally independent: evidence of a differential effect of smooth pursuit eye movements. Experiment Brain Res 234(9):2491–2504. https://doi.org/10.1007/s00221-016-4654-0
56. Meier B, Robinson M (2004) Why the sunny side is up: associations between affect and vertical position. Psychol Sci 15(4):243–247. https://doi.org/10.1111/j.0956-7976.2004.00659.x
57. Meier B, Moller A, Chen J, Riemer-Peltz M (2011) Spatial metaphor and real estate. Soc Psychol Personal Sci 2(5):547–553. https://doi.org/10.1177/1948550611401042
58. Montoro P, Contreras M, Elosúa M, Marmolejo-Ramos F (2015) Cross-modal metaphorical mapping of spoken emotion words onto vertical space. Front Psychol 6. https://doi.org/10.3389/fpsyg.2015.01205

59. Nelson L, Simmons J (2009) On Southbound ease and northbound fees: literal consequences of the metaphoric link between vertical position and cardinal direction. J Market Res 46(6):715–724. https://doi.org/10.1509/jmkr.46.6.715
60. Saarinen T, Parton M, Billberg R (1996) Relative size of continents on world sketch maps. Cartographica Int J Geogr Informat Geovisual 33(2):37–48. https://doi.org/10.3138/f981-783n-123m-446r
61. Novitski BJ (1998) Rendering real and imagined buildings: the art of computer modeling from the Palace of Kublai Khan to Le Corbusiers villas. Rockport Publishers, Glouster
62. Nadasdy Z, Nguyen TP, Török Á, Shen JY, Briggs DE, Modur PN, Buchanan RJ (2017) Context-dependent spatially periodic activity in the human entorhinal cortex. Proc Nat Acad Sci 114(17):E3516–E3525
63. Friston K, Holmes A, Worsley K, Poline J, Frith C, Frackowiak R (1994) Statistical parametric maps in functional imaging: a general linear approach. Human Brain Map 2(4):189–210. https://doi.org/10.1002/hbm.460020402
64. Huettel SA, Song AW, McCarthy G (2004) Functional magnetic resonance imaging. Sinauer, Sunderland MA
65. King F, Jayender J, Bhagavatula S, Shyn P, Pieper S, Kapur T et al (2016) An immersive virtual reality environment for diagnostic imaging. J Med Robot Res 01(01):1640003. https://doi.org/10.1142/s2424905x16400031
66. Zhang S, Demiralp C, Keefe D, DaSilva M, Laidlaw D, Greenberg B et al (2001)An immersive virtual environment for DT-MRI volume visualization applications: a case study. Proceedings Visualization, VIS '01., 437–583. https://doi.org/10.1109/visual.2001.964545
67. Alaraj A, Luciano C, Bailey D, Elsenousi A, Roitberg B, Bernardo A et al (2015) Virtual reality cerebral aneurysm clipping simulation with real-time haptic feedback. Neurosurgery 11(0–2):52–58. https://doi.org/10.1227/neu.0000000000000583
68. Barsom E, Graafland M, Schijven M (2016) Systematic review on the effectiveness of augmented reality applications in medical training. Surg Endosc 30(10):4174–4183. https://doi.org/10.1007/s00464-016-4800-6
69. Nigicser I, Szabó B, Jaksa L, Nagy Á D, Garamvölgyi T, Barcza Sz, Galambos P, Heidegger T (2016) Anatomically relevant pelvic phantom for surgical simulation. In: 2016 7th IEEE international conference on cognitive infocommunications (CogInfoCom), Wroclaw, pp 427–432. https://doi.org/10.1109/CogInfoCom.2016.7804587
70. Török Á, Kóbor A, Persa G, et al. (2017). Temporal dynamics of object location processing in allocentric reference frame. Psychophysiology 1–13. https://doi.org/10.1111/psyp.12886
71. Erdös P, Harary F, Tutte W (1965) On the dimension of a graph. Mathematika 12(02):118. https://doi.org/10.1112/s0025579300005222
72. Erdös P, Simonovits M (1980) On the chromatic number of geometric graphs. Ars Combinator 9:229–246
73. Schnapp B, Warren W (2010) Wormholes in virtual reality: what spatial knowledge is learned for navigation? J Vis 7(9):758–758. https://doi.org/10.1167/7.9.758
74. Vass L, Copara M, Seyal M, Shahlaie K, Farias S, Shen P, Ekstrom A (2016) Oscillations go the distance: low-frequency human hippocampal oscillations code spatial distance in the absence of sensory cues during teleportation. Neuron 89(6):1180–1186. https://doi.org/10.1016/j.neuron.2016.01.045
75. Verhaeghen P, Marcoen A (1996) On the mechanisms of plasticity in young and older adults after instruction in the method of loci: Evidence for an amplification model. Psychol Aging 11(1):164–178. https://doi.org/10.1037//0882-7974.11.1.164
76. Török Á, Nguyen T, Kolozsvári O, Buchanan R, Nádasdy Z (2014) Reference frames in virtual spatial navigation are viewpoint dependent. Front Human Neurosci 8:1–17. https://doi.org/10.3389/fnhum.2014.00646
77. Spelke E, Lee S, Izard V (2010) Beyond Core Knowledge: Natural Geometry. Cognitive Science 34(5):863–884. https://doi.org/10.1111/j.1551-6709.2010.01110.x
78. Peters RA (1969) Dynamics of the vestibular system and their relation to motion perception, spatial disorientation, and illusions. NASA CR 1309

79. Finkelstein A, Derdikman D, Rubin A, Foerster J, Las L, Ulanovsky N (2014) Three-dimensional head-direction coding in the bat brain. Nature 517(7533):159–164. https://doi.org/10.1038/nature14031

80. Yartsev M, Ulanovsky N (2013) Representation of three-dimensional space in the hippocampus of flying bats. Science 340(6130):367–372. https://doi.org/10.1126/science.1235338

81. Geva-Sagiv M, Las L, Yovel Y, Ulanovsky N (2015) Spatial cognition in bats and rats: from sensory acquisition to multiscale maps and navigation. Nature Rev Neurosci 16(2):94–108. https://doi.org/10.1038/nrn3888

82. Bostock M, Carter S, Tse A (2014) Is it better to rent or buy?. The New York Times

83. Weldon M (2015) The future X network: a bell labs perspective (1st ed.). CRC Press

84. Di Cesare SC, Sarlegna F, Bourdin C, Mestre D, Bringoux L (2014) Combined influence of visual scene and body tilt on arm pointing movements: gravity matters!. PLoS ONE 9(6):e99866. https://doi.org/10.1371/journal.pone.0099866

Chapter 4
Executive Functions and Personality from a Systemic-Ecological Perspective

A Quantitative Analysis of the Interactions Among Emotional, Cognitive and Aware Dimensions of Personality

Raffaele Sperandeo, Mauro Maldonato, Enrico Moretto and Silvia Dell'Orco

Abstract This paper intends to clarify the complexity of the concept of Executive Functions and to show its relationships with cognitive, emotional and awareness dimensions. In this sense, Executive Functions cannot be considered preconditions of normal physiological functioning, when they are undamaged or, on the contrary, the cause of pathological phenomena, when they are in deficit. Although the small size of our sample does not allow us to explore completely this hypothesis, our data show that there is probably no direct relationship between Executive Functions deficiencies and psychopathological manifestations. In the light of these empirical evidences and theoretical implications, it is plausible to move from a linear interpretative model to a circular explanatory hypothesis.

R. Sperandeo (✉) · M. Maldonato · S. Dell'Orco
Department of Human Sciences, DISU University of Basilicata, Potenza, Italy
e-mail: raffaele.sperandeo@gmail.com

M. Maldonato
e-mail: mauro.maldonato@unibas.it

S. Dell'Orco
e-mail: silvia.dellorco@unibas.it

E. Moretto
SiPGI Postgraduate School of Integrated Gestalt Psychotherapy,
Torre Annunziata, Naples, Italy
e-mail: enrico.more@gmail.com

© Springer International Publishing AG, part of Springer Nature 2019
R. Klempous et al. (eds.), *Cognitive Infocommunications, Theory and Applications*,
Topics in Intelligent Engineering and Informatics 13,
https://doi.org/10.1007/978-3-319-95996-2_4

4.1   Theoretical Framework

A fundamental aspect of the mind is the allocation of cognitive resources and the modulation of affective impulses in relation to personal goals and to environmental, social and relational requirements. The Executive Functions (EF) expression refers to these regulatory mechanisms that allow people to organize creatively complex patterns of behaviour and thought. So the EF do not correspond to distinguished affective and cognitive processes, but control these processes when environmental situations cannot be faced on the basis of schemes already learned by the subjects [1]. Executive control is necessary to tackle new tasks, to formulate goals and plans and finally to choose, execute and correct behaviour sequences oriented towards an aim. The EF are conscious and aware processes and allow to reinterpret the past and to plan the future; to stop inappropriate automatic behaviours; to direct, allocate and support attention strategically [2].

4.2   Executive Functions and Neurobiological Systems

The Executive Functions are considered allocated in the frontal cortex, mainly in the dorsolateral cortex that is connected to the orbitofrontal cortex, to the cingulate gyrus and to parietal lobes. They adapt the thought and the behaviour to environmental circumstances by the information processing, and they harmonize the needs of the subject with reality, by modulating affectivity. The orbitofrontal cortex seems to be the neurobiological basis of pre-verbal consciousness that regulates impulses, attention and affectivity, while the dorsolateral cortex appears to be the support of awareness, able to formulate strategies and planning voluntary actions [3]. This neuronal circuit, characterized by high levels of plasticity, is the last to mature, evolves throughout life, ensures the efficiency of selective attention, of decision making and of conflict resolution [4]. This circuit is connected to the cranial nerve nuclei that produce dopamine and norepinephrine and to the motivational nuclei of the basal ganglia; the dopaminergic circuits activate the search for pleasure, the noradrenergic circuits inhibit it, allowing the adjustment of the alert and attention [5]. There is a general consent on the presence of two attention systems: a low one, based on reflections and an evolutional one, based on awareness [6]. The circuit is connected with the motor systems through which organizes behavioural responses and receives inputs from the limbic system, by which it attributes emotional values to experience; this set of connections allows the reading of other people's social intentions and an efficient emotional and behavioural adjustment [7].

4.3 Executive Functions and Personality

The executive functions have an adaptation task. This explains their central role in the relationship life but makes evident the difficulty in approaching the study of executive functions with the classic neuropsychological approach based on the symptom [8]. Contrary to what happens in the study of classic cognitive functions, it is not easy to establish that a deficit in an executive task corresponds to a pathological condition, both for the transverse variability and instability in the time of the deficiency, and for the intrinsic difficulties to systems of evaluation of the EF (eg the construct validity problems of tests and task impurity of the same EF) so the measures are never specific and pure [9]. In our research, the first of the problems described, that is, the instability and variability of the disorder, is the most important one: a person may fall into some trials and not in others related to the same FE and also he may have variables performances to the same test in a short time away; finally, some patients may fail a task in an evaluative clinical context but not in everyday life situations and vice versa [10]. These findings can be interpreted as intrinsic problems to the evaluation of executive functions, but more acutely as distortions caused by the theoretical reference model on which the selection of patients to be tested is based. A clear example of this complex aspect to understand, is the automatic use behaviour described by Lhermitte [11]: the presentation of a common object leads, automatically, the patient to take it and use it. According to Goldberg [12], the phenomenon can be interpreted as a "motor release" based on the imbalance between the activity of the medial motor (volunteers) and lateral (automatic) systems. This takes place in severe impairment of the nervous system conditions (as in SMA) but also in environmental stress and imaginative ab-sorption situations. Finally, paradoxically, the automatic motor behaviour emerges in conditions in which a need for environmental hypercontrol is involved; in this case the person grasps the examiner's hands who touches the palm of his hands (task planned in Frontal Assessment Battery FAB) and he asks to understand the task to be performed because he does not tolerate to participate in an activity whose meaning he doesn't understand [13]. A correct interpretation of these facts requires that the EF are not seen as independent processes but as interrelated phenomena with the personality and the environment, which can be studied properly in a systemic and ecological vision [14]. The following study is proposed as a systemic and ecological approach to executive functions, which are interpreted within the narrow circular dynamic with personality traits and not as neuropsychological autonomous processes from whose dysfunction pathological phenomena could result.

4.4 Introduction to the Study

Executive functions (EF), coherent with recent studies, can be defined as a group of specific cognitive processes that organize emotions, motor activities, and are associated with the ability to fix, organize, and adjust behaviors which are focused on achieving a specific goal [15]. The fact that they are not confined solely to the frontal areas, thanks to the extensive connections with virtually all other regions of the brain [16], they integrate information from all the other areas of the brain with which they are connected. EFs are fundamental to various psychological skills such as controlling and regulating emotions, social commitment and starting from this assumption, many neuropsychological researches link them to emotional development and pro-social behaviors [17]. Further evidence of the close connection between executive functions and personality development comes from studies of personality disorders. Specifically, it is possible to identify the temperamental basis of self-regulation [18] and to note that traumatic experiences [19, 20] and personality disorders [21] have a significant impact on memory, direction of attention, and ability to control emotions. Most studies that relate executive functions to personality have been conducted on people with personality disorders, but the evidence of the prevalence of prefrontal areas involved in self-regulation of environmental responses [22] allows to generalize and identify the presence of these processes throughout the population. Psychological disorders are distributed in a continuum, that is, they are described in continuity with the adaptive personality, requiring a dimensional rather than a categorical evaluation, which is, on the other hand, a kind of approach that tends to polarize disturbances and aspects of the personality according to a dichotomous system (extroversion vs. introversion, emotional stability versus emotional disregulation) [23], moving away from the complexity present in real clinical practice. Among the various dimensional approaches, Cloninger's Temperament and Character Inventory (TCI) model describes the personality as a complex adaptive system supported by seven major dimensions that interact with each other in a composite system that self-organizes and self-regulates itself, to then find behavioral patterns that allow to address all the internal and external demands that are required during the life cycle [24]. This hierarchical system is formed by the interaction of the two distinct psychobiological dimensions of temperament and character. Temperament describes the individual differences of unplanned response to heterogeneous environmental stimuli (such as danger, novelty, frustration) and include response patterns that involve behaviours which derive from primary emotions [25, 26] such as fear, aggression, sociability, and attachment [27, 28]. Temperament can therefore be conceptualized as the biological core of personality, as it reflects the biological variability of personality that remains relatively stable throughout the course of life, since it derives from hereditary biological characteristics that are the basis of the neurobiological systems of activation/inhibition of specific behaviors. In this model, temperament is divided into four dimensions: the novelty seeking, avoidance, reward dependence and persistence, evaluated on a bipolar continuum that goes from the low to high expression of a behavioral trait. Characteristical dimensions, on the other hand, are considered to be socio-culturally determined, non-linear functions that mature with time in a non-gradual way, through the relationship between genetic features and social and relational learning. The Character is defined by 3 dimensions that include the Self-Directedness dimension, Cooperativeness and Self-Transcendence. Temperament involves the basic sensory processes, while character regards the higher cognitive functions, they are therefore functionally and developmentally correlated, but develop independently through differentiated processes. Research programs such as those dealing with "affective computing" can benefit from research that correlates personalities and executive functions. Affective computing is based on the assumption that it is possible to create machines that can perceive, influence, and show emotions or other human-like affective phenomena [29]. This type of research is aimed at increasing the human affective experience through new technologies, at the same time increasing knowledge of the correlation between affectivity and physical and mental health, placing itself as a bridge of conjunction between the study of I.A., psychology and biomedical engineering. From this point of view, it is possible to describe cerebral and personological characteristics as embodied processes that emerge from the bi-directional interaction that occurs between an organism and the environment. Pathological development follows the same self-defined organizational logic described for the normal personality, even though it is influenced by other elements, such as pathogenic environments, which have an impact on the individual's adaptation skills. Ultimately, it is possible to say that the most relevant organismic functions are the temperamental and characterial traits because of which character derives from temperament and at the same time intervenes by modifying the significance and relevance of sensory perceptions, which, in turn, are involved in a complex interaction with brain and with interpersonal relationships.

4.4.1 Aim of the Study

This ecological vision allows to see how fundamental the relationship between executive functions and the individual-environment adaptive system is and how this process is essential in temperament and character structure. It therefore seems crucial to estimate the quantitative link between specific executive functions and personality components according to the seven-factor Cloninger model. Given the neuroplasticity of the neural networks of the prefrontal systems, where they do not have pathological psychological characteristics, they meet the individual's bio-psycho-social specifications and participate in establishing that individual's experience of by defining his personality. This factor, which is, the result of the continuous circularity between the environment and the individual, appears again in the definition of his temperament and character components and resumes the circle again, converting it into an activation experience in the environment. The aim of our study is to explore, in a sample of psychiatric outpatients, the complex network of links between the components involved in individual-environment adaptive processes and to trace, as far as possible, predictive models between the areas of temperament, character, and executive functions.

Table 4.1 Academic status, civil status and career status of the subjects in the sample

Status	Frequency	Percentage (%)
Elementary	2	1.4
Junior high school	19	13.7
High school	19	13.7
University graduate	37	26.6
Single	80	58.0
Married	47	34.1
Separated	7	5.1
Widow/Widower	4	2.9
Unemployed	50	35.2
Employed	91	64.1
Retired	1	0.7

4.4.2 Methods

The sample

Since it is common knowledge that subjects with borderline personality disorders frequently present deficiencies in the executive functions, subjects who had this type of disorder at the beginning of the treatment were excluded from the study. In addition, subjects with psychotic disorders, cognitive deficits, brain degenerative processes and a history of cranial traumas were also excluded. Furthermore, all the patients included in the study requested psychotherapy for non-clinical problems. All subjects were informed and agreed that the data collected for clinical evaluations would be used in a scientific study in compliance with the laws on anonymity and the confidentiality of sensitive information. The sample consists of 142 subjects with an average age of 34 years (SD = 12.48) with a minimum of 17 and a maximum of 66 years. The socio-demographic characteristics of the sample are described in Table 4.1.

4.4.3 Instruments

In this sample of outpatients admitted to a mental health clinic in the years 2015–2016, executive functions were evaluated using the frontal assessment battery (FAB), while the character temperament traits were evaluated using Cloninges's temperament and character (TCI). The FAB is a short test that evaluates frontal functions extensively. It consists of 6 subtests that assess the following executive functions [30, 31]:

- Conceptualization (explored through a similarity test);
- Mental flexibility (explored with a phonemic fluency test);

- Motor programming (explored through the Luria series);
- Susceptibility to interference (assessed by testing with conflicting instructions);
- The Inhibitory control (assessed with a go-no-go test);
- Autonomy from the environment (assessed by grasping behavior).

The scores of the individual subtests are between 0 and 3, where 3 is the value that expresses the proper functioning of the full efficiency of the single function. In addition, you can get a total score for age and education level with a cut-off of 13.5.

The TCI is a self-administered questionnaire of 240 questions [26] that measure 4 dimensions of temperament and 3 of character. The main functions of the TCI are described in Table 4.2.

Table 4.2 Dimension of temperament and character

Temperament	Character
NS-40 items The first dimension of the temperament is novelty seeking: high scores on this dimension are characteristic of persons with a tendency to explore, prodiga to be extravagant, impulsive and irritable; low scores characterize subjects who are uncommunicative, conscientious, frugal, and stoic	SD-44 items The first character dimension is Self-Directedness: higher scores indicate maturity, strength, responsibility, reliability, constructive, good individual integration, self-esteem and self-confidence; lower scores indicate immaturity, fragility, tendency to blame others, unreliability and poor ability to integrate
HA-35 items The second dimension of the temperament is harm avoidance	C-42 items The second character dimension is Cooperativeness: high scores characterize empathetic subjects, tolerant, compassionate, supportive and loyal; low scores characterize self-centered subjects, intolerant, opportunistic and vindictive
RD-24 items The third dimension of temperament is reward dependence	ST-33 items The third character dimension is Self-Transcendence: high scores describe individuals who are patient, happy, spiritual, selfless and unassuming, able to tolerate uncertainty; low scores describen individuals who are proud, arrogant, impatient, materialistic and dissatisfied
P-8 items The fourth dimension of temperament is persistence: higher scores indicate diligence and reliability, despite fatigue and frustration; low scores describe people who give up easily	

4.4.4 Statistical Analysis

Descriptive analyzes were performed on categorical and dimensional variables. The FAB subscales have been transformed into dichotomous variables in which subjects with scores below 3 were assigned a value of 0, while subjects with value 3 were assigned a value 1. Finally, a logistic regression was performed to obtain a model explanatory of the effect of the character temperament dimension on the 6 subtests of the FAB. In the logistic regression test, the TCI temperament dimensions were used as explanatory variables, while the 6 tasks of the FAB, converted into dichotomous variables, were used as dependent variables.

4.4.5 Results and Discussion

Although the subjects of the sample do not show pathological conditions typically related to the deficiency of the executive functions, many of them have FAB's tasks characterized by reduced efficiency as shown in Table 4.3.

This fact points out that there is no specific relationship between a deficiency in the executive functions and psychic symptoms, rather that ineffective executive functions are common among subjects without psychic or neuropsychological disorders and can be attributed at least to some extent to the impact on the frontal areas of normal dimensions of temperament and character. In support of this hypothesis the evidence in Tables 4.4 and 4.5 show the existence of small, but significant causal relationships between the dimension of the personality and the executive functions. Specifically, Table 4.4 highlights how environmental autonomy reduces efficiency in people with high scores in novelty seeking and harm avoidance, and becomes more efficient as scores increase in the reward dependence dimension. In contrast, interference sensitivity is less efficient in subjects with high scores in the reward dependence dimension. Finally, the efficiency of mental flexibility increases as persistence and novelty seeking grow, and it decreases as the scores in harm avoidance increase.

Table 4.3 Distribution of FAB's tasks

Tasks	Subject with deficits tasks N (%)	Subject with efficient tasks N (%)
Conceptualization	98 (69)	44 (31)
Mental flexibility	57 (40)	85 (60)
Motor programming	36 (25.4)	106 (74.6)
University graduate	37	26.6
Sensibility to interference	34 (23.9)	108 (76.1)
Inibitory control	52 (36.6)	90 (63.4)
Autonomy from environment	9 (6.3)	133 (93.7)

Table 4.4 Logistic regression: temperament versus tasks of FAB

Dependent variables	Explicative variables	B	Sign	Exp (B)	95% C.I. Exp (B)	R2 di Cox
Autonomy from environment	Novelty seeking	−0.326	0.031	0.722	0.537–0.97	0.91
	Harm avoidance	−0.24	0.037	0.787	0.628–0.985	
	Reward dependence	0.28	0.036	1.3236	1.019–1.719	
Susceptibility to interference	Reward dependence	−0.035	0.016	0.966	0.939–0.993	0.589
Mental flexibility	Persistence	0.024	0.046	1.024	1–1.049	0.644
	Novelty seeking	−0.027	0.025	1.027	1.003–1.051	
	Harm avoidance	−0.022	0.047	0.979	0.958–1	

Table 4.5 Logistic regression: character versus tasks of FAB

Dependent variables	Explicative variables	B	Sign	Exp (B)	95% C.I. Exp (B)	R2 di Cox
Autonomy from environment	Cooperativeness	−0.073	0.013	0.93	0.878–0.985	0.885
Inibitory control	Self directiveness	0.039	0.039	1.04	1.002–1.08	0.020
	Cooperativeness	−0.047	0.005	0.954	0.923–0.985	0.555
Motor programming	Cooperativeness	−0.05	0.014	0.952	0.915–0.99	0.668
Mental flexibility	Cooperativeness	−0.023	0.043	0.977	0.955–0.999	0.607

Table 4.5 shows the negative effect of the dimension of cooperativeness on the FAB tasks regarding autonomy from the environment, inibitory control, programming, mental flexibility and the positive effect of the dimension of self-directiveness on inhibitory control. Although the inhibition or activation effects on specific executive functions for each unit variation of the score of the personality dimension are very small, the effect of high scores on the dimensions of the TCI on frontal functions can be significant. In addition, there is no relationship between inefficiency of executive functions and symptoms because even extreme personality scores do not represent pathological traits, but express normal adaptive modes. Finally, in the case of cooperativeness, high scores are the index of optimal adaptive functioning and are, in any case, associated with the inefficiency of 4 executive tasks.

4.5 Conclusions

Our sample data shows that there is probably no direct relationship between executive function deficiencies and psychopathological manifestations. The usual hypothesis that considers starting from a neurological alteration of the frontal areas from which can result a deficiency of the executive functions and therefore a psychopathological symptom, according to a linear and reductive explanatory model, should be abandoned in favor of a circular explanatory hypothesis [32, 33]. According to this model, executive functions are not at the origin of psychopathological phenomena, but interact in a bidirectional fashion with adaptive personality traits that are habitual attitudes and habits deeply rooted in an individual's lifestyle. As a consequence, the dimensions of the adaptive personality trigger the activation or inhibition of the executive functions because they push the subject to exercise or to abandon them; as a consequence, the executive functions, circularly, enhanced or inhibited, accentuate and stabilize the personality dimension. For example, cooperativeness favoring social interaction inhibits the exercise of environmental autonomy and inhibition control of emotional expressions, reducing the efficiency of their respective executive functions related to such processes; reciprocally, the inefficiency of the mechanisms of inhibitory control and environmental autonomy tend to favor the stabilization of cooperativeness. However, the small size of the sample on which our study is based does not allow us to explore this explanatory hypothesis, which is at present a possible alternative to the ordinary interpretation of the activity of the frontal areas. It is essential to broaden the number of subjects and to highlight studies aimed at falsifying the circular hypothesis because this type of approach to the study of brain interaction and relational processes can deeply modify the computational patterns of the mind [34].

References

1. Burgess PW (2003) Assessment of executive function. In: Handbook of clinical neuropsychology, pp 302–321
2. Rabbitt P (1997) Introduction: methodologies and models in the study of executive function. In: Methodology of frontal and executive function, pp 1–38
3. Cozolino L (2014) The neuroscience of human relationships: attachment and the developing social brain (Norton series on interpersonal neurobiology). WW Norton & Company
4. Repa JC, Muller J, Apergis J, Desrochers TM, Zhou Y, LeDoux JE (2001) Two different lateral amygdala cell populations contribute to the initiation and storage of memory. Nature Neurosci 4(7):724–731
5. Schore AN (2003) Affect dysregulation and disorders of the self (Norton series on interpersonal neurobiology). WW Norton & Company
6. Maldonato M, Dell'Orco S (2015) Making decisions under uncertainty emotions, risk and biases. Smart Innov Syst Technol Springer 37:293–302
7. Hart S (2008) Brain, attachment, personality: An introduction to neuroaffective development. Karnac Books
8. Burgess PW (1997) Theory and methodology in executive function research. In: Methodology of frontal and executive function, pp 81–116

9. Maldonato M, Oliverio A, Esposito A (2017) Neuronal symphonies: musical improvisation and the centrencephalic space of functional integration. World Fut J New Paradig Res 1–20

10. Eslinger PJ, Damasio AR (1985) Severe disturbance of higher cognition after bilateral frontal lobe ablation patient EVR. Neurology 35(12):1731–1731

11. Lhermitte F (1983) Utilization behaviour and its relation to lesions of the frontal lobes. Brain 106(2):237–255

12. Goldberg G (1985) Supplementary motor area structure and function: review and hypotheses. Behavior Brain Sci 8(04):567–588

13. Sperandeo R, Maldonato M, Baldo G, Dell'Orco S (2016) Executive functions, temperament and character traits: a quantitative analysis of the relationship between personality and prefrontal functions. In 7th IEEE International Conference on Cognitive Infocommunications (CogInfoCom), pp. 000043–000048. IEEE

14. Brien LO' (2007) Achieving a successfull and stainable return to workforce after ABI: a client approach. Brain Inj 21:465–478

15. McCloskey G, Perkins LA, Van Divner B (2009) Assessment and intervention for executive function difficulties. Routledge, School-based practice in action series, New York

16. Stuss DT, Alexander MP (2007) Is there a dysexecutive system. Philos Trans Royal Soc LondonSer B Biol Sci 362(1481):901–915

17. Bush G, Luu P, Posner MI (2000) Cognitive and emotional influences in the anterior cingulate cortex. Trends Cogn Sci 4:215–222

18. Hoermann S, Clarkin JF, Hull JW, Levy KN (2005) The construct of effortful control: an approach to borderline personality. Psychopathology 38:82–86

19. Cloitre M, Cancienne J, Brodsky B, Dulit R, Perry SW (1996) Memory performance among women with parental abuse histories: enhanced directed forgetting or directed remembering. J Abnorm Psychol 105(2):204

20. Svrakic DM, Draganic S, Hill K, Bayon C, Przybeck TR, Cloninger CR (2002) Temperament, character, and personality disorders: Etiologic, diagnostic, treatment issues. Acta Psychiatrica Scandinavica 106:189–195

21. Arntz A, Appels C, Sieswerda S (2000) Hypervigilance in borderline disorder: a test with the emotional Stroop paradigm. J Person Disord 14(4):366–373

22. Meares R (2012) A dissociation model of borderline personality disorder. Norton Series on Interpersonal Neurobiology. WW Norton & Company, New York

23. Diamond A (2013) Executive functions. Ann Rev Psychol 64:135–68

24. Fossati A, Cloninger CR, Villa D, Borroni S, Grazioli F, Giarolli L, Battaglia M, Maffei C (2007) Reliability and validity of the Italian version of the temperament and character inventory-revised in an outpatient sample. Comprehen Psychiatr 48(4):380–387

25. Cloninger CR (1987) A systematic method for clinical description and classification of personality variants. Propos Arch Gen Psychiatr 44:573–588

26. Cloninger CR, Svrakic DM, Przybeck TR (1993) A psychobiological model of temperament and character. Arch Gen Psychiatr 50(12):975–990

27. Svrakic M, Dragan R, Cloninger C (2013) Psychobiological model of personality: guidelines for pharmacotherapy of personality disorder. Curr Psychopharmacol 2.3:190–203

28. Gillespie A, Nathan et al (2017) The genetic and environmental relationship between Cloninger's Dimensions of Temperament and Character. Personality and individual differences 35.8, 2003, pp 1931–1946. PMC. Web. 24 Sept 2017

29. Picard R (1997) Affective computing. The MIT Press, Massachusetts

30. Fogassi L, Ferrari PF, Gesierich B, Rozzi S, Chersi F, Rizzolatti G (2005) Parietal lobe: from action organization to intention understanding. Science 308(5722):662–667

31. Appollonio I, Leone M, Isella V, Piamarta F, Consoli T, Villa ML, Nichelli P (2005) The frontal assessment battery (FAB): normative values in an Italian population sample. Neurol Sci 26(2):108–116

32. Sperandeo R et al (2016) executive functions, temperament and character traits: a quantitative analysis of the relationship between personality and prefrontal functions. In: 7th IEEE International Conference on Cognitive Infocommunications(CogInfoCom). IEEE

33. Cantone D, Sperandeo R, Maldonato M (2012) A dimensional approach to personality disorders in a model of juvenile offenders. Revista Latinoamericana de Psicopatologia Fundament 15(1):42–57
34. Baranyi P (2012) Csap, definition and synergies of cognitive infocommunications. Acta Polytechnica Hungarica 9(1):67–83

Chapter 5
Mirroring and Prediction of Gestures from Interlocutor's Behavior

Costanza Navarretta

Abstract Mirroring and synchronization of non-verbal behavior is an important characteristics of human conduct also in communication. The aims of this paper are to analyze the occurrences of mirroring gestures, which comprise head movements, facial expressions, body postures and hand gestures, in first encounters and to determine whether information about the gestures of an individual can be used to predict the presence and the class of the gestures of the interlocutor. The contribution of related speech token is also investigated. The analysis of the encounters shows that 20–30% of the head movements, facial expressions and body postures are mirrored in the corpus, while there are only few occurrences of mirrored hand gestures. The latter are therefore not included in the prediction experiments. The results of the experiments, in which various machine learning algorithms have been applied, show that information about the shape and duration of the gestures of one participant contributes to the prediction of the presence and class of the gestures of the other participant, and that adding information about the related speech tokens in some cases improves the prediction performance. These results indicate that it is useful to take mirroring into account when designing and implementing cognitive aware info-communicative devices.

5.1 Introduction

Humans adapt to their surroundings and particularly to the behavior of other humans when they interact with each other. The mirror system has been judged to be a central factor in the process of learning from the actions of others and the ability of

C. Navarretta (✉)
Department of Nordic Studies and Linguistics, University of Copenhagen,
2300, Copenhagen, Denmark
e-mail: costanza@hum.ku.dk

© Springer International Publishing AG, part of Springer Nature 2019   91
R. Klempous et al. (eds.), *Cognitive Infocommunications, Theory and Applications*,
Topics in Intelligent Engineering and Informatics 13,
https://doi.org/10.1007/978-3-319-95996-2_5

reacting to these in an appropriate way [24, 26]. The mirror system, which was first discovered in apes and then found in humans, consists of a number of neurons, called the mirror neurons, which are activated a) when individuals perform an action and b) when they see other individuals perform the same action. More specifically, mirror neurons fire when subjects observe other individuals perform meaningful actions, and these comprise facial expressions, which show emotions and attitudinal states, [28, 29] and goal-directed gestures such as eating and handling objects [20]. Mirror neurons are also activated when subjects identify an action through sounds [21] or look at actions performed by e.g. robotic arms [17]. Mirror neurons play also an important role in the development of gesture and language [12, 23] and of social skills [22, 25].

Speech and gestures are the main components of face-to-face communication and are also referred to as auditory and visual modalities, respectively. Gesture is used in this paper as a general term for referring to co-speech non-verbal behavior, and it comprises inter alia body posture, facial expressions, head movements and hand gestures.

Since mirroring seems to be an important cognitive mechanism in human-human communication, it is worthy to determine to what extent it occurs in everyday interactions, and whether its influence on human behavior can be formalized and therefore included in advanced cogninfocommunicative technologies as those described in [2].

In this paper, we build upon previous research which addressed adaptive behavior in communication, and present a study of co-occurring and mirroring gestures in first encounters. Furthermore, we describe machine learning experiments in which classifiers are trained on information of the gestures of one individual in order to predict the behavior of the interlocutor extending previous studies that focused on facial expressions [15, 16]. Our main hypothesis is that mirroring effects are frequent in face-to-face communication, and they involve all gestures, and that mirroring should and can be accounted for in advanced and cognitively aware info-communicative technologies.

The paper's structure is the following. First in Sect. 5.2, we account for related studies, then in Sect. 5.3 we describe the corpus and the annotations. In Sect. 5.4, the extraction of mirroring gestures is described and the analysis of mirroring gestures is presented. In the following Sect. 5.5, machine learning experiments aimed to determine whether information about the gestures of one subject contributes to the prediction of the presence and type of gestures produced by the interlocutor are presented. In Sect. 5.6, we discuss the results of the experiments, and finally in Sect. 5.7, we conclude and present directions for future work.

5.2 Related Studies

The fact that humans align their verbal behavior in conversations has been observed in numerous studies [5, 7, 9]. In particular, [5] noticed that speech overlaps increased during long telephone conversations, and they interpreted these overlaps as a sign of synchronization between speakers. The importance of mirroring gestures, and in particular of mirroring facial expressions has also been addressed in neurological studies [8], but also in interactions between humans and software agents. More specifically, facial mirroring behavior has been tested in meetings between humans and embodied software agents [11]. The results of these experiments show that subjects smiled more often during meetings in which the software agent smiled frequently than in interactions in which the agent's smiles were less frequent, even though the subjects were not conscious of the difference between the two smiling conditions.

In previous studies of Danish first encounter conversations, speech overlaps and overlapping facial expressions were found to be frequent [13, 14]. However [14] also detected differences between the two modalities. More specifically, speech overlaps increase during the encounters confirming the study of [5], while the frequency of overlapping facial expressions does not change during the conversations. This indicates that mirroring the interlocutors' facial expressions is a natural phenomenon which is independent from the content of the conversation. In another study [15], we found that co-occurring facial expressions in the first encounters were often mirrored, that is they had the same shape features as the facial expressions produced by the interlocutor. The analysis of these expressions showed that especially smile and laughter co-occur. Furthermore, there is a strong association between smile occurrences by a participant and laughter occurrences by the other participant. The analysis of the emotions connected with the mirroring facial expressions showed that mirroring expressions do not always convey the same emotion as that conveyed by the mirrored facial expressions [15]. Amusement and friendliness are the emotion which both subjects showed at the same time most frequently. Emotions which indicate individual affective states such as certainty, self-confidence, hesitancy, embarrassment and uncertainty are usually not shared by the two participants, and some emotions co-occur with complementary emotions, for example hesitancy co-occurs with support, and uncertainty co-occurs with friendliness. Finally, machine learning experiments have shown that information about the shape of facial expressions and head movements by one subject can be used to predict a similar behavior by the interlocutor in first encounters [16]. The present research builds upon [16] and extends the analysis and the prediction experiment of mirroring behaviors to other gesture types. Eyebrows data have been added to the analysis of facial expressions, and mirrored body postures have been addressed. Finally, more machine learning algorithms, trainings and evaluation methods have been applied to the data.

5.3 The Corpus

The data are the Danish NOMCO first meetings which were collected and annotated by researchers at the University of Copenhagen. The data can be obtained at http://cst.ku.dk/projekter/projekter_slut/nomco/. The data are the Danish part of a number of multimodal corpora which were collected in a Nordic research project in order

Fig. 5.1 Snapshot from the first encounters video: mirrored scowl faces

to study and compare language and gestures in first encounters in different cultures [10, 18, 27]. The corpus is approximately one hour long, and it comprises twelve five-minutes dyadic encounters involving six male and six female participants. Each subject participated in two meetings, one with a female and one with a male. The participants were between 21 and 36 years old and did not know each other in advance. They talked freely during the meetings.

The Danish meetings were audio- and video-recorded by three cameras in a university studio. Figure 5.1 shows a situation from the data in which one participant mirrors the scowling facial expression of the other participant. In Fig. 5.2 is another snapshot in which one participant mirrors both the laughing facial expression and the both-hand raising gestures of the other participant. The annotations of the NOMCO corpus are in XML-format, and they were checked and agreed upon by three or four coders. They comprise the time aligned speech transcriptions and gesture annotations. Speech tokens consist of words, pauses, filled pauses and hesitations. The communicative gesture descriptions are expressed via pre-defined shape and function attribute-value pairs as accounted for in the MUMIN annotation scheme [1]. Multilinks connect a gesture to the speech tokens to which they have been judged to be connected via a semantic or pragmatic relation. The speech segments can have been uttered by the gesturer and/or by the interlocutor. The inter-coder agreement was measured as Cohen's kappa scores [6] and was found to be between 0.6 and 0.8 for head movements and facial expressions [19].

The shape features which are relevant to the present study are in Table 5.1.

The shape features of head movements describe the type of movement and whether the movement is single or repeated. Facial expressions are characterized by a general expression feature and a feature characterizing eyebrow position. The direction of the

Fig. 5.2 Snapshot from the first encounters videos: mirrored laughter and arm gestures

Table 5.1 Shape features

Behavior attribute	Behavior value
HeadMovement	Nod, Jerk (Up-nod), HeadBackward, HeadForward, Tilt, SideTurn, Shake, Waggle, HeadOther
HeadRepetition	Single, Repeated
General face	Smile, Laugh, Scowl, FaceOther
Eyebrows	Frown, Raise, BrowsOther
BodyDirection	BodyForward, BodyBackward, BodyUp, BodyDown, BodySide, BodyTurn, BodyDirectionOther
Shoulders	Shrug, ShouldersOther
Handedness	BothHandsSym, BothHandsAsym, RightSingleHand, LeftSingleHand
HandRepetition	Single, Repeated
TrajectoryLeftHand	LeftHandForward, LeftHandBackward, LeftHandSide, LeftHandUp, LeftHandDown, LeftHandComplex, LeftHandOther
TrajectoryRightHand	RightHandForward, RightHandBackward, RightHandSide, RightHandUp, RightHandDown, RightHandComplex, RightHandOther

body and the position of the shoulders define body postures, and hand gestures are described by the hand(s) involved in the movement, their trajectory and information about whether the hand gesture is repeated.

5.4 Analysis

In the first encounters, there are 3117 head movements, 1449 facial expressions, 982 body postures and 570 hand gestures. The most frequent gesture types are head movements while the less common are hand gestures. This is because some of the participants keep their hands in their pockets during the encounters.

We have extracted co-occurring gestures from the XML-annotation files with a perl script. A gesture of a subject is considered as mirroring if it co-occurs with but starts after the mirrored gesture by the other participant. Furthermore, the two gestures must have the same shape features. No restrictions are posed on the duration of the co-occurrence. The minimal overlap of two gestures is a frame, which in Anvil lasts 41.67 ms and a mirroring gesture starts at least 41.67 ms after the mirrored one.

883 head movements out of 3117 precede and co-occur with an head movement by the other participant and the correlation between head movements of a participant and the mirrored head movements of the other is positive and the Pearson 2-tailed correlation is statistically significant ($r(3117) < 0.0001$). A detailed analysis of the data shows that the head movements which are more often mirrored are nods, shakes and tilts. This is not surprising since they are the most frequent head movements in the encounters. Moreover, the types of head movement which co-occur significantly more often than expected [1] are Shake, Tilt, HeadBackward, HeadForward and Upnod. Nods and up-nods, and nods and shakes also co-occur more often than expected.

509 facial expressions (383 general face and 126 eyebrows) out of the 1449 facial expressions co-occur with and precede a facial expression produced by the interlocutor and their correlation is positive and its level is statistically significant (Pearson t-tailed correlation's $r(1449) < 0.0001$). The correlation between the duration of the mirrored and mirroring facial expressions is equally significant. The most common mirrored facial expressions are smiles, laughs and raised eyebrows. Smiles and laughs co-occur also with each other more frequently than expected.

199 out of the 982 body postures in the corpus co-occur with and precede a body posture of the other participant. The Pearson 2-tailed correlation between mirrored and mirroring body postures is positive and its level is still statistically significant ($r(982) = 0.049 < 0.05$). The body postures which co-occur particularly often are BodyBackward, BodyDown, BodySide and BodyTurn. Also the correlation between shoulders' shrugs is strong. Only 30 out of 530 occurrences of hand gestures co-occur with and precede a hand gesture of the other participant. The hand gestures which co-occur most often involve both hands or the right hand gestures. Left hand gestures are not mirrored at all in these data.

---

[1]Correlation $r(3117) < 0.001$.

5.5 Machine Learning Experiments

Given the frequency and correlation of co-occurring gestures by the two participants in the encounters, we aim to determine to what degree information about the gesture of one participant can be useful to predict the presence and then the shape of a gesture of the second participant. In a previous study [16], a similar experiment was performed on facial expressions and head movements, and the results showed, not surprisingly, that the best results were obtained when predicting the presence of a gesture by an individual from the presence of the same type of gesture by the interlocutor. In this study, we extend the experiments to body posture and test more machine learning algorithms and evaluation methods on the data. Moreover, we present an experiment in which we penalize the assignment of the prediction of a gesture to the null category, which is the most common case, in order to improve the identification of co-occurring gestures. We do not include hand gestures in the prediction experiments because the number of hand gestures and co-occurring hand gestures in these data is too small.

The algorithms which we tested are the WEKA's implementations [30] of a multilayer perceptron using DeepLearning 4J (DL4J), Naive Bayes, Simple logistics, Ada boost, support vector, and the WEKA implementation of Elman RNN by John Salatas.[2] The best results were obtained with Naive Bayes and the second best results were achieved by the Deep learning algorithm. In what follows, we only report the results from these two algorithms.

We trained all the prediction algorithms and used three training and validation methods. The first method was splitting the data-set randomly in two parts and using 66% of the data for training and and 34% for testing. The second method consisted of WEKA's 10-fold classification, in which the original dataset is randomly partitioned into 10 subsets. Nine subsets are used for training and the remaining subset is used for testing the model. Cross-validation is repeated 10 times, using each data subset exactly one time as validation data. The final results are obtained averaging the results obtained in each test run. Finally, we used 12-fold cross-validation, but this time the data were not divided randomly. Instead the annotations from each encounter formed a sub-sample, following a strategy which is often applied on multimodal data. The best results were achieved applying the latter training and testing method, and only the results obtained with this method are reported in what follows. We use as baseline the results of a majority classifier because it is the hardest baseline to beat, since the NONE class occurs in all data-sets in approximately two third of the cases and its prior is much higher than for the co-occurrence class. The results of each experiment are reported with Precision (P), Recall (R), and F-score figures. F-score is calculated as follows:

$$F - score = \frac{2 \times Precision \times Recall}{Precision + Recall}. \tag{5.1}$$

---

[2]http://jsalatas.ictpro.gr/weka/.

In the first experiments, we determine to what extent information about head movements, facial expressions and body postures of one participant contribute to the prediction of a head movement, facial-expression or body posture of the interlocutor, respectively. In the training data, we include the duration and shape features of each gesture. The shape features describing each gesture are those shown in Table 5.1 and they were extracted from the annotations. In a following step, we added to the training data the speech tokens to which gestures are connected semantically. In the tables showing the results of the machine learning experiments, *selfspeech* refers to the speech tokens uttered by the gesturer, *otherspeech* indicates the speech tokens uttered by the other participant, and *bothspeech* refers to all the speech tokens to which a gesture was connected to independently from which participant uttered them.

In Table 5.2, the results obtained by the ZeroR algorithm and by the Naive Bayes and DL4j algorithm for each type of gesture with the different training features are shown.

The table shows that features describing head movements, facial expressions and body postures produced by one conversation participant are useful for predicting the presence of the same type of gesture produced by the other participant in the Danish data. Significance of results with respect to the baseline was calculated with a corrected paired t-test and the results are significant if $p < 0.05$.

The results also indicate that information about the speech tokens uttered by the producer of the gestures improves the prediction of the mirrored gestures, and that this information is especially useful in the case of facial expressions and body postures.

The confusion matrices for the best results achieved by the Naive Bayes algorithm on head movements, facial expressions and body postures are in Tables 5.3, 5.4 and 5.5, respectively.

The confusion matrices show how the results of the prediction algorithm reflect the skewness of the data. In the following group of experiments, we trained the best performing algorithm on the same data-sets as in the preceding experiment, but we weighted the two classes penalizing the prediction of the non gestural class according to its prior probability. The results of these experiments are in Table 5.6.

The table shows that penalizing the choice of the non gestural category results in an improvement of the Naive Bayes prediction with respect to the majority baseline and the confusion matrices for the head movements, facial expressions and body postures in the best runs (Tables 5.7, 5.8, and 5.9) show that the number of true positives for the less frequent classes, the mirrored gestures, is significantly higher than in the preceding experiments.

In the second groups of experiments we aimed to determine to what degree the information about the duration and shape of gestures by a subject, as well as of the related speech tokens, contributes to the prediction of the presence and class of co-occurring head movements, facial expressions and body postures by the other participant. We expected the prediction to work less well in these experiments than in the former experiments since not all gestural subclasses are co-occurring, and some of the gestural subclasses only seldom occur in the first encounters [16]. The results

Table 5.2 Predicting the presence of a co-occurring gesture by the interlocutor

Dataset	Algorithm	Precision	Recall	F-measure
Head	Mayority	0.51	0.72	0.6
Head	NaiveBayes	0.67	0.71	0.67
Head+dur	NaiveBayes	0.7	0.73	**0.7**
Head+dur+selfspeech	NaiveBayes	0.69	0.72	0.69
Head+dur+otherspeech	NaiveBayes	0.7	0.7	0.69
Head+dur+allspeech	NaiveBayes	0.7	0.73	0.7
Head	DL4j	0.62	0.65	0.63
Head+dur	DL4j	0.66	0.7	067
Head+dur+selfspeech	DL4j	0.67	0.69	0.66
Head+dur+otherspeech	DL4j	0.66	0.7	0.67
Head+dur+bothspeech	DL4j	0.65	0.7	0.65
Face	Majority	0.42	0.65	0.51
Face	NaiveBayes	0.61	0.65	0.6
Face+dur	NaiveBayes	0.7	0.71	0.68
Face+dur+selfspeech	NaiveBayes	0.7	0.71	**0.69**
Face+dur+otherspeech	NaiveBayes	0.7	0.71	0.68
Face+dur+bothspeech	NaiveBayes	0.63	0.71	0.69
Face	DL4j	0.58	0.61	0.59
Face+dur	DL4j	0.67	0.68	0.67
Face+dur+selfspeech	DL4j	0.64	0.66	0.62
Face+dur+otherspeech	DL4j	0.65	0.67	0.65
Face+dur+bothspeech	DL4j	0.64	0.66	0.61
Body	Majority	0.62	0.79	0.69
Body	NaiveBayes	0.8	0.81	0.74
Body+dur	NaiveBayes	0.75	0.79	0.75
Body+dur+selfspeech	NaiveBayes	0.75	0.79	**0.76**
Body+dur+otherspeech	NaiveBayes	0.74	0.79	0.75
Body+dur+allspeech	NaiveBayes	0.7	0.77	0.72
Body	DL4j	0.7	0.78	0.72
Body+dur	DL4j	0.73	0.77	0.74
Body+dur+selfspeech	DL4j	0.72	0.78	0.74
Body+dur+otherspeech	DL4j	0.71	0.75	0.73
Body+dur+allspeech	DL4j	0.67	0.79	0.72

Table 5.3 Confusion matrix for prediction of the presence of head movements

a	b	Classified as
225	658	a = Head movement
211	2023	b = No head movement

Table 5.4 Confusion matrix for prediction of the presence of facial expressions

a	b	Classified as
203	306	a = Facial expression
109	831	b = No facial expression

Table 5.5 Confusion matrix for prediction of the presence of body postures

a	b	Classified as
733	50	a = No body posture
162	37	b= Body posture

Table 5.6 Weighted prediction of the presence of a co-occurring gesture by the interlocutor

Dataset	Algorithm	Precision	Recall	F-measure
Head	Mayority	0.29	0.54	0.38
Head	NaiveBayes	0.58	0.56	0.56
Head+dur	NaiveBayes	0.61	0.61	0.61
Head+dur+selfspeech	NaiveBayes	0.64	0.64	**0.64**
Head+dur+allspeech	NaiveBayes	0.64	0.64	0.64
Face	Majority	0.3	0.54	0.38
Face	NaiveBayes	0.63	0.63	0.63
Face+dur	NaiveBayes	0.64	0.63	0.63
Face+dur+selfspeech	NaiveBayes	0.63	0.63	**0.63**
Face+dur+otherspeech	NaiveBayes	0.64	0.63	0.63
Face+dur+bothspeech	NaiveBayes	0.63	0.63	0.63
Body	Majority	0.48	0.7	0.57
Body	NaiveBayes	0.7	0.72	0.64
Body+dur	NaiveBayes	0.7	0.7	0.7
Body+dur+selfspeech	NaiveBayes	0.68	0.68	**0.68**
Body+dur+otherspeech	NaiveBayes	0.66	0.7	0.66
Body+dur+allspeech	NaiveBayes	0.69	0.69	0.68

Table 5.7 Confusion matrix for weighed prediction of the presence of head movements

a	b	Classified as
783	541	a = Head movement
366	751	b = No head movement

Table 5.8 Confusion matrix for weighted prediction of the presence of facial expressions

a	b	Classified as
411	164	a = Facial expression
245	288	b = No facial expression

Table 5.9 Confusion matrix for the weighted prediction of the presence of body postures

a	b	Classified as
75	82	a = No body posture
81	277	b = Body posture

of these experiments are shown in Table 5.10 which is organized as the Tables 5.2 and 5.6.

The results of these prediction experiments show that information about the shape and duration of the gestures of one participant and of the speech tokens uttered by the same participant improves the performance of the algorithms in predicting the subclass of the mirrored gestures. However, the results also confirm the fact that the prediction algorithms only partially improve the results of the baseline algorithm since only some gestural subclasses are frequent in the data and are mirrored or co-occur significantly often with other gestural subclasses. Not surprisingly, the confusion matrices show that the machine learning algorithms only to some extent predict the co-occurrence of the most frequent classes for each gesture type, which in the data are *Nod*, *Tilt* and *SideTurn* for head movements, *Smile* for facial expressions, *BodyDirectionOther* and *BodyBackward* for body postures. The confusion matrix of the results obtained by the Naive Bayes classifier trained on facial expression data is given as an example in Table 5.11.

Table 5.11 also shows that the class Smile is often predicted instead of the class Laughter, and this is not surprising since the two classes often co-occur. In the future, the weighted strategy applied to the data in the preceding experiment could be extended also for predicting the occurrence of the gestural subclasses, but it would be useful to do it on larger datasets to avoid tuning the weights to the occurrences of subclasses in a specific dataset.

5.6 Discussion

The analysis of co-occurring gestures in the Danish first encounters show that not only facial expressions and head movements, but also body postures and single facial treats such as eyebrow positions often co-occur and in many cases the co-occurring gestures of one participant mirror a gesture of their interlocutor. The high presence of co-occurring and mirroring gestures is not surprising given the set-up of these data in

Table 5.10 Prediction of the presence and class of co-occurring head movements, facial expressions, body postures

Dataset	Algorithm	Precision	Recall	F-measure
Head	Majority	0.51	0.72	0.6
Head	NaiveBayes	0.51	0.72	0.6
Head+dur	NaiveBayes	0.54	0.71	0.61
Head+dur+selfspeech	NaiveBayes	0.56	0.73	**0.62**
Head+dur+otherspeech	NaiveBayes	0.55	0.72	0.61
Head+dur+allspeech	NaiveBayes	0.54	0.71	0.61
Head	DJ4	0.53	0.69	0.6
Head+dur	DJ4	0.54	0.69	0.6
Head+dur+selfspeech	DJ4	0.54	0.71	0.61
Head+dur+otherspeech	DJ4	0.54	0.7	0.6
Head+dur+allspeech	DJ4	0.54	0.71	0.61
Face	Majority	0.54	0.74	0.62
Face	NaiveBayes	0.63	0.73	0.66
Face+dur	NaiveBayes	0.63	0.73	0.66
Face+dur+selfspeech	NaiveBayes	0.65	0.73	**0.68**
Face+dur+otherspeech	NaiveBayes	0.64	0.72	0.66
Face+dur+allspeech	NaiveBayes	0.65	0.73	0.68
Face	DL4j	0.59	0.7	0.62
Face+dur	DL4j	0.62	0.71	0.65
Face+dur+selfspeech	DL4j	0.63	0.7	0.65
Face+dur+otherspeech	DL4j	0.61	0.69	0.63
Face+dur+allspeech	DL4j	0.64	0.71	0.66
Body	Majority	0.62	0.79	0.69
Body	NaiveBayes	0.62	0.78	0.69
Body+dur	NaiveBayes	0.65	0.77	0.71
Body+dur+selfspeech	NaiveBayes	0.65	0.78	**0.71**
Body+dur+otherspeech	NaiveBayes	0.65	0.77	0.7
Body+dur+allspeech	NaiveBayes	0.65	0.78	0.71
Body	DL4j	0.64	0.76	0.69
Body+dur	DL4j	0.66	0.76	0.7
Body+dur+selfspeech	DL4j	0.64	0.75	0.69
Body+dur+otherspeech	DL4j	0.66	0.76	0.7
Body+dur+allspeech	DL4j	0.63	0.75	0.69

Table 5.11 Confusion matrix for the prediction of the presence and subclass of facial expressions

a	b	c	d	e	Classified as
59	194	0	2	0	a = Smile
65	996	0	4	0	b = No facial expression
3	19	0	0	0	c = FaceOther
26	78	0	2	0	d = Laughter
0	1	0	0	0	e = Scowl

which the two participants stand in front of each other. The behavior of the participants in the first encounters confirms that mirrored gestures are common in face-to-face communication, as suggested by the mirroring theory [3, 4, 23]. Mirroring hand gestures are seldom in these data, and in general the participants only produced few hand gestures, partly because some of them kept their hands in their pockets during large part of the encounters. The hand gestures which are mirrored involved often right hand gestures or both-hand gestures, as the one in Fig. 5.2.

The head movements which are often mirrored are the most frequent occurring movements in these data, that is nods, shakes and tilts. The co-occurrence is statistically significant for shake, tilt, head backward, head forward and up-nod, while it is not significant for nod. However, nods and up-nods also co-occur significantly often, and they can be considered variations of the same movement and their function is also often related to feedback. In the future, the two classes could be collapsed. Surprisingly, we found that nods and shakes co-occur more often than expected. The majority of the mirroring shakes and nods have the function of feedback giving, while the mirrored shakes and nods have the function of self feedback or feedback eliciting.

The body postures which co-occur significantly often in the encounters are BodyBackward, BodyDown, BodySide, BodyTurn, BodyOther and shrugs of the shoulders.

Facial expressions, as also reported in [15], co-occur very often and they are frequently mirrored. The most common mirrored facial expressions are smiles and laughs as well as raised eyebrows.

Since head movements, body postures and facial expressions co-occur significantly often and in many case are mirrored, we trained machine learning algorithms in order to predict the gesture of a participant from the gestures of the interlocutor. We trained and tested the prediction algorithms with three different methods, and the best results were achieved training the algorithms on 11 encounters and testing them on the twelfth left out encounter, and then repeating the process twelve times and averaging all the twelve evaluation results in order to obtain a single evaluation measure. We also applied different classifiers on the data, and the classifier that produced the best prediction results was Naive Bayes, while the second best prediction was made by a deep learning implementation.

The results of the machine learning experiments act to predict whether mirroring gestures are produced and what is their class show that shape and duration information about the gestures of one individual improves the prediction of the presence of a co-occurring gesture of the same type by the interlocutor and also to the identification of its class. Therefore, the starting hypothesis that mirroring can contribute to model gestural behavior is confirmed.

Information about speech tokens semantically related to the mirrored gesture contributes to the identification of the mirroring gestures in nearly all cases, however the impact on the prediction of speech tokens is not very strong, the most important features being the shape and duration of gestures.

The results of the prediction algorithms corroborate the observation in [3] that co-occurring behaviors are not always produced consciously, and therefore they are not necessarily related to speech content. In conclusion, this study shows that, at least in dyadic face-to-face conversations, features describing the body behavior of a conversation participant are useful for training classifiers to predict the behaviors of their interlocutors, and that this information can be implemented in intelligent ICT.

Since our analysis and machine learning experiments only concern first encounters in Danish, the obtained results should be tested on more and different interaction types.

Some classes of gestures only occur seldom in the first encounters and some gesture classes never co-occur with the interlocutor's gestures. The skewness of the data is clearly reflected in the results of the prediction algorithms. One way to partly obviate to the skewness of the data is assigning higher weights to classes which do not occur frequently in the training dataset. We tested this in the first group of experiments aimed to predict the presence of a gesture produced by a participant from the gestures produced by the second participant. The results of these experiments show that the number of true positives belonging to the less frequent class augments significantly when weights are used. In the future, we will extend the weighting strategy to the prediction of the subclasses of the gestures produced by a participant from the gestures produced by the interlocutor, but we need more training data in order to confirm the impact of mirroring on the implementation of cognitively aware information and communication devices.

Mirroring is important because it concerns the way in which humans relate to their interlocutors, synchronize their behaviors to the others' gestures and show empathy. Including mirroring as a factor improves the prediction of the gestures of conversation participants with respect to the majority baseline. However, information about the interlocutor's gestures is not sufficient to understand and model gesture production and many other factors must be taken into account since the production of co-speech gestures is an extremely complex phenomenon. Therefore, it is necessary to include many more factors in order to model gesture production. These factors comprise prosodic features, discourse content and context, the communicative settings, the language, the personality of the individuals and their relation to the interlocutors, as well as the physical and cultural environments.

5.7   Conclusion

In this paper, we have taken our departure from the mirroring theory according to which a number of neurons are activated in the brain not only when humans produce an action, but also when they see an other individual produce the same action. Researchers have proposed that mirroring is a central factor in the development of social skills, such as the ability of recognizing the intention and goals of other individuals and showing empathy, and that it is also crucial in learning language and gestures [12, 22, 23, 25]. One of the effects of mirroring in communication is the synchronization of verbal and non-verbal behavior by conversation participants, and mirroring of gestures. In this work, we have extended previous research on co-occurring gestures and their impact on the prediction of the gestures which are produced in face-to-face conversations. The analysis of the data showed that head movements, facial expressions and body postures are often synchronized, while hand gestures are not frequent in these data, and they seldom are mirrored. For these reasons, we excluded hand gestures from the prediction experiments.

We trained classifiers on features describing the shape and duration of gestures and the speech tokens semantically related to them in order to predict from the gestures of a participant the presence and then the subclass of the gestures produced by the other participant in the Danish first encounters. The machine learning experiments were applied not only to facial expressions and head movements as in previous work, but also to body postures. Furthermore, we tested different algorithms and training/testing methods, and tested the effect of penalizing the prediction of the frequent non gestural class in the experiments aimed to predict the presence of a gesture by a participant from the gestures by the other participant.

The results of the prediction experiments show that information about the gesture of a participant improves the prediction of the presence and class of gestures produced by the interlocutor with respect to the majority baseline which always predicts that the interlocutor does not produce a gesture. This baseline covers the majority of the data. Speech tokens related to the gesture and uttered by the gesturer contribute to the prediction in most cases. Penalizing the prediction of the frequent non gestural class according to its higher prior significantly improves the results of the best working Naive Bayes algorithm. Even though the results which we have obtained on the first encounters are promising, more factors must be taken into account when modeling gestures in communication. Moreover, our work has only addressed a type of data in one language. Therefore, our results most be tested on more data types and languages.

In the future, we will extend our research to more corpora, and we will investigate individual differences in mirroring. Our analysis has also shows that some types of gestures are mirrored more frequently than other ones, and therefore it should be investigated whether it would be better to predict them individually and extend the weighting strategy to the prediction of the subclass of the mirrored gestures.

References

1. Allwood J, Cerrato L, Jokinen K, Navarretta C, Paggio P (2007) The MUMIN coding scheme for the annotation of feedback, turn management and sequencing. Multimodal Corpora for Modelling Human Multimodal Behaviour. Spec Issue Int J Lang Resour Eval 41(3–4):273–287
2. Baranyi P, Csapo A, Sallai G (2015) Cognitive Infocommunications (CogInfoCom). Springer
3. Bernieri F (1988) Coordinated movement and rapport in teacher-student interactions. J Nonverbal Behav 12:120–138
4. Bernieri F, Davis J, Rosenthal R, Knee CR (1994) Interactional synchrony and rapport: measuring synchrony in displays devoid of sound and facial affect. Pers Soc Psychol Bull 20:303–311
5. Campbell N, Scherer S (2010) Comparing measures of synchrony and alignment in dialogue speech timing with respect to turn-taking activity. In: Proceedings of Interspeech. pp 2546–2549
6. Cohen J (1960) A coefficient of agreement for nominal scales. Educ Psychol Meas 20(1):37–46
7. Condon W, Sander L (1974) Synchrony demonstrated between movements of the neonate and adult speech. Child Dev 45(2):456–462
8. Dimberg U, Thunberg M, Elmehed K (2000) Unconscious facial reactions to emotional facial expressions. Psychol Sci 11(1):86–89
9. Esposito A, Marinaro M (2007) What pauses can tell us about speech and gesture partnership. In: Fundamentals of verbal and nonverbal communication and the biometric issue. NATO publishing series sub-series E: human and societal dynamics, vol 18. IOS Press, pp 45–57
10. Jokinen K (2013) Multimodal feedback in first encounter interactions. In: Kurosi M (ed) Human-computer interaction. In: Proceedings of 15th international conference on interaction modalities and techniques, HCI International 2013, Las Vegas, NV, USA, 21–26 July, 2013, Part IV. Springer, Berlin, Heidelberg, pp 262–271
11. Krämer N, Kopp S, Becker-Asano C, Sommer N (2013) Smile and the world will smile with you–the effects of a virtual agent's smile on users' evaluation and behavior. Int J Hum Comput Stud 71(3):335–349
12. McNeill D (2005) Gesture and thought. University of Chicago Press
13. Navarretta C (2013) Transfer learning in multimodal corpora. In: IEEE (ed). Proceedings of the 4th IEEE international conference on cognitive infocommunications (CogInfoCom2013). Budapest, Hungary, pp 195–200
14. Navarretta C (2014) Alignment of speech and facial expressions and familiarity in first encounters. In: Proceedings of the 5th IEEE international conference on cognitive infocommunications (CogInfoCom2014). IEEE Signal Processing Society, Vietri, Italy, pp 185–190
15. Navarretta C (2016) Mirroring facial expressions and emotions in dyadic conversations. In: N.C.C. (Chair), Choukri K, Declerck T, Goggi S, Grobelnik M, Maegaard B, Mariani J, Mazo H, Moreno A, Odijk J, Piperidis S (eds) Proceedings of the tenth international conference on language resources and evaluation (LREC 2016). European Language Resources Association (ELRA), Paris, France, pp 469–474
16. Navarretta C (2016) Predicting an individual's gestures from the interlocutor's co-occurring gestures and related speech. In: 2016 7th IEEE international conference on cognitive infocommunications (CogInfoCom). pp 233–238. https://doi.org/10.1109/CogInfoCom.2016.7804554
17. Nelissen K, Luppino G, Vanduffel W, Rizzolatti G, Orban GA (2005) Observing others: multiple action representation in the frontal lobe. Science 310(5746):332–336
18. Paggio P, Ahlsén E, Allwood J, Jokinen K, Navarretta C (2010) The NOMCO multimodal Nordic resource—goals and characteristics. In: Proceedings of LREC 2010. Malta, pp 2968–2973
19. Paggio P, Navarretta C (2011) Head movements, facial expressions and feedback in danish first encounters interactions: a culture-specific analysis. In: Stephanidis C (ed) 6th international conference on universal access in human-computer interaction—users diversity, UAHCI 2011, Held as Part of HCI International 2011, no. 6766 in LNCS. Springer, Orlando, Florida, pp 583–690

20. di Pellegrino G, Fadiga L, Fogassi L, Gallese V, Rizzolatti G (1992) Understanding motor events: a neurophysiological study. Exp Brain Res 91(1):176–180
21. Ricciardi E, Bonino D, Sani L, Vecchi T, Guazzelli M, Haxby JV, Fadiga L, Pietrini P (2009) Do we really need vision? How blind people "See" the actions of others. J Neurosci 29(31):9719–9724
22. Rizzolatti G (2005) The mirror neuron system and its function in humans. Anat Embryol 210:419–421
23. Rizzolatti G, Arbib M (1998) Language within our grasp. Trends Neurosci 21(5):188–194
24. Rizzolatti G, Craighero L (2004) The mirror-neuron system. Annu Rev Neurosci 27:169–192
25. Rizzolatti G, Fabbri-Destro M (2008) The mirror system and its role in social cognition. Curr Opin Neurobiol 18:179–184
26. Rizzolatti G, Fadiga L, Gallese V, Fogassi L (1996) Premotor cortex and the recognition of motor actions. Cognit Brain Res 3(2):131–141
27. Särg D, Jokinen K (2015) Nodding in Estonian first encounters. In: Proceedings of the 2nd European and the 5th Nordic symposium on multimodal communication, pp 87–95
28. Singer T, Seymour B, O'Doherty J, Kaube H, Dolan RJ, Frith CD (2004) Empathy for pain involves the affective but not sensory components of pain. Science 303(5661):1157–1162
29. Wicker B, Keysers C, Plailly J, Royet JP, Gallese V, Rizzolatti G (2003) Both of us disgusted in my insula: the common neural basis of seeing and feeling disgust. Neuron 40(3):655–664
30. Witten IH, Frank E (2005) Data mining: practical machine learning tools and techniques, 2nd edn. Morgan Kaufmann, San Francisco

Chapter 6
Automatic Labeling Affective Scenes in Spoken Conversations

Firoj Alam, Morena Danieli and Giuseppe Riccardi

Abstract Research in affective computing has mainly focused on analyzing human emotional states as perceivable within limited contexts such as speech utterances. In our study, we focus on the dynamic transitions of the emotional states that are appearing throughout the conversations and investigate computational models to automatically label emotional states using the proposed *affective scene framework*. An affective scene includes a complete sequence of emotional states in a conversation from its start to its end. Affective scene instances include different patterns of behavior such as *who* manifests an emotional state, *when* it is manifested, and *which* kinds of changes occur due to the influence of one's emotion onto another interlocutor. In this paper, we present the design and training of an automatic affective scene segmentation and classification system for spoken conversations. We comparatively evaluate the contributions of different feature types in the acoustic, lexical and psycholinguistic space and their correlations and combination.

F. Alam (✉) · M. Danieli · G. Riccardi
Department of Information Engineering and Computer Science, University of Trento, Trento, Italy
e-mail: firoj.alam@unitn.it; firoj.alam@alumni.unitn.it

M. Danieli
e-mail: morena.danieli@unitn.it

G. Riccardi
e-mail: giuseppe.riccardi@unitn.it

© Springer International Publishing AG, part of Springer Nature 2019
R. Klempous et al. (eds.), *Cognitive Infocommunications, Theory and Applications*,
Topics in Intelligent Engineering and Informatics 13,
https://doi.org/10.1007/978-3-319-95996-2_6

109

6.1   Introduction

In social interactions, the emotional states of the interlocutors[1] continuously change over time. The flow of emotional states is influenced by the speaker's internal state, by the interlocutor's response, and by the surrounding environmental stimuli in the situational context [21]. For the development of behavioral analytics systems, there is a need to automatically recognize the flow of speakers' emotions in socially regulated conversations.

Behavioral analytics systems that can analyze the emotional flow may facilitate the automatic analysis of naturally-occurring conversations from different areas of human relationships, including therapist versus patient, teacher versus student, and agent versus customer interactions.

State of the art literature provides evidence that the understanding of emotions is important in human-machine dialogue systems [40] and call center interactions [18].

During the past century, the study of human emotions was a critical field of investigation in humanities. While most of the studies in this area focused on the classification of emotions into basic and complex categories, some psychological studies focused on the emotional process itself. For example, the appraisal theory [43] illustrated the sequential organization of the emotional process. According to this theory, an emotional state can change because of the underlying appraisals and reaction processes. Moreover, the sequential organization model assumes that the appraisal process is always active, and it continuously evaluates an event or situation to update its organism's information. Based on the appraisal phenomenon, Gross proposed the *modal model* of emotion [25]. This model illustrates the emotional process in term of the situation $\rightarrow$ attention $\rightarrow$ appraisal $\rightarrow$ response sequence. Gross's theory states that the emotion-arousal process is induced by a situation, i.e., a physical or virtual space that can be objectively defined. The situation compels an attention of the subject and it elicits the subject's appraisal process and the related emotional response. The response may generate actions that in turn modify the initial situation. Based on this model, Danieli et al. [17] proposed a concept, named *affective scene*. The purpose was to model the complex interplay between the expression of emotions and its impact in different conversational interactions.

In the field of automatic analysis, the research on affective computing mainly focused on analyzing low-level signals such as utterances, turns in spoken conversation or images. However, the present studies lack the modeling of the entire emotional space at the higher level, such as the flow of an emotion sequence throughout the conversation. There could be several reasons such as lack of operationalized concepts and the difficulty in representing them in conversational interactions. The study of Lee et al. [33] suggests that modeling the conditional dependency between two interlocutors' emotional states in sequence improves the automatic classification performance. In [21], Filipowicz et al. studied the how emotional transitions influence the social interactions. They observed that it could lead to different interpersonal

---

[1] In this chapter, the word 'interlocutor' encompasses the person speaking and expressing (*speaker*) the emotion and the person listening and perceiving (*listener*) that emotion.

behaviors. Regarding the research for designing automatic emotion recognition systems, several behavioral cues have been explored using various modalities such as audio and video, and multi-modal [27, 30, 31, 44].

The importance of summarizing spoken conversations, in terms of emotional manifestations, long-term traits, and conversational discourse, is presented in [48]. Authors demonstrated that such a summary can be useful in different application domains.

The research related to the use of paralinguistics for behavioral analysis from spoken conversations include personality [2–4], overlap discourse [12, 14], overlap in context [15], turn-taking dynamics [11], user satisfaction [16], and interlocutors' coordination in emotional segment [5].

In our study, we focus on designing the *affective scene framework* to automatically label affective scenes. The *motivation* of designing an affective scene framework is also practical because from the automatically labeled affective scenes one can interpret the different patterns of the speakers' affective behavior in situated contexts. For the experiment and evaluation described in this chapter, we utilized call center's dyadic spoken conversations. We investigate the call center agent and customer's affective behaviors such as *customer manifested anger or frustration, however, the agent was not empathic* (see Sect. 6.4). This kind of understanding can help to pinpoint the customer's problem, and it can also help in reducing phenomena like the churn rate.[2] Such an understanding can help to provide a better customer service. We propose the computational models that can automatically classify emotional segments occurring in conversations. For the design and evaluation of the model, we utilize *verbal* and *vocal non-verbal* cues in terms of acoustic, lexical, and psycholinguistic features. The work described in this chapter may contribute to the cognitive infocommunications research field by providing a model of affective scenes that may be adopted to investigate intra-cognitive communications [6] issues.

This chapter is organized as follows. In Sect. 6.2, we briefly review the terminologies used in the affective computing research, which are relevant for this work. In Sect. 6.3, we give a brief overview of the corpus that we used to investigate the affective scene (Sect. 6.4) and design the framework. We present the experimental details in Sect. 6.5, and discuss the results in Sect. 6.6. Finally, we provide conclusions in Sect. 6.7.

6.2   Terminology

In this section, we review terminologies from affective behavior research that are relevant for the context of this work.

---

[2]The *churn rate* is "the percentage of customers who stop buying the products or services of a particular company." In the telecommunication industry, some studies found that the approximate annual churn rate is 30% [24, 49].

Behavior: It is defined as "... quite broadly to include anything an individual does when interacting with the physical environment, including crying, speaking, listening, running, jumping, shifting attention, and even thinking" [22].

Behavioral Signals: Signs that are direct manifestations of individual's internal states being affected by the situation, the task and the context. Signals can be overt and/or covert. Examples of overt signals are changes in the speaking rate or lips getting stiff. Examples of covert cues are changes in the heart-rate, galvanic skin response or skin temperature.

Affect: It is an umbrella term that covers a variety of phenomena that we experience such as emotion, stress, empathy, mood, and interpersonal stance [29, 41]. All of these states share a special affective quality that sets them apart from the neutral states. In order to distinguish between each of them, Scherer [42] defined a design-feature approach, which consists of seven distinguished dimensions, including intensity, duration, synchronization, event-focus, appraisal elicitation, rapidity of change, and behavioral-impact.

Affective Behavior: The component of the behavior that can be explained in terms of affect analysis.

Emotion: There is a variety of definitions of this concept. A few of them are reported below. According to Scherer [42], emotion is a relatively brief and synchronized response, by all or most organismic subsystems, to the evaluation of an external or internal stimulus.

Gross's [26] definition of emotion refers to its *modal model*, which is based on three core features such as (1) what gives rise to emotions (when an individual attend and evaluate a situation), (2) what makes up an emotion (subjective experience, behavior, and peripheral physiology), and (3) malleability of emotion.

According to Frijda "emotions are intense feelings that are directed to someone or something" [23, 41].

Emotional State: In the psychological and psychiatric literature the concept "emotional state" is often used as being coextensive with some given emotion, and it is not explicitly defined from itself. However, studies that analyze human emotions in a more situated perspective, have proposed that emotional states are conditions of the psychological and physiological processes that generate an emotional response, and that contextualize, regulate, or otherwise alter such a response [7].

Empathy: According to Hoffman [28], "Empathy can be defined as an emotional state triggered by another's emotional state or situation, in which one feels what the other feels or would normally be expected to feel in his situation". For this work, we follow this definition of empathy.

By McCall and Singer [34], empathy is defined based on four key components: "*First*, empathy refers to an affective state. *Second*, that state is elicited by the inference or imagination of another person's state. *Third*, that state is isomorphic with the other person's state. *Fourth*, the empathizer knows that the other person is the source of the state. In other words, empathy is the experience of vicariously feeling what another person is feeling without confounding the feeling with one's own direct experience".

Perry and Shamay-Tsoory [37] "... denotes empathy as our ability to identify with or to feel what the other is feeling."

6.3   The Corpus and Its Annotations

The corpus of Italian conversations analyzed in this study consists of 1894 customer-agent phone dialogues amounting ~210 h. It was recorded on two separate channels with 16 bits per sample, and a sampling rate of 8 kHz. The length of the conversations is 406 ± 193 (mean ± standard deviation) seconds. The corpus was annotated with *empathy (Emp)*, basic and complex emotions. The basic emotions include *anger (Ang)*, and complex emotions include *frustration (Fru), satisfaction (Sat), dissatisfaction (Dis)*. We also introduced *neutral (Neu)* state tag as a relative concept to support annotators while identifying the emotions in the conversational context. The annotation protocols were defined based on the Gross's *modal model* of emotion. More details of the annotation process can be found in [17]. The annotators' task was to identify the occurrences of emotional segments in the continuous speech signal where they can perceive a transition in the emotional state of the speaker. They identified the onsets of the variations and assigned an emotional label. Hence, the emotional segment may consist of one or more turns.[3] Moreover, the annotators were also instructed to focus both on their perception of speech variations, such as acoustic and prosodic quality of pairs of speech segments, and on the linguistic content of the utterances.

For the evaluation of the annotation model, we randomly selected 64 (~448 min) spoken conversations. The annotators were hired for the annotation task. The demographic information of the annotators consists of Italian native speakers, similar age, and ethnicity but different gender. The goal was to assess whether the annotators can perceive the speech variations at the same onset position, as well as their agreement of assigning the emotional labels on speech segments. The comparison showed that 0.31 of the annotated speech segments were exactly tagged by the two annotators at the same onset positions, while 0.53 was the percentage of cases where the two annotators perceived the emergence of an emotional attitude of the speaker occurring at different, yet contiguous, time frames of the same dialog turns.

To measure the quality of the annotations, we calculated inter-annotator agreement using the kappa measure [8]. It is widely used to assess the degree of agreement among the annotators. The kappa coefficient ranges between 0 (i.e., agreement is due to chance) and 1 (i.e., perfect agreement). Values above 0.6 suggests an acceptable agreement. We found reliable kappa results as shown in Table 6.1. The annotation task was complex, which was also taking more time for the annotation. From our observation, we found that on average the annotation time of a conversation was ~18 min. The analysis of annotators' disagreement showed that the inevitable portion of the subjectivity of this task had a greater impact in cases where the differences of the emotional labels were more nuanced, like in case of frustration and dissatisfaction.

---

[3]Turn refers to the spoken content of a speaker at a time. For example, speaker A says something, which is speaker A's turn, then, speaker B says something, which is speaker B's turn.

Table 6.1 Kappa results of the annotation

Emotional state	Agent/Customer	Kappa
Empathy	Agent	0.74
Anger	Customer	0.75
Frustration	Customer	0.67
Satisfaction	Customer	0.69
Dissatisfaction	Customer	0.71

6.4 Affective Scene

In [17], Danieli et al. defined the *affective scene* on the basis of the Gross's *modal model* of emotion [25, 26]. The concept of **affective scene** is defined as "*an emotional episode where an individual is interacting, in an environment, with a second individual and is affected by an emotion-arousing process that (a) generates a variation in their emotional state, and (b) triggers a behavioral and linguistic response.*" The affective scene extends from the event-triggering of the 'unfolding of emotions' throughout the closure event when individuals disengage themselves from the communicative context. To describe the affective scenes, we have considered three factors:

- *who* showed the variation of their emotional state *when*,
- *how* the induced emotion affected the other interlocutor's emotional appraisal and response, and
- *which* modifications occurred by such a response with respect to the state that triggered the scene.

In Fig. 6.1, we present an example of an emotional sequence, which shows an affective scene. In the example, we see there is a sequence of three events.

1. *who:* customer 'first' manifested frustration.
2. *how:* agent appraised customer's emotion and responded with empathic behaviors.
3. *which:* customer manifested satisfaction at the end of the conversations.

The lack of empathic response from the agents may cause different patterns of emotional response from the customer. It can lead to the manifestation of customer's anger, dissatisfaction, or frustration.

From the manual annotations of our corpus, we selected a subset containing 566 conversations where we investigated different patterns of affective behaviors. Some of the selected patterns are presented below.

- agents were neutral when customers manifested either anger, frustration or dissatisfaction.
- agents were neutral when customers manifested either anger, frustration or dissatisfaction followed by satisfaction at the end of the conversations.

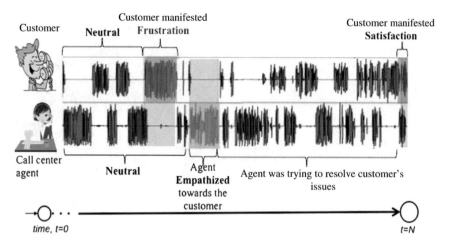

Fig. 6.1 An annotated example of a call center conversation. The interaction between call center's agent and customer shows that customer first manifested frustration, then agent empathized towards the customer, and finally customer manifested satisfaction at the end of the conversation

- agents were empathic, and customer manifested either anger, frustration or dissatisfaction.
- agents were empathic, and customer manifested either anger, frustration or dissatisfaction followed by satisfaction at the end of the conversations.
- agents were empathic, and customer manifested satisfaction.

In Fig. 6.2, we depict the distribution of the behavioral patterns between speakers that we mentioned above. In the figure, *A:Emp-C:Neg* represents agents (A) were empathic (Emp), and customers manifested negative emotions and *C:Neg,Sat* rep-

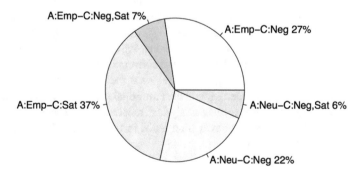

Fig. 6.2 A set of conversations consisting of different patterns of affective behavior analyzed from emotional state sequence. For instance, the pattern (A:Emp-C:Neg,Sat) indicates that, in this corpus, there were 7% of conversations where agents showed empathy and the customer negative emotions, however, at the end they were satisfied. This pattern can explain the success of the operator in handling the state of the customer

Fig. 6.3 The proportion of conversations, in which either agent (AF) or customer (CF) expressed emotion at the start of the conversation, and the manifestation of customer's emotional state at the end of the conversation. *Pos* and *Neg* represent customer manifested either positive or negative emotion

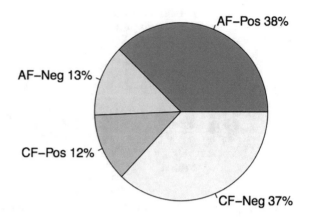

resents customer manifested negative emotions followed by satisfaction at the end of the conversations. For the sake of this analysis, we grouped anger, frustration, and dissatisfaction into negative. From this analysis, we see that in 22% conversations customers manifested negative emotions whereas agents were neutral. In 27% conversations, even if the agent were empathic, but customers manifested negative emotions. These two scenarios might be an important factor for call center managers. Manual intervention might be necessary for these conversations. Other scenarios are also interesting to look into, for example, the pattern *A:Emp-C:Neg,Sat* occurred in 7% conversations, where customers' manifestation of satisfaction followed customers' negative emotions. It may be agents were empathic, and customer's emotional arousal reduced over the course of the conversations.

The other patterns that we wanted to investigate are the manifestation of emotions at the start, and at the end of the conversations. It includes *who* is manifesting emotion at the start of the conversations, i.e., agent first (AF) or customer first (CF), and *which* types (positive (Pos) or negative (Neg)) of emotions customer were manifesting at the end of the conversations. In Fig. 6.3, we present such an analysis. The critical part is that in 13% (AF-Neg) and 37% (CF-Neg) conversations customer manifested negative emotions at the end of the conversations. We observed that customer manifestations of negative emotion at the start of the conversations might lead to the higher chance of the manifestations of negative emotions at the end of the conversations. In this analysis, we did not consider what is the emotional manifestations of the agents at the end of the conversations. Such analytical results indicate that more insights can be found from affective scenes, which leads to the fact that it is necessary to automatically detect affective scenes.

6.5 Affective Scene Framework

In Fig. 6.4, we present the proposed affective scene framework. Several architectural decisions can be made, however, we found this approach is a good starting point towards the automation of the affective scenes in conversations. The whole architectural pipeline can also be useful to process visual signals too. As shown in Fig. 6.4, at first we separated the conversation channel-wise. This process can be done in two different manners. Either two channels will input independently to the system, or a classification module can be designed to separate the two speakers' speech channels. After that, an automatic segmenter was used to separate speech and non-speech segments (see Sect. 6.5.1). Then emotion segment classifiers (Sect. 6.5.2) were employed, which automatically assign each speech segment with an emotion label. After that, using the emotion sequence labeler module (see Sect. 6.5.3), we combined the segment classifiers' output and segment meta information, i.e., start and end time boundary of the segment, to design emotional sequence for the whole conversation.

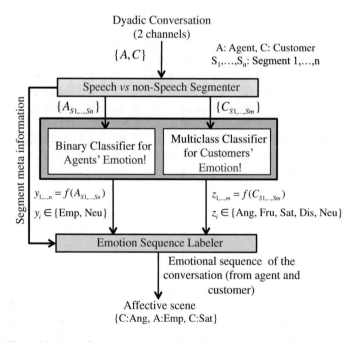

Fig. 6.4 The architecture of the proposed automatic affective scene classification

6.5.1   Classification of Speech and Non-speech Segments

To classify speech and non-speech segments within a conversational channel we used an off-the-shelf speech and non-speech segment classifier. It has been designed using forced aligned transcriptions and trained the model using Hidden Markov Models (HMMs). The transcriptions were obtained from 150 conversations, which consists of approximately 100 h of speech. To train the model, Mel Frequency Cepstral Coefficient (MFCC) features were extracted from the conversations with 25 ms per frame and 100 frames per second with a step size of 10 ms. The model was trained using 32 Gaussian mixtures with a beam size 50, and utilized Kaldi [39] speech recognition framework. The F-measure of the segment classifier is 66.42% [1].

6.5.2   Segment Classifiers

We designed separate emotion classifiers to deal with customer and agent channels' speech segments. For the agent channel, we designed a binary classifier in which the class labels include *empathy (Emp)*, and *neutral (Neu)*. Whereas for the customer channel we grouped anger and frustration into negative, then designed a multiclass classifier in which class labels are *negative (Neg)*, *dissatisfaction (Dis)*, *satisfaction (Sat)*, and *neutral (Neu)*. We investigated each system using acoustic, lexical, and psycholinguistic features, also with their decision-level combination. For the experiment, we separated the conversations into the training, development, and test set and their proportion was 0.70, 0.15, and 0.15, respectively. We separated them at the conversation level while we maintained speaker independence for the agent conversations. Due to the unavailability of the customer information in the corpus, we assumed that each conversation is from an independent speaker.

6.5.2.1   Feature Extraction

Acoustic features

We extracted acoustic features from speech signal using openSMILE [19] and the configurations, which we made publicly available.[4] The approach of the feature extraction process is that we first extract low-level features and then project them onto statistical functionals. The success of this approach has been reported in the literature in which studies have been conducted in different paralinguistic tasks [3, 46, 47].

We extracted low-level acoustic features with 100 frames per second. The frame size for voice-quality features was 60 ms with a Gaussian window function and $\sigma = 0.4$. For the other low-level features, the frame size was 25 ms with a Hamming

---

[4]https://github.com/firojalam/openSMILE-configuration.

window function. The low-level features include zero crossing rate, MFCC (1–12), root mean square frame energy, fundamental frequency, pitch, harmonics-to-noise ratio, spectral features, voice probability, Mel-spectrum band 1–26, centroid, max, min, and flux.

We apply many statistical functionals, here, we report a few of them. More details can be found in [1]. The set includes range, max, min, the geometric and quadratic mean, linear and quadratic regression coefficients, arithmetic and quadratic mean, quartiles and interquartile range and percentile.

After applying statistical functionals onto low-level features, the resulted total number of acoustic features consists of 6861.

Lexical features

We extracted lexical features from automatic transcriptions, and we used an in-house developed ASR system [13] to get the transcriptions. The word error rate of the ASR system is 31.78% on the test set. To understand how the results differ from training data we also evaluated the system on the training set and obtained an WER of 20.87%. To design the classification model, the textual information needs to be converted into vector form and the widely used approach is bag-of-words or bag-of-ngrams. For this work, we converted the transcriptions of each segment into bag-of-words vectors. The applied logarithmic term frequencies (tf) and inverse document frequencies (idf) method. Since contextual information provides better classification results, therefore, we extracted trigram features. The n-gram approach results in a large dictionary, which increase computational cost and also introduce overfitting. To avoid such drawbacks, we removed the stop-words and filtered out lower frequent words and only kept the 10 K most frequent n-grams.

Psycholinguistic features

We used LIWC [36] to extract the psycholinguistic features from the automatic transcriptions. It is a knowledge-based system containing dictionaries for several languages in which word categories are associated with a set of a lexicon. The system computes frequency or relative frequency for each word category by matching the transcriptions. The word categories are used as features for the machine learning tasks. The word categories include linguistic, psychological, paralinguistic, personal concern and punctuations. Since the transcriptions were Italian, therefore, we used Italian dictionary, which contains 85 word categories. Moreover, the LIWC system extracts 6 general and 12 punctuation categories, which results in 103 features. We removed features that are not found in our dataset. For example, punctuations are not available with transcriptions. Hence, we extracted 89 psycholinguistic features.

6.5.2.2 Feature Selection

To reduce the dimensionality, we applied *Relief* feature selection algorithm [32]. The motivation of using this technique is that it showed improved classification performance and at the same time it reduced computational cost. We observed that in previous paralinguistic tasks [2]. Our feature selection process is as follows. We rank the feature set according to Relief feature selection scores. Then, we generate learning curves by incrementally adding batches (e.g. we added 200 features each time) of ranked features. After that, we select the optimal set of features [2]. To obtain a better performance during the feature selection process, we discretize the feature values into 10 equal-frequency bins [50]. The main reason is that the feature selection algorithm does not work well with the continuous-valued features. The approach of equal-frequency binning is that it divides data into k bins, where each bin contains an equal number of values. For our experiment, we have chosen to use the value of k as 10. To get an unbiased estimate on the classification results, we experimented feature selection and discretization approaches on the development set.

6.5.2.3   Classification Experiments and Evaluation

Classification method

We designed the classification models using SVM [38] and applied linear kernel for lexical and acoustic features. For the psycholinguistic features, we applied Gaussian kernel of the SVM. We optimized the parameters C and G by tuning it on the development set. The range of values that we experiment for C and G is $[10^{-5}, 10^{-4}, ..., 10]$. In order to get the results on the test set, we first combined the training and development dataset. After that, we trained the models using the optimized parameters. Since we have three classifier's decisions, therefore, to get a single decision we applied *majority voting*.

Undersampling and oversampling

One of the important problems in designing machine learning model is the imbalance class label distribution. For example, the original distribution of the Emp versus Neu segments was 6% and 94% respectively. The original distribution of customer's channel emotions such as Neg, Dis, Sat, and Neu were 1%, 1%, 2% and 96%, respectively. To deal with such a problem, we undersampled the instances of majority classes at the data level[5] and oversampled the instances of minority classes at the feature level.[6] Our undersampling approach is very intuitive in a sense that we capture the variation of segment length. To do that, we defined a set of bins, which varies segment lengths. After that, we randomly selected N segments from each bin. The size of N is experimental and problem specific. For this study, we used the size of N as 1. However, we found that this is an optimal number, which we obtained by running experiments on the development set. The pseudocode of undersampling approach is presented in Algorithm 1. The boundaries i.e., start and end, of each

---

[5]By data level, we refer to the data preparation phase, i.e., before feature extraction we select segments of the majority class, which is neutral in this case.

[6]By feature level, we refer to the over-sampling process on feature vector for minority classes.

bin in this study are set based on the experimental observation. It can also be done based on percentiles of segments' length *or* applying a supervised approach such as multi-interval discretization by Fayyad and Irani [20]. With such an approach we are providing the variation of segment length as well as reducing the samples and influence of majority class.

Algorithm 1 Undersampling algorithm, C- Conversation containing segments, N-number of segment from each bin of segment duration, l-class label to be under-sampled. It returns randomly selected segments. Bin can be designed in several way, such as equal frequency or equal interval.

```
 1: procedure DOWN- SAMPLING(C, N, l)
 2:     newSegmentList ← List
 3:     segmentLengthBin ← {0.1 − 4.0, 4.0 − 7.0, 7.0 − 10.0, 10.0 − 15.0, > 15.0}
 4:     segIndex = 0
 5:     for all seg ∈ C do
 6:         if seg.tag = l then
 7:             dur ← duration(seg)
 8:             segmentLengthBin[dur] ← seg
 9:         else
10:             newSegmentList[segIndex] ← seg
11:             segIndex = segIndex + 1
12:         end if
13:     end for
14:     for all bin ∈ segmentLengthBin do
15:         bin ← shuffle(bin)                  ▷ Shuffles the bin elements for randomization
16:         for i = 1 : N do
17:             newSegmentList[segIndex] ← bin[i]
18:             segIndex = segIndex + 1
19:         end for
20:     end for
21:     nC ← newSegmentList
22:     return nC                             ▷ Set of segments randomly selected from C's segments.
23: end procedure
```

For the oversampling, we used SMOTE algorithm [10] and applied it on the training set. After applying the sampling techniques, the distribution for empathy versus neutral segments becomes 30% and 70%, respectively, and the distribution for Neg, Dis, Sat, and Neu becomes 7.5%, 7.5%, 12% and 73%, respectively. Such distributions are still very skewed, however, more balancing does not help in classification performance, which we experimented with the development set. The reason is with too much under-sampling of the majority class we are reducing the variations of the training instances and classifier does not capture much information. On the other hand too much synthetic generation of oversampling, we are generating similar instances and classifier does not gain much information about the minority classes.

Evaluation metric

To measure the performance of the system we used an well-known metric Un-weighted Average (UA) recall. It is a widely accepted metric for the paralinguistic tasks [45]. We have extended this measure to take into account the segmentation errors. Similar type of evaluation is typically done by NIST in diarization tasks [9, 35]. UA is the average recall of class labels, which we computed from a weighted confusion matrix, as shown in Eq. 6.1, where the weight for each segment is the corresponding segment length.

$$C(f) = \left\{ c_{i,j}(f) = \sum_{s \in S_T} [((y = i) \wedge (f(s) = j)) \times length(s)] \right\} \tag{6.1}$$

where $C(f)$ is the $n \times n$ confusion matrix of the classifier f, s is the segment, $length(s)$ is the duration of s, y is the reference label of s, $f(s)$ is the automatic label for s. The indices i and j are the reference and automatic/classified class labels of the confusion matrix.

6.5.3 Affective Scene Labeling

Using the *emotion sequence labeler* module as shown in Fig. 6.4, we combined the emotion sequence from both agent and customers' channels by utilizing the output of the emotion segment classifier and the meta information of segments from the speech versus non-speech segmenter. An example of such a sequence is as follows, *C: Fru* → *A: Emp* → *C: Sat* where customer 'first' manifested emotion, then agent empathized towards the customer, after that customer's emotion regulated towards satisfaction, which is a similar representation we presented in Fig. 6.1.

6.6 Results and Discussion

In Table 6.2, we present the performances of the emotion segment classification systems, which we investigated using different types of features such as acoustic, lexical (automatic transcriptions), and psycholinguistic, and also their decision level combination. To compute the baseline, we have randomly selected the class labels based on the prior class distribution. In addition, we also present average results (Avg) to get an understanding of the performance of the whole system.

   We obtained the best results with *majority voting* for both segment classification systems. To understand whether the obtained best results are statistically significant or not we run the significant test. For which we used McNemar's test. From the results, we observed that the results of majority voting are statistically significant with $p < 0.05$ compared to any other classifier's results. Moreover, compared to the baseline, a relative improvement of the system with agent's emotions is 42.2%, and

Table 6.2 Segment classification results in terms of weighted average recall using acoustic, lexical, and psycholinguistic features together with decision level fusion. Ac—acoustic features; Lex—lexical features; LIWC—psycholinguistic features; Maj—majority voting. Values inside parenthesis represent feature dimension

Exp	Agent	Customer	Avg
	Binary {Emp, Neu}	Multiclass {Neg, Dis, Sat, Neu}	
Baseline	49.3	24.4	36.9
Ac	(2400) 68.1	(4600) 47.4	57.7
Lex	(8000) 65.6	(5200) 56.5	61.0
LIWC	(89) 67.3	(89) 51.9	59.6
Ac+Lex	(2600) 68.3	(5800) 49.2	58.8
Maj(Ac+Lex+LIWC)	70.1	56.9	63.5

the system with the customer's emotions is 61.51%. The performance of the system with customer's emotions is lower compared to the other system, which is due to the complexity of the task in terms of class distributions, and multiclass classification task.

6.6.1 Empathy

From the Table 6.2, we see that for individual feature-based models, the model using acoustic features provides the best performance compared to models using lexical and psycholinguistic features, respectively. The results of the acoustic features are significantly better than random baseline with a p value less than $2.2E-16$. When there are no transcriptions (i.e., either manual or automatic) available then the use of acoustic features is the ideal situation as it provides a useful and low-computation classification model. The performance of the LIWC's features is better than lexical features. The advantage of this feature set is that its size is very small, i.e., a number of features is 89, which is computationally less expensive. Compared to the baseline results, the performances of all classification systems are higher and statistically significant with a p value less than $2.2E-16$.

In terms of feature and decision level combination, we obtained the best results using *majority voting*. From the statistical significance test, we observe that the results of the majority voting are statistically significant with a p value equal to 0.0004 compared to any other classification models' results. We also conducted experiments how linear combination of acoustic and lexical features performs based on the previous findings in other paralinguistic tasks [3]. We observed that it has not improved performance as shown in Table 6.2.

In Figs. 6.5 and 6.6, we present class-wise correlation analysis on acoustic and lexical features, which we performed correlation analysis on top-ranked features. To

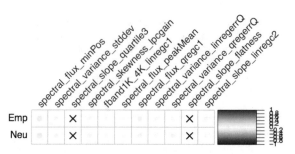

Fig. 6.5 Correlation analysis of acoustic features. Cells without asteric "×" are statistically significant. The color in each cell represents the correlation coefficients (r) and its magnitude is represented by the depth of the color. The "×" symbol represents the corresponding r is not significant. Description of the features is presented in Table 6.3

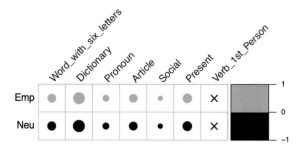

Fig. 6.6 Correlation analysis of LIWC features. Cells without asteric "×', are statistically significant. The color in each cell represents the correlation coefficients (r) and its magnitude is represented by the depth of the color. The "×" symbol represents the corresponding r is not significant

rank the features we used a feature selection algorithm discussed in Sect. 6.5.2.2. From the feature selection and correlation analysis, our findings suggest that spectral features contribute most to the classification decision. The top-ranked low-level acoustic feature includes spectral, energy, and mfcc in ranked order and the statistical functionals include minimum segment length, standard deviation, quadratic mean and quadratic regression coefficient.

From the Fig. 6.5, we observe that spectral-flux feature is positively correlated with empathy and is negatively correlated with neutral. In the figure, the "×" symbol represents they are not statistically significant. From the figure, we see that even if spectral-slope features appeared in top 10 features, however, they are not highly correlated with any class label. From our analysis of psycholinguistic features, we observed that character length, dictionary word, pronoun, article, social connotation, words convey present tense are positively correlated with empathy and negatively correlated with neutral.

Our findings from lexical feature analysis suggest some words and phrases are positively correlated with empathy and negatively correlated with neutral such as

Table 6.3 Selected acoustic features and their description obtained from empathy segment

Feature	Description
spectral_flux_minPos	Absolute position of the minimum value of the spectral flux feature
spectral_variance_stddev	The standard deviation of the spectral variance
spectral_slope_quartile3	The third quartile (75% percentile) of the spectral slope
spectral_skewness_lpcgain	The linear predictive coding gain of spectral skewness
fband1K_4K_linregc1	The slope of a linear approximation of the fband 1000–4000 contour
spectral_flux_peakMean	The arithmetic mean of peak spectral flux
spectral_flux_qregc1	The quadratic regression coefficient 1 of the spectral flux
spectral_variance_linregerrQ	The quadratic error of the spectral variance, computed as the difference of the linear approximation and the actual contour
spectral_variance_qregerrQ	The quadratic error of the spectral variance, computed between contour and quadratic regression line
spectral_slope_flatness	The contour flatness of spectral slope, which is a ratio of geometric mean and absolute value of arithmetic mean
spectral_slope_linregc2	The offset of a linear approximation of the contour of spectral_slope

"posso aiutarla/can I help you", "vediamo un po/let's see". An example of a positive correlation with neutral is "cosa posso esserle/what can I do".

In another study, we have done an experiment with manual segments, which shows that we can reach above 90% UA with a better segmentation approach.

6.6.2 Basic and Complex Emotions

The classification task for basic and complex emotions is much more complex than empathy classification task due to the multi-class classification task. From the results, as shown in Table 6.2, we observed that the performance of LIWC features is lower than lexical features. However, it is higher than acoustic features, and the number of features is very low for this set. The decision level combination has not improved the performance for this case, whereas it shows higher improvement for the classification of empathy. For the linear combination of acoustic and lexical features performance is also lower compared to the lexical features. The feature dimension for the lexical feature set is comparatively higher than the acoustic and LIWC feature sets.

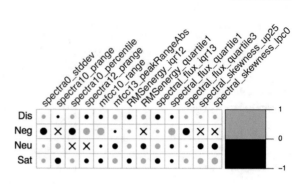

Fig. 6.7 Correlation analysis of acoustic features Cells without asteric "×" are statistically significant. The color in each cell represents the correlation coefficients (r) and its magnitude is represented by the depth of the color. The "×" symbol represents the corresponding r is not significant. Description of the features is presented in Table 6.4

We observed that the recall of dissatisfaction is comparatively lower than other emotional categories. It also confuses with satisfaction due to the fact the manifestation of both satisfaction and dissatisfaction appear at the end of the conversation. For this reason, there is an overlap of the linguistic content, which also effects the paralinguistic properties of the spoken content. Using acoustic features, we obtained better performance for negative and neutral, whereas using the lexical feature we obtained better performance for negative and satisfaction. In terms of discriminative characteristics, spectral, voice-quality, pitch, energy, and mfcc features are highly important. The statistical functionals include the arithmetic mean of the peak, quadratic regression, the gain of linear predictive coefficients, flatness, quartile, and percentiles.

We have analyzed them in terms of class-wise correlation analysis as presented in Fig. 6.7. The number in each cell in the figure represents the correlation value

Table 6.4 Selected acoustic features and their description obtained from the segments of basic and complex emotion

Feature	Description
spectra0_stddev	The standard deviation of the rasta style auditory spectra band 0
spectra10_prange	The percentile range of the rasta style auditory spectra band 10
spectra10_percentile	The percentile of the rasta style auditory spectra band 10
spectra12_prange	The percentile range of the rasta style auditory spectra band 12
mfcc10_range	The range of the derivative of the mfcc 10
mfcc13_peakRangeAbs	The absolute peak range of the derivative of the mfcc 13
RMSenergy_iqr12	The inter-quartile range: quartile2-quartile1 of the root-mean square energy 12
RMSenergy_quartile1	The first quartile (25% percentile) of the root-mean square energy
spectral_flux_iqr13	The inter-quartile range: quartile3-quartile1of the spectral flux
spectral_flux_quartile1	The first quartile (25% percentile) of the spectral flux
spectral_flux_quartile3	The third quartile (75% percentile) of the spectral flux
spectral_skewness_up25	The percentage of time the signal is above (25% * range + min) of the spectral skewness
spectral_skewness_lpc0	The linear predictive coding of spectral skewness

Fig. 6.8 Correlation analysis of LIWC features. Cells without asteric "×" are statistically significant. The color in each cell represents the correlation coefficients (r) and its magnitude is represented by the depth of the color. The "×" symbol represents the corresponding r is not significant

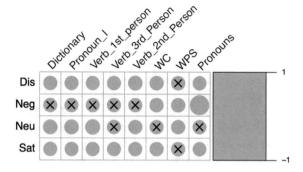

between class-label and a feature. The color in each cell represents the positive and negative association. The "×" symbol represent the association is not significant with $p = 0.05$. Even though the correlation value is very low, close to zero, however, most of them are statistically significant. Spectral and rasta style auditory spectrum features are positively associated with satisfaction. For neutral, spectral features are negatively associated. MFCC and rasta style auditory spectrum features are positively correlated with negative emotion. Satisfaction and dissatisfaction are mostly similar; the only dissimilarity exists in the strength of positive and negative association in some features.

The correlation analysis of LIWC features is presented in Fig. 6.8. The highly discriminative LIWC features include personal pronouns, words associated with emotion and verb. Similar to the acoustic features, satisfaction and dissatisfaction are quite similar in their correlation with LIWC features. However, there exists a disassociation in the strength of the correlation. First three features, such as words containing in the dictionary, pronoun (I), and 1st person verb, are negatively correlated with neutral, and these features are positively correlated with the negative emotion.

The correlation analysis of lexical features shows that negative emotion is highly associated with negative words whereas satisfaction represents mostly positive words such as "grazie mille/thank you so much/", "benissimo/very well/" and "perfetto/perfect/". The difference between satisfaction and dissatisfaction is that dissatisfaction represents some negativity, however, there is not much lexical difference. It might be because the annotators mostly focused on the tone of the voice, which distinguished the annotation of satisfaction and dissatisfaction. It is needed to mention that the LIWC and lexical feature analysis has been done based on ASR transcription.

To understand the upper-bound of the classification system, we designed a system by exploiting manual segments with which we obtained UA 70.9% using acoustic features.

Note that, we do not present any figure of the correlation analysis of lexical features as it is very difficult to make a general conclusion from such graphical representation.

6.7 Conclusions

In this chapter, we address the design of an affective scene system to automatically label the sequence of emotional manifestations in a dyadic conversation. Our presented framework shows the complete pipeline from speech segmentation to sequence labeling, which might be useful in analyzing affective and behavioral patterns in different applicative scenarios. The results of the automatic classification system using decision combination on call center conversations are significantly better than random baseline. We investigated different feature sets such as acoustic, lexical, and psycholinguistic features. Moreover, we also investigated feature and decision level combinations for designing emotion segment classifiers. The investigation of the feature sets suggests that lexical and psycholinguistic features can be useful for the automatic classification task. In both emotion segment classification systems, we obtained better results using decision combination. In future, as an extension of this research, one can explore the study of affective scenes in multi-party contexts and social domains.

References

1. Alam F (2016) Computational models for analyzing affective behaviors and personality from speech and text. PhD thesis, University of Trento
2. Alam F, Riccardi G (2013) Comparative study of speaker personality traits recognition in conversational and broadcast news speech. In: Proceedings of interspeech, ISCA, pp 2851–2855
3. Alam F, Riccardi G (2014) Fusion of acoustic, linguistic and psycholinguistic features for speaker personality traits recognition. In: Proceedings of international conference on acoustics, speech and signal processing (ICASSP), pp 955–959
4. Alam F, Riccardi G (2014) Predicting personality traits using multimodal information. In: Proceedings of the 2014 ACM multi media on workshop on computational personality recognition, ACM, pp 15–18
5. Alam F, Chowdhury SA, Danieli M, Riccardi G (2016) How interlocutors coordinate with each other within emotional segments? In: COLING: international conference on computational linguistics
6. Baranyi P, Csapó Á (2012) Definition and synergies of cognitive infocommunications. Acta Polytech Hung 9(1):67–83
7. Barrett LF, Lewis M, Haviland-Jones JM (2016) Handbook of emotions. Guilford Publications
8. Carletta J (1996) Assessing agreement on classification tasks: the kappa statistic. Comput Linguist 22(2):249–254
9. Castán D, Ortega A, Miguel A (2014) Lleida E (2014) Audio segmentation-by-classification approach based on factor analysis in broadcast news domain. EURASIP J Audio Speech Music Process 1:1–13
10. Chawla NV, Bowyer KW, Hall LO, Kegelmeyer WP (2002) Smote: synthetic minority over-sampling technique. J Artif Intell Res 321–357
11. Chowdhury SA (2017) Computational modeling of turn-taking dynamics in spoken conversations. PhD thesis, University of Trento
12. Chowdhury SA, Riccardi G (2017) A deep learning approach to modeling competitiveness in spoken conversation. In: Proceedings of international conference on acoustics, speech and signal processing (ICASSP), IEEE

13. Chowdhury SA, Riccardi G, Alam F (2014) Unsupervised recognition and clustering of speech overlaps in spoken conversations. In: Proceedings of workshop on speech, language and audio in multimedia—SLAM2014. pp 62–66
14. Chowdhury SA, Danieli M, Riccardi G (2015) Annotating and categorizing competition in overlap speech. In: Proceedings of ICASSP. IEEE
15. Chowdhury SA, Danieli M, Riccardi G (2015) The role of speakers and context in classifying competition in overlapping speech. In: Sixteenth annual conference of the international speech communication association
16. Chowdhury SA, Stepanov E, Riccardi G (2016) Predicting user satisfaction from turn-taking in spoken conversations. In: Proceedings of Interspeech
17. Danieli M, Riccardi G, Alam F (2015) Emotion unfolding and affective scenes: a case study in spoken conversations. In: Proceedings of emotion representations and modelling for companion systems (ERM4CT) 2015. ICMI
18. Devillers L, Vidrascu L (2006) Real-life emotions detection with lexical and paralinguistic cues on human-human call center dialogs. In: Proceedings of Interspeech. pp 801–804
19. Eyben F, Weninger F, Gross F, Schuller B (2013) Recent developments in opensmile, the munich open-source multimedia feature extractor. In: Proceedings of the 21st ACM international conference on Multimedia (ACMM). ACM, pp 835–838
20. Fayyad UM, Irani KB (1993) Multi-interval discretization of continuousvalued attributes for classification learning. Thirteenth international joint conference on articial intelligence, vol 2. Morgan Kaufmann Publishers, pp 1022–1027
21. Filipowicz A, Barsade S, Melwani S (2011) Understanding emotional transitions: the interpersonal consequences of changing emotions in negotiations. J Pers Soc Psychol 101(3):541
22. Fisher W, Groff R, Roane H (2011) Applied behavior analysis: history, philosophy, principles, and basic methods. In: Handbook of applied behavior analysis, pp 3–13
23. Frijda NH (1993) Moods, emotion episodes, and emotions
24. Galanis D, Karabetsos S, Koutsombogera M, Papageorgiou H, Esposito A, Riviello MT (2013) Classification of emotional speech units in call centre interactions. In: 2013 IEEE 4th international conference on cognitive infocommunications (CogInfoCom). IEEE, pp 403–406
25. Gross JJ (1998) The emerging field of emotion regulation: an integrative review. Rev Gen Psychol 2(3):271
26. Gross JJ, Thompson RA (2007) Emotion regulation: conceptual foundations. In: Handbook of emotion regulation, vol 3, p 24
27. Harrigan J, Rosenthal R (2008) New handbook of methods in nonverbal behavior research. Oxford University Press
28. Hoffman ML (2008) Empathy and prosocial behavior. Handb Emot 3:440–455
29. Juslin PN, Scherer KR (2005) Vocal expression of affect. In: The new handbook of methods in nonverbal behavior research. pp 65–135
30. Kim S, Georgiou PG, Lee S, Narayanan S (2007) Real-time emotion detection system using speech: Multi-modal fusion of different timescale features. In: Proceedings of multimedia signal processing, 2007 (MMSP 2007). pp 48–51
31. Konar A, Chakraborty A (2014) Emotion recognition: a pattern analysis approach. Wiley
32. Kononenko I (1994) Estimating attributes: analysis and extensions of relief. In: Proceedings of machine learning: European conference on machine learning (ECML). Springer, pp 171–182
33. Lee CC, Busso C, Lee S, Narayanan SS (2009) Modeling mutual influence of interlocutor emotion states in dyadic spoken interactions. In: Proceedings of Interspeech. pp 1983–1986
34. McCall C, Singer T (2013) Empathy and the brain. In: Understanding other minds: Perspectives from developmental social neuroscience. pp 195–214
35. NIST (2009) The 2009 RT-09 RIch transcription meeting recognition evaluation plan. NIST
36. Pennebaker JW, Francis ME, Booth RJ (2001) Linguistic inquiry and word count: Liwc 2001. Lawrence Erlbaum Associates, Mahway, p 71
37. Perry A, Shamay-Tsoory S (2013) Understanding emotional and cognitive empathy: a neuropsychological. In: Understanding other minds: Perspectives from developmental social neuroscience. Oup Oxford, p 178

38. Platt J (1998) Fast training of support vector machines using sequential minimal optimization. MIT Press. http://research.microsoft.com/~jplatt/smo.html
39. Povey D, Ghoshal A, Boulianne G, Burget L, Glembek O, Goel N, Hannemann M, Motlicek P, Qian Y, Schwarz P, et al (2011) The kaldi speech recognition toolkit. In: Proceedings of automatic speech recognition and understanding workshop (ASRU). pp 1–4
40. Riccardi G, Hakkani-Tür D (2005) Grounding emotions in human-machine conversational systems. In: Lecture notes in computer science. Springer, pp 144–154
41. Robbins S, Judge TA, Millett B, Boyle M (2013) Organisational behaviour. Pearson Higher Education AU
42. Scherer KR (2000) Psychological models of emotion. Neuropsychol Emot 137(3):137–162
43. Scherer KR (2001) Appraisal considered as a process of multilevel sequential checking. Theory Methods Res Apprais Process Emot 92–120
44. Schuller B, Batliner A (2013) Computational paralinguistics: emotion, affect and personality in speech and language processing. Wiley
45. Schuller B, Steidl S, Batliner A (2009a) The interspeech 2009 emotion challenge. In: Proceedings of Interspeech. pp 312–315
46. Schuller B, Vlasenko B, Eyben F, Rigoll G, Wendemuth A (2009b) Acoustic emotion recognition: a benchmark comparison of performances. In: Proceedings of automatic speech recognition and understanding workshop (ASRU). pp 552–557
47. Schuller B, Steidl S, Batliner A, Burkhardt F, Devillers L, Müller C, Narayanan S (2013) Paralinguistics in speech and language state-of-the-art and the challenge. Comput Speech Lang 27(1):4–39
48. Stepanov E, Favre B, Alam F, Chowdhury S, Singla K, Trione J, Béchet F, Riccardi G (2015) Automatic summarization of call-center conversations. In: In Proceedings of the IEEE automatic speech recognition and understanding workshop (ASRU 2015)
49. Tamaddoni Jahromi A, Sepehri MM, Teimourpour B, Choobdar S (2010) Modeling customer churn in a non-contractual setting: the case of telecommunications service providers. J Strateg Mark 18(7):587–598
50. Witten IH, Frank E (2005) Data mining: practical machine learning tools and techniques. Morgan Kaufmann

Chapter 7
Tracking the Expression of Annoyance in Call Centers

Jon Irastorza and María Inés Torres

Abstract Machine learning researchers have dealt with the identification of emotional cues from speech since it is research domain showing a large number of potential applications. Many acoustic parameters have been analyzed when searching for cues to identify emotional categories. Then classical classifiers and also outstanding computational approaches have been developed. Experiments have been carried out mainly over induced emotions, even if recently research is shifting to work over spontaneous emotions. In such a framework, it is worth mentioning that the expression of spontaneous emotions depends on cultural factors, on the particular individual and also on the specific situation. In this work, we were interested in the emotional shifts during conversation. In particular we were aimed to track the annoyance shifts appearing in phone conversations to complaint services. To this end we analyzed a set of audio files showing different ways to express annoyance. The call center operators found disappointment, impotence or anger as expression of annoyance. However, our experiments showed that variations of parameters derived from intensity combined with some spectral information and suprasegmental features are very robust for each speaker and annoyance rate. The work also discussed the annotation problem arising when dealing with human labelling of subjective events. In this work we proposed an extended rating scale in order to include annotators disagreements. Our frame classification results validated the chosen annotation procedure. Experimental results also showed that shifts in customer annoyance rates could be potentially tracked during phone calls.

J. Irastorza (✉) · M. Inés Torres (✉)
Speech Interactive Research Group, Universidad el País Vasco UPV/EHU, Leioa, Spain
e-mail: jon.irastorza@ehu.es

M. Inés Torres
e-mail: manes.torres@ehu.es

© Springer International Publishing AG, part of Springer Nature 2019
R. Klempous et al. (eds.), *Cognitive Infocommunications, Theory and Applications*,
Topics in Intelligent Engineering and Informatics 13,
https://doi.org/10.1007/978-3-319-95996-2_7

7.1 Introduction

In recent year, the machine learning community has paid increasing attention to model the emotional status based on parameters derived from the analysis of the voice, the language, the face, the gestures or the ECG [9]. But data-driven approaches need corpora of human spontaneous behavior annotated with emotional labels [9, 26], which is a challenging requirement mainly due to the subjectivity of emotion perception by humans [9, 10]. As a consequence much research is being carried out over corpora that cover simulated or induced emotional behavior [10, 16]. It is worth mentioning that the selection of the situation where spontaneous emotions can be collected strongly depends on the goals of the research to be carried out. In particular the emotion identification from speech shows a wide range of potential applications and research objectives [8, 24]. Some examples are early detection of Alzheimer's disease [18], the detection of valency onsets in medical emergency calls [26] or in Stock Exchange Customer Service Centres [9]. It is clear that different sponteneous emotions arise in each of the previous situations.

When we deal with the recognition of emotions from speech signals we find a number of short-term features such as pitch, excitation signals, vocal tract features such as formants [25, 27], prosodic features [6] such as pitch loudness, speaking rate, rhythm, voice quality and articulation [12, 26], latency to speak, pauses [11, 15], features derived from energy [16] as well as feature combinations, etc.

Some surveys dealing with databases, classifiers, features and also with the set of categories to be identified in the analysis of emotional speech can also be found in the literature [1, 3, 23, 25]. Regarding methodology, classical classifiers such as the Bayesian or SVM have been proposed to analyze the feature distributions. The model of continuous affective dimensions is also an emerging challenge when dealing with continuous rating of emotion labelled during real interaction [19]. In this approach recurrent neural networks have been proposed to integrate contextual information and then predict emotion in continuous time to just deal with arousal and valence [21, 28].

However, once again further performance of any set parameters chosen or each identification method proposed depends of the research question to be addressed. In this work we addressed a research question proposed by a Spanish call-center providing customer assistance [15]. The goal of the company was to automatically detect annoyance rates during customer calls for further analysis, which resulted in a very challenge and novel goal. Their motivation was to verify if the policies that the company proposed to be implemented by operators when they have to deal with annoyed and angry customers really lead to shifts in the customer behavior. Thus an automatic procedure to detect those shifts would allow the company to evaluate their policies through the analysis of the recorded audios. Moreover their proposed to provide the operators with this information in real time, i.e., during the conversation. As a consequence our final work was aimed at tracking shifts in customer annoyance during conversations to complain services. To begin with, a short number of subjects showing very different ways to express their annoyance when complaining about a service were analyzed in this work. It is worth mentioninig that the call center operators found disappointment, impotence or anger as expression of annoyance. We also discussed the annotation problem arising when dealing with human labelling of such subjective events. In this work we proposed an extended rating scale in order to include annotators disagreements. Then a certain amount of features were evaluated as potencial hints to track annoyance degrees through parametric and geometric classifiers. Intensity values and derivatives along with their corresponding suprasegmental values combined with some spectral analysis have demonstrated to be very robust in the experiments carried out over all the expressions of annoyance analyzed. Our results validated the proposed annotation procedure. The experimental results also showed that shifts in customer annoyance rates could be potentially tracked during phone conversations. In summary, the problem addressed in this work is the analysis of the intra-cognitive communication [4, 5] between customers and operators of a customer assistant service. In this framework the analysis carried out through the feature discussed can be seen as an artificial cognitive capability that actually measures a Human cognitive load.[1]

7.2 Expression of Annoyance and Human Annotation

In this section, we describe the difficulties in get a consensus in both the human expression and the human perception of emotions. In particular we analyze the topics over a set of phone conversations aimed to complain about services provided by phone and internet companies.

7.2.1 Annotation Perception Difficulties

Due to the perceptual differences between people, the annotation of emotions is a very challenging problem to be addressed. Perceptual difficulties are caused by factors related to the culture [17] and the environment where we live as well as beliefs and experiences in our mental and maturity growth stage. In addition, people with psychiatric disorders like autism have more problems to understand and recognize emotion [2, 22]. These challenges are present primarily when we try to distinguish emotions with similar valence and arousal shown in Fig. 7.1. Nevertheless, we find more problems in distinguishing degrees within an emotional pure state. For instance, if we take annoyance as emotional pure state, low angry, medium angry and high angry could be three degrees to distinguish. In Sect. 7.2.3.3 we can observe real difficulties annotating degrees of annoyance.

---

[1]This paper is a revised and extended version of a paper that was presented in [13].

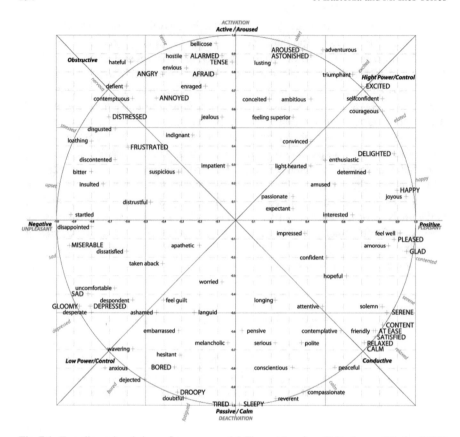

Fig. 7.1 Two-dimensional circumflex space model. Figure taken from Paltoglou and Thelwall [20]

7.2.2   First Annotation Procedure

7.2.2.1   Expressing Annoyance During Phone Calls

Often, we find irate customers who call Telecommunications services for a variety
of reasons because of slower broadband, landline fault or bill problem, among other
things. The Spanish call center company involved in this work offers customer assistance for several phone, tv and internet service providers. The customer complaint
services of these companies receive a certain number of phone-calls from angry or
annoyed customers. But the way of expressing annoyance is not the same for all the
customers. Some of them are furious and shout; others speak quickly with frequent
and very short micropauses but do not shout [15], others seems to be more fed-up
than angry; others feel impotent after a number of service failures and calls to the
customer service.

Table 7.1 Expert annotations labels and duration of each audio analyzed

Expert annotation labels	Duration	Customer speech duration
Disappointed	00:00:42	00:00:31
Angry 1	00:00:42	00:00:10
Angry 2	00:00:35	00:00:24
Extremely angry	00:16:20	00:07:50
Fed-up	00:01:08	00:00:20
Impotent	00:01:02	00:00:24
Annoyed in disagreement	00:01:35	00:00:29

7.2.2.2   Call Center Annotation Procedure

Due to the high number of recordings and issues regarding privacy, among other things, many audios were analyzed by experienced Call Center operators. These latter drew up a dataset and seven conversations with angry situations were selected and annotated as very annoyed customers. The customers show discomfort and disagreement because of service failures. In a second step, each conversation was named with the specific level of annoyance showed by the customer. Thus call center operators qualified the seven subjects in conversations as follows: *Disappointed*, *Angry* (2 records), *Extremely angry*, *Fed-up*, *Impotent* and *Annoyed in disagreement*. All these feelings correspond to the different ways the customer in the study expressed their annoyance with the service provided. More specifically they correspond to the way the human operators perceived customer feelings. Table 7.1 shows the duration of the conversations.

7.2.3   Second Annotation Procedure

Call center operators annotated each full audio file with one label. However we wanted to annotate segments inside each audio file since our final goal is to detect shifts in emotional degree appearing during the conversation. Thus we carried out the following segment-level annotation process.

7.2.3.1   Annotation Tool

The second annotation procedure was carried out using the *Praat* [7] software tool, which is a package for the scientific analysis of speech in phonetics designed to record, analyze synthesize and manipulate speech signals. Figure 7.2 shows a screenshot of the annotation procedure using *Praat* where from the top down we can observe

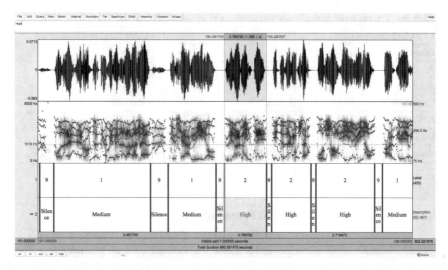

Fig. 7.2 Annotation example of 20 s interval of *Very Angry* audio using *Praat*. From the top down oscillogram tier, spectrogram tier, label tier and label-description tier are represented respectively

four tiers representing respectively the audio waveform, the spectrogram joined with acoustic features, the annoyance numbered scale and the annoyance worded scale.

7.2.3.2   Annotating Degrees of Annoyance

For this second annotation procedure, two Spanish male members of the research group [14, 15] annotated manually the segments. The task was done separately by each annotator in order to analyze and join annotations later. The age of each participant was 25 and 26 years respectively. Both Spanish annotators were graduate students.

First of all, audios were divided marking time steps into discrete segments such as customer speech-valid for these experiments-, operator speech, silence and over-lapping. Then, annotators were asked to locate the changes of annoyance from the customer giving just one instruction: they had to identify up to three different levels of annoyance: zero for neutral or very low, one for medium and two for high degree. Table 7.2 shows the annotation labels chosen.

7.2.3.3   Annotation Difficulties

In the annotation procedure, a high level of disagreement between both annotators was obtained. Annotating and classifying emotions is a complex task, where usually a significant level of disagreement is present due to context, individual and cultural factors [10] that create an environment of subjectivity. Moreover, in this labeling

Table 7.2 Segment annotation levels

Label	Description
0	Neutral or very low
1	Medium
2	High
	Silence
	Operator or overlapping segments

Table 7.3 Level scale defined

Ann. 1 label	Ann. 2 label	Final category
0	0	Very low
0	1	Low
1	1	Medium
1	2	High
2	2	Very high

process the annotators had to identify between different levels of anger, which is a further difficulty because of the different ways to expressed anger by each customer analyzed. The behavior of each annotator to annotate, resulted many disagreements. Second annotator, tended to perceive one lower degree than the first one. For this reason, we can observe many segments annotated as zero/one and one/two, while one/zero and two/one were rare. Figure 7.3 shows graphically disagreements between annotators. This issue between annotators appeared because the second-one paid more attention to the words and the first one paid more attention to the tone of voice. These reasons could explain why some number of zero/two were present in *Very*

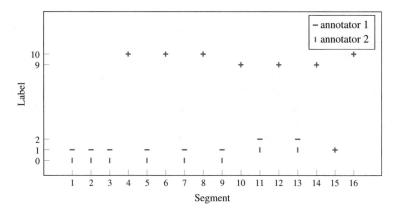

Fig. 7.3 Disagreements between annotators in *Disappointed* audio

Angry audio file. However, the annotator agreement was high in the identification of the time steps where they perceived changes in the degree of expression. Then, just one of them was chosen to fix segments bounds.

7.2.3.4 Defining Level Scale

At this point we needed to arrange disagreements defining an extended scale of annoyance expression. On this scale we have three agreed degrees corresponding to *very low*, *medium* and *very high*. The latter degree is only present in the audio entitled as *very angry*. Table 7.3 shows segment annotation label. Furthermore, we have two disagreed degrees corresponding to *low* and *high*, which correspond to *low-medium* and *medium-high* disagreements respectively. Table 7.4 shows the final number of segments and frames for each audio file and annoyance level. In this Table, the audio *annoyed in disagreement* does not appear because of not changes on the customer rate of annoyance during the conversation.

7.3 Features

In this section, we describe the primary features used as well as the procedure and the features package selected. Primary features extraction, as well as the annotation procedure described in Sect. 7.2.3, was carried out using the *Praat* [7] software tool. In this case, scripts were written and used to extract the features desired. Furthermore, we wrote scripts to calculate normalized and derivatives values.

7.3.1 Primary Features

For this work, we used four primary features like Pitch, Intensity, Mel-Frequency Cepstral Coefficients (MFCC) and Linear Prediction Coefficients (LPC). The Pitch represents the fundamental frequency (f_0) of speech signal (Fig. 7.4). The intensity is defined as the energy per unit area transmitted by the acoustic wave (Fig. 7.5). The MFCC is based on frequency domain using the Mel scale and is a representation of the real cepstral of a windowed short-time signal derived from the Fast Fourier Transform (FFT) of that signal. The LPC coefficients are calculated by LPC estimation which describes the inverse transfer function of the human vocal tract.

Table 7.4 Number of segments and frames for each audio file

	Activation level category											
	v_Low		Low		Medium		High		v_High		All	
	Segments	Frames	Segments	Frames	Segments	Frames	Segments	Frames	Segments	Frames	Segments	Frames
Disappointed	0	0	5	2095	1	456	1	3495	0	0	7	6046
Angry 1	0	0	0	0	5	1545	1	346	0	0	6	1891
Angry 2	1	412	2	1471	1	499	1	2445	0	0	5	4827
Very angry	0	0	27	9421	40	15,922	97	4,3026	45	1,9221	209	87590
Fed-up	0	0	3	783	11	3018	1	334	0	0	15	4135
Impotent	4	2600	4	1432	2	732	0	0	0	0	9	4764
All	5	3012	41	15,202	59	22,172	101	4,9646	45	1,9221	251	10,9253

Fig. 7.4 Pitch feature
represented in *Praat*

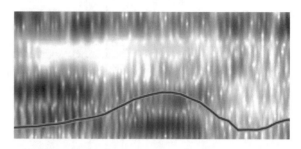

Fig. 7.5 Intensity feature
represented in *Praat*

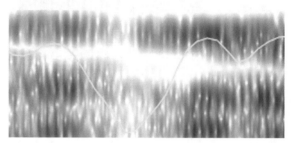

7.3.2   Procedure

In these experiments, we chose a short-term and a long-term analysis of speech signal
using 20 ms overlapping window and 330 ms window [14] respectively. This period
roughly corresponds to the word utterance level for Spanish tongue. In our work it
includes 66 overlapped frames each 5 ms. Overall, we extracted 350 values per frame
such as primary, normalized, derivative and suprasegmental parameters.

On the one hand, as primary local features, we calculated 35 parameters: Pitch,
Intensity, 5 Formants, 12 Mel-Frequency Cepstral Coefficients (MFCC) and 16 Linear Prediction Coefficients (LPC). Then we also calculated 35 normalized values
using the mean from all the speaker frames and 70 derivative values -first and second
derivatives- by comparing the local value with the following frame.

On the other hand, as suprasegmental features, we calculated 210 parameters: the
smoothed value, i.e., mean value over the suprasegmental analysis window, smoothed
first and second derivatives, the standard deviation and the standard deviation of the
first and second derivatives.

7.4   Experimental Evaluation

The experiments carried-out had a two-fold objective: one was to analyze the validity
of the assumptions made in the annotation procedure, i.e., the defined set of classes,
the other was to select a set of discriminative parameters for further frame classification. To this end two sets of experiments were carried out. In the first one we got

speaker dependent results, whereas in the second one all the audios were included in global experiments.

7.4.1 Experimental Framework

Several tools are used to analyse and experience such as Accord Framework, CUDA or Scikit-Learn. This latter was used in these experiments. More specifically, we used a parametric Naïve Bayes Classifier (NB) and the Support Vector Machine (SVM), which is a no parametric learning model. We also used k-Nearest Neighbors (k-NN) to compare with SVM due to the results of the latter.

7.4.1.1 Classifiers

Naïve Bayes Classifier assumes that features are conditionally independent but they are distributed according with some parametric distribution whose parameters have to be estimated during the learning procedure, typically using a Maximum Likelihood Estimate (MLE). Bayes theorem states the following relationship:

$$P(y|x_1, \ldots, x_n) = \frac{P(y)P(x_1, \ldots, x_n|y)}{P(x_1, \ldots, x_n)},$$

where y represents a class variable and x_1 through x_n a dependent feature vector. In our use case, we have chosen to use the Gaussian Naive Bayes algorithm for classification. The likelihood of the features is assumed to be Gaussian:

$$P(x_i|y) = \frac{1}{\sqrt{2\pi\sigma_y^2}} exp\left(-\frac{(x_i - \mu_y)^2}{2\sigma_y^2}\right).$$

On the other hand geometric classifiers do not assume any distribution of data and focus on searching boundaries. In particular Support Vector Machine (SVM) looks for hyperplanes separating data in different classes in high dimensional spaces, whereas K-NN classification is based on distance between class prototypes. The disadvantage of the SVM classifier with respect both NB and k-NN classifiers is the time complexity. An SVM is a quadratic programming problem that can scale between:

$$O(n_{features} \times n_{samples}^2) \text{ and } O(n_{features} \times n_{samples}^3).$$

For this reason, the compute and storage requirements increase depending on the dataset.

7.4.1.2 Experiments

Table 7.4 show us the number of segments and set of frames used. Categorizing and segmenting speech segments in different degrees of annoyance are difficult and subjective tasks, making the frame classification process which is our goal in this work even more complex.

Then two series of classification experiments were carried out using two different sets of features. To this end we shuffled the set of frames of each audio file and then split this set into a training and a test set that included 70% and 30% of frames, respectively. In both series of experiments we used the frame classification accuracy as evaluation metric. Then we also calculated precision, recall and F-measure values for each category defined in Sect. 7.2.3. The metrics have been selected to evaluate the quality of the frame classification. In both Sects. 7.4.2.1 and 7.4.2.2 we used accurary and F-measure as most representative evaluation metrics of each experiments. F-measure is defined as follows:

$$F = 2\frac{recall \times precision}{recall + precision}$$

7.4.2 Speaker Dependent Experiments

7.4.2.1 First Series of Experiments

The goal of this series of experiments was to analyze the performance from the previous cited three classifiers using the selected sets of features and classifying annoyance degrees for each speaker.

Firstly, SVM and NB models were used to classify frames in order to evaluate the following set of features: LP and MFC coefficients, Formants, Pitch, Intensity and two more sets adding the normalized value, the first and second derivatives and the suprasegmental values to Pitch and Intensity. The latter two sets resulting in a total of 10 values and are referenced in Figures and Tables as dPitch and dIntensity. We also tested the set Primary of the 35 primary local parameters. In Table 7.5 we can see the frame classification accuracy for each set of features, speaker and classifier.

We can see the best results achieved when the SVM classifier and the set dIntensity were used. The highest and the lowest accuracies were obtained in *Angry 1* and *Very Angry* audio files with 0.87 and 0.56 respectively. Suprasegmental features seem to be fairly reliable when annoyance degrees are classified. The other sets of features showed worse performance except for the *Angry 2* audio file whose results we included here can be considered only a little worse. Furthermore, LPC show slightly better performance than MFCC and dPitch unimproved Pitch performance.

Then again, the NB classifier show worse performance and the best classification accuracies for all audios were achieved by MFCC, suggesting that these spectral features matches better the parametric distribution assumption of NB classifier.

Table 7.5 Frame classification accuracy for each set of features and customer audio file, using SVM and NB

		LPC	MFCC	Formants	Intensity	dIntensity	Pitch	d_Pitch	Primary
SVM	Disappointed	0.61	0.60	0.58	0.60	**0.84**	0.60	0.61	0.57
	Angry 1	0.81	0.81	0.80	0.82	**0.87**	0.81	0.82	0.80
	Angry 2	0.60	0.51	0.50	0.50	**0.80**	0.56	0.54	0.50
	Very angry	0.47	0.47	0.47	0.47	**0.56**	0.47	0.48	0.47
	Fed-up	0.74	0.72	0.72	0.71	**0.83**	0.73	0.75	0.74
	Impotent	0.58	0.55	0.54	0.57	**0.76**	0.55	0.59	0.54
NB	Disappointed	0.56	**0.62**	0.59	0.59	0.60	0.57	0.59	0.55
	Angry 1	0.65	**0.82**	0.78	0.79	0.81	0.80	0.85	0.72
	Angry 2	0.35	**0.56**	0.54	0.51	0.49	0.52	0.36	0.50
	Very angry	0.44	**0.48**	0.47	0.46	0.45	0.47	0.37	0.40
	Fed-up	0.63	**0.75**	0.71	0.72	0.65	0.73	0.37	0.69
	Impotent	0.51	**0.56**	0.55	0.53	0.55	0.54	0.38	0.52

Finally, we can se in Table 7.5 highest accuracy results for customer entitled as *Angry 1* and lowest accuracy results for customer entitled as *Very angry*.

7.4.2.2 Second Series of Experiments

Secondly, SVM, NB and k-NN classifiers were evaluated using the following set of features: LP and MFC coefficients, dIntensity and dIntensity combined with LPC and MFCC. We compared both SVM and NB and both SVM and k-NN classifiers. The results obtained are represented in Figs. 7.6 and 7.7 where bar graphs show

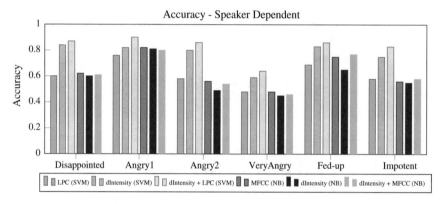

Fig. 7.6 Comparison of SVM and NB frame classification approaches. The three bar graphs on the left side of each audio file (blue, red and brown) correspond to results obtained by SVM classifier whereas the ones of the right side (grey, purple and green) corresponds to the results obtained by NB

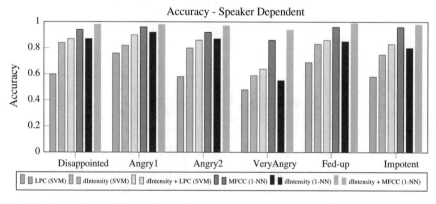

Fig. 7.7 Comparison of SVM and k-NN (k = 1) frame classification approaches. The three bar graphs on the left side of each audio file (blue, red and brown) correspond to results obtained by SVM classifier whereas the ones of the right side (grey, purple and green) corresponds to the results obtained by k-NN

the frame identification accuracies per each of features selected for each audio file. Figure 7.6 shows the frame classification accuracy for each audio and annoyance rate, when both SVM and NB models were used to classify the selected sets of features. It confirms that classification performance achieve better results classifying with SVM. The highest accuracy results were obtained combining spectral information from LP coefficients and suprasegmental information—LPC plus dIntensity features- for all audio files analyzed. The worst accuracy results were obtained with *Very Angry* achieving 0.64. In the rest of the audios we obtained an accuracy higher than 0.85 for all the speakers. The overall accuracy of this classifier with dIntensity features was 0.68 versus 0.45 of a majority guess baseline. The annoyance degree classification in this model can be measured to an accuracy better than 20% for the *Angry 1* audio file and 53% for the *Angry 2* one.

In order to extend our study, we decided to compare SVM results with a distance-based classifier. Thus, we chose K-Nearest Neighbors Classifier wich is a nonparametric classification algorithm that find a predefined number of k training samples closest in distance to the new point, and assigns their predominant label. Figure 7.7 shows highest accuracies values with k-NN for k = 1 when MFCC features were added to the dIntensity ones. K-NN classifier obtained an accuracy of 0.91 for *very angry* audio file and an accuracy around 0.98 for al the others customers. It should be noted that we are trying to classify segments at frame-level. For this reason, it is easy to find the closest prototype and we found the SVM classifier more robust and reliable.

Figures 7.8, 7.9, 7.10, 7.11 and 7.12 show the frame classification F-measure obtained by each annoyance rate for each audio file, when SVM model was used to classify the selected sets of features. Bar graphs in the Figures show the frame identification F-measure for each class, based on each of the sets of features selected for these experiments. They confirm that the set of dIntensity along with the LPC values led to the best frame classification results for every activation rate. Figures 7.9

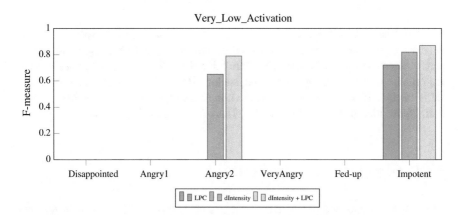

Fig. 7.8 Comparison of SVM frame classification approach for *Very Low* activation value

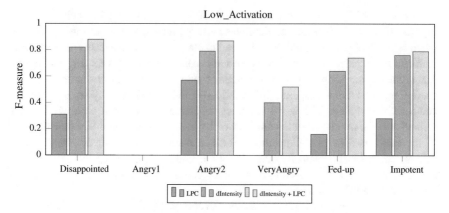

Fig. 7.9 Comparison of SVM frame classification approach for *Low* activation value

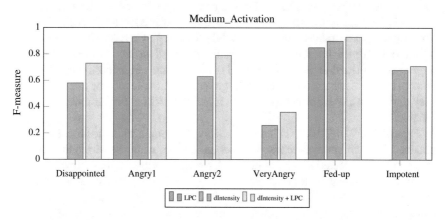

Fig. 7.10 Comparison of SVM frame classification approach for *Medium* activation value

and 7.10 also show poor classification rates for *Low* and *Medium* activation rates in *very angry* audio file, which explains the lower overall frame classification accuracy obtained in this audio file (see Table 7.5 and Fig. 7.6). Finally, we can see the importance of the suprasegmental analysis included in the intensity analysis when shifts have to be detected. However the combination with spectral features led to even higher identification rates both in terms of Accuracy and F-measure.

7.4.3 Global Experiments Results

Global experiments was aimed at analyzing the behavior of the best sets of features in previous experiments (see Fig. 7.6) when all the frames in audio files were shuffled to be considered in a global experiment. To this end we split the whole set of frames

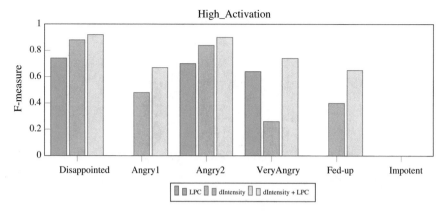

Fig. 7.11 Comparison of SVM frame classification approach for *High* activation value

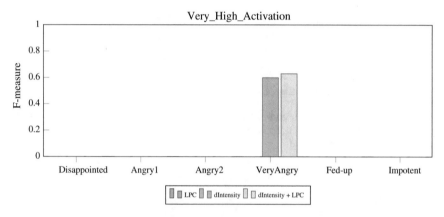

Fig. 7.12 Comparison of SVM frame classification approach for *Very High* activation value

into ten folders to carry out a 10-Cross Validation evaluation procedure. The whole sample set included 109253 frames. Silence frames were not considered in this task.

Figure 7.13 shows lower classification rates than the previous speaker dependent experiments in Fig. 7.6. However, similar behavior of the set of features and classifiers is observed in this case. The highest frame classification accuracy, 0.95, was obtained when K-NN classifier was used for *dIntensity + MFCC* feature vector. K-NN classifier continues giving excellent results due to the reasons explained in Sect. 7.4.2.2. Anyway, in Fig. 7.14 we can observe F-measure scores around 0.7 and 0.8 using SVM and NB classifiers respectively for *Very Low*.

Fig. 7.14 shows the F-measure obtained by both SVM and K-NN classifiers for each category representing the global anger degree defined in Sect. 7.2 through a cross-validation evaluation procedure of the sets of features in Fig. 7.6.

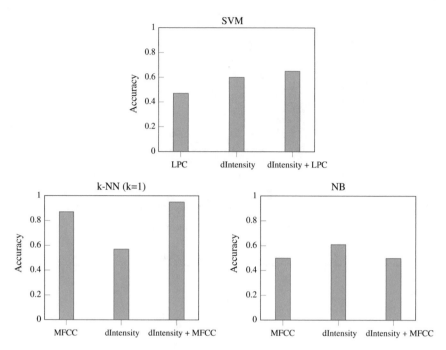

Fig. 7.13 Frame identification accuracy for each category and feature vector when SVM, k-NN and NB classifiers were used in a global cross validation experiment where all frames extracted from all audio files were included

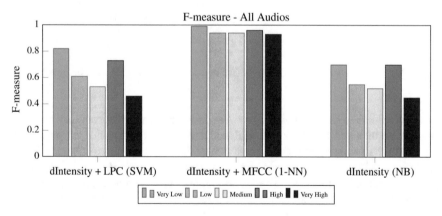

Fig. 7.14 F-measure values computed for each category in this second series of experiments by SVM, k-NN and NB classifiers through the feature sets in Fig. 7.6 and a cross validation procedure

7.5 Conclusions

In this work we addressed the detection and the of the spontaneous expression of annoyance during real conversations between customers and operators of phone customer services, which is a really novel and challenging goal. Our final work has been to identify and tracking shifts in customer annoyance during the phone-calls, in Spanish. To this end we have analyzed a set of audio files showing different ways to express the annoyance or anger of the customer. Some of them were furious and shouted; others spoke quick with frequent and very short micropauses but did not shout [15], others seemed to be more fed-up than angry; others felt impotent after a number of service failures and calls to the customer service. A number of seven conversations were analyzed in the work. In them the call center operators found disappointment, impotence or anger as expression of annoyance in these audio files. The work has also discussed the annotation problem arising when dealing with human labelling of subjective events. In this work an extended rating scale has been proposed in order to include annotators disagreements. Then a certain amount of features were evaluated as potencial hints to track annoyance degrees through parametric and geometric classifiers. Local features including acoustic and prosodic parameters, their normalized values, derivatives and a set of suprasegmental parameters have been extracted. Intensity and intensity-based suprasegmental features has shown to be very robust to identify class boundaries in every audio file analyzed. A combination of intensity-based suprasegmental features with LPC coefficients led to the best frame classification accuracies for all the expressions of annoyance analyzed for SVM classifier, whereas the combination with MFCC coefficients got the best results when K-NN classifier was used.

The obtained frame classification results validated the chosen annotation procedure. However it should be extended to a higher number of both conversations and annotators so that the procedure could be adjusted and improved. Experimental results also showed that shifts in customer annoyance rates could be potentially tracked during phone calls.

One of the main goals of this study was design and implement a tool capable of monitoring the different customer annoyance degrees. The annotation process is not an easy task because of the different issues mentioned in Sect. 7.2.1. These factors appear when annotators begin to annotate data as can be proven by Sect. 7.2.3.3. Thus, the subjectivity of this process is directly reflected in recognition performance because the input subjectivity will generate output subjectivity. We can try to solve this problem in some environment cases or cases where the subject analyzed is known thanks to the analyst cognitive capacity. For this reason, at this early stage, we can consider the synergy between the Decision Support System (DSS) and the call center in order to gain objectivity and later adjust the DSS.

The problem addressed in this work is a good example to show synergies between humans and artificial cognitive systems, between engineering and cognitive sciences [5]. In particular the behavior of the classifiers to identify different activation levels at frame level and their ability to identify shifts in annoyance level can be interpreted as a measure of the human cognitive load when dealing with the same problem.

Acknowledgements This work has been partially funded by the Spanish Science Ministry under grant TIN2014-54288-C4-4-R and by the EU H2020 project EMPATHIC grant N 769872.

References

1. Anagnostopoulos CN, Iliou T, Giannoukos I (2015) Features and classifiers for emotion recognition from speech: a survey from 2000 to 2011. Artif Intell Rev 43(2):155–177
2. Ashwin C, Chapman E, Colle L, Baron-Cohen S (2006) Impaired recognition of negative basic emotions in autism: a test of the amygdala theory. Social neuroscience 1(3–4):349–363
3. Ayadi ME, Kamel MS, Karray F (2011) Survey on speech emotion recognition: features, classification schemes, and databases. Pattern Recognit 44(3):572–587
4. Baranyi P, Csapó A (2012) Definition and synergies of cognitive infocommunications. Acta Polytech Hung 9(1):67–83
5. Baranyi P, Csapó A, Sallai G (2015) Cognitive Infocommunications (CogInfoCom). Springer International
6. Ben-David BM, Multani N, Shakuf V, Rudzicz F, van Lieshout PHHM (2016) Prosody and semantics are separate but not separable channels in the perception of emotional speech: test for rating of emotions in speech. J Speech Lang Hear Res 59(1):72–89
7. Boersma P, Weenink D (2016) Praat: doing phonetics by computer. Software tool, University of Amsterdam, version 6. 0.15. http://www.fon.hum.uva.nl/praat/
8. Clavel C, Callejas Z (2016) Sentiment analysis: from opinion mining to human-agent interaction. IEEE Trans Affect Comput 7(1):74–93
9. Devillers L, Vidrascu L, Lamel L (2005) Challenges in real-life emotion annotation and machine learning based detection. Neural Netw 18(4):407–422
10. Eskimez SE, Imade K, Yang N, Sturge-Apple M, Duan Z, Heinzelman W (2016) Emotion classification: how does an automated system compare to naive human coders? In: Proceedings of the IEEE international conference on acoustics, speech and signal processing (ICASSP 2016), pp 2274–2278. https://doi.org/10.1109/ICASSP.2016.7472082
11. Esposito A, Esposito AM, Likforman-Sulem L, Maldonato MN, Vinciarelli A (2016) Recent advances in nonlinear speech processing, chap on the significance of speech pauses in depressive disorders. In: Results on read and spontaneous narratives. Springer International Publishing, Cham, pp 73–82
12. Girard JM, Cohn JF (2016) Automated audiovisual depression analysis. Curr Opin Psychol 4:75–79. https://doi.org/10.1016/j.copsyc.2014.12.010
13. Irastorza J, Torres MI (2016) Analyzing the expression of annoyance during phone calls to complaint services. In: 2016 7th IEEE international conference on cognitive info communications (CogInfoCom). IEEE, pp 103–106
14. Iturriza M (2015) Identificacin de activacin emocional adaptada a cada locutor. Graduation thesis Universidad del País Vasco
15. Justo R, Horno O, Serras M, Torres MI (2014) Tracking emotional hints in spoken interaction. In: Proceedings of VIII Jornadas en Tecnología del Habla and IV Iberian SLTech Workshop (IberSpeech 2014), pp 216–226
16. Kim JC, Clements MA (2015) Multimodal affect classification at various temporal lengths. IEEE Trans Affect Comput 6(4):371–384
17. Koeda M, Belin P, Hama T, Masuda T, Matsuura M, Okubo Y (2013) Cross-cultural differences in the processing of non-verbal affective vocalizations by Japanese and canadian listeners

18. Meilán JJG, Martínez-Sácnhez F, Carro J, López DE, Millian-Morell L, Arana JM (2014) Speech in alzheimer's disease: can temporal and acoustic parameters discriminate dementia? Dement Geriatr Cognit Disord 37(5–6):327–334

19. Mencattini A, Martinelli E, Ringeval F, Schuller B, Natlae CD (2016) Continuous estimation of emotions in speech by dynamic cooperative speaker models. IEEE Trans Affect Comput PP(99):1–1. https://doi.org/10.1109/TAFFC.2016.2531664

20. Paltoglou G, Thelwall M (2013) Seeing stars of valence and arousal in blog posts. IEEE Trans Affect Comput 4(1):116–123

21. Ringeval F, Eyben F, Kroupi E, Yuce A, Thiran JP, Ebrahimi T, Lalanne D, Schuller B (2015) Prediction of asynchronous dimensional emotion ratings from audiovisual and physiological data. Pattern Recognit Lett 66:22–30

22. Rump KM, Giovannelli JL, Minshew NJ, Strauss MS (2009) The development of emotion recognition in individuals with autism. Child Dev 80(5):1434–1447

23. Schuller B, Batliner A, Steidl S, Seppi D (2011) Recognising realistic emotions and affect in speech: state of the art and lessons learnt from the first challenge. Speech Commun 53(9–10):1062–1087

24. Valstar M, Schuller B, Smith K, Almaev T, Eyben F, Krajewski J, Cowie R, Pantic M (2014) Avec 2014: 3D dimensional affect and depression recognition challenge. In: Proceedings of the 4th international workshop on audio/visual emotion challenge, ACM, New York, NY, USA, AVEC '14, pp 3–10

25. Ververidis D, Kotropoulos C (2006) Emotional speech recognition: resources, features, and methods. Speech Commun 48(9):1162–1181

26. Vidrascu L, Devillers L (2005) detection of real-life emotions in call centers. In: Proceedings of interspeech'05: the 6th annual conference of the international speech communication association, ISCA. Lisbon, Portugal, pp 1841–1844

27. Wang K, An N, Li BN, Zhang Y, Li L (2015) Speech emotion recognition using fourier parameters. IEEE Trans Affect Comput 6(1):69–75

28. Wollmer M, Eyben F, Reiter S, Schuller B, Cox C, Douglas-Cowie E, Cowie R (2008) Abandoning emotion classes—towards continuous emotion recognition with modelling of long-range dependencies, pp 597–600

Chapter 8
Modeling of Filled Pauses and Prolongations to Improve Slovak Spontaneous Speech Recognition

Ján Staš, Daniel Hládek and Jozef Juhár

Abstract This chapter summarizes the results of modeling filled pauses and prolongations to improve Slovak spontaneous speech recognition, by introducing them into the language model and speech recognition dictionary. We propose a hybrid method of robust statistical language modeling that combines hidden-event filled pause modeling with the random distribution of prolongations in a corpus of the Slovak written texts. We decided to use existing triphone context-dependent acoustic models designed for regular words for implicit acoustic modeling of selected types of hesitation fillers. Filled pauses are modeled and represented in a recognition dictionary by a small set of phonetic classes with similar acoustic-phonetic properties. We significantly improved the robustness for individual speakers in the task of transcription of the Slovak TEDx talks and speech recognition performance up to 4.56% for prolongations and 7.90% for filled pauses, relatively on average.

8.1 Introduction

The performance of automatic speech recognition depends on the task and availability of spoken and written language resources for the given language. The complexity of speech recognition system is influenced by the target speaker's speaking style, the size of vocabulary and language model (LM) used in the system.

J. Staš (✉) · D. Hládek · J. Juhár
Faculty of Electrical Engineering and Informatics, Department of Electronics and Multimedia Communications, Technical University of Košice, Park Komenského 13, 04200 Košice, Slovakia
e-mail: jan.stas@tuke.sk

D. Hládek
e-mail: daniel.hladek@tuke.sk

J. Juhár
e-mail: jozef.juhar@tuke.sk

© Springer International Publishing AG, part of Springer Nature 2019
R. Klempous et al. (eds.), *Cognitive Infocommunications, Theory and Applications*,
Topics in Intelligent Engineering and Informatics 13,
https://doi.org/10.1007/978-3-319-95996-2_8

Moreover, each speech recognition system, besides complexity, is strongly language dependent. Especially, for highly inflective languages like Slovak, it is also necessary to deal with the problems such as data sparseness, free word order, rich morphology and variability of pronunciation in spontaneous speech.

Spontaneous speech is the most challenging task in the design and development of large vocabulary continuous speech recognition (LVCSR) system for any language. The variability of natural speech is significantly larger when compared to the prepared speech [24] due to the increasing number of speech disruptions and a lot of hesitation fillers in conversations. Hesitation fillers such as filled and silent pauses, word prolongations, repetitions and other artifacts occur within the flow of otherwise fluent speech. These phenomena cannot be properly evaluated during the decoding of speech utterances and various errors arise in recognized hypotheses. Moreover, the most widely used language models are trained on written texts [9], where the equal probability is often assigned to the words or phrases with similar pronunciation but different meaning. Such language models can produce errors because the assigned probability is often different from the true probability [17].

The problem of frequently appearing errors in spontaneous speech recognition can be eliminated by introducing the most frequent hesitation fillers (such as reflexes, filled pauses, word lengthenings, repeating words, etc.) into the speech recognition dictionary and the language model training data. The relative-frequency of occurrence and distribution of these hesitation phenomena can be obtained from manually prepared annotations of speech recordings in the given language that are used in the acoustic model (AM) training. The same problems with processing and modeling of prosodic events have to be solved in the speech synthesis [1], spoken dialogue systems [16], human-computer interaction systems [18], and many others.

In the context of Cognitive Infocommunications [2], modeling of hesitation fillers in spontaneous speech can be seen as a problem of investigation of a new set of artificial cognitive capabilities to enhance speech-oriented technologies and interaction between human subjects and artificial systems. In other words, this problem can be viewed as a transition from more general cognitive capabilities (general speech recognition systems) to a specific set of cognitive capabilities (user-aware systems) [3]. In the design and development of the speech-oriented technologies, filled pause modeling extend speech recognition capabilities and the user's perceptual and cognitive capabilities. It is a typical example of how users do not "simply" interact with speech technologies because of imperfections of human speech.

This chapter describes the results of introducing selected types of hesitation fillers such as filled pauses and prolongations into the language model and speech recognition dictionary to improve the Slovak spontaneous speech recognition system. We propose a hybrid method of robust statistical language modeling that combines hidden-event filled pause modeling with random distribution of prolongations in a corpus of Slovak written texts. We decided to use triphone context-dependent acoustic models (designed for regular words) for implicit modeling of the most frequent filled pauses and prolongations. Moreover, filled pauses are modeled and represented in the speech recognition dictionary by a small set of phonetic classes with similar acoustic-phonetic properties, in similar way as it was reported in [40].

Preliminary results show a slight improvement in the recognition accuracy for individual speakers in the task of transcription of the Slovak TEDx talks.

The chapter is organized as follows. Section 8.2 discusses some related works that are relevant to our research objectives. The collection of manually prepared annotations of speech recordings that was used for the analysis of hesitation fillers in spontaneous Slovak speech is summarized in the Sect. 8.3. The investigated hesitation fillers and the proposed approach to handle them are presented in the Sect. 8.5. The next section introduces the speech recognition setup used in the experimental part of this paper. The results of modeling filled pauses and prolongations in the automatic transcription and subtitling of the Slovak TEDx talks are discussed in the Sect. 8.7. Finally, the Sect. 8.8 summarizes the contribution of this research and concludes the chapter with future directions.

8.2 Related Works

There are many research works to enhance speech recognition quality of using additional knowledge resources or different models of hesitation fillers.

Andreas Stolcke and Elizabeth Shriberg introduced the first LM that expresses the frequency distribution of hesitation fillers and repetitions in spontaneous speech. Elizabeth Shriberg shows in her Ph.D. thesis [27] that number of filled pauses occur more often at the beginning of the sentences and grow exponentially with the sentence length. Based on this observation, they created a representative set of hesitation fillers by considering surrounding context into account and specific LM that model their frequency distribution in a corpus of spontaneous speech [34]. Gauvain et al. [8] showed that the error rates increase if too many filled pauses are inserted into language model training data.

Authors in [12, 25] examined places where the hesitation fillers and repetitions frequently occur. Beňuš et al. [5] note that important sentence, clause or phrase boundaries can be followed by conversational fillers. Ohtsuki et al. [17] found that major filled pauses appear at the beginning of sentences or just after commas. To cope with the number of recognition errors caused by conversational fillers, authors introduced filled pauses into language models and reduced word error rate by 4.40% in Japanese broadcast news transcription task.

Pakhomov and Savova [19] showed that a language model based on a training corpus with uniformly distributed filled pauses, significantly improves speech recognition performance over the language model without filled pauses. Moreover, they concluded that a stochastic method with random insertion uniformly distributed filled pauses around the average sentence length yield better results, compared to random insertion at other range.

Liu et al. [12] introduced three types of LMs: 1. hidden-event word-based LM that describes the relation of keywords and disfluent events in spontaneous speech; 2. hidden-event part-of-speech (POS) LM that captures syntactically generalized patterns; and 3. repetition-pattern LM for detection of repeating words [10].

Peters [20] compared three different approaches of filled pauses modeling: 1. treat filled pauses as regular words; 2. use the LM for both words and filled pauses, but discard all filled pauses from the history; and 3. use fixed, context-independent probability for filled pauses. The author concludes that discarding filled pauses from the history is helpful if the sentence is continued after the interruption.

Schramm et al. [28] reported that using merged counts and discarding filled pauses from the language model history reduced the error rate on medical transcriptions by 2.20%, while perplexity was reduced by 7% (relative).

Watanabe made two hypotheses [38]:

Hypothesis 1 *Filled pause tends to occur at the linguistic break such as phrase, clause or sentence boundaries, where the frequency of filled pauses increases according to the strength of break.*

Hypothesis 2 *The frequency of filled pauses increases when the succeeding phrase or clause has complex structure.*

These hypotheses were supported by the comprehensive study of a corpus of academic and simulated lectures in Japanese. He also investigated the effects of filled and silent pauses on listener's comprehension.

Somiya et al. [29] reported how filled pauses used in a lecture speech influenced the listening and understanding ability of the audience. It was shown how filled and silent pauses affect the ability to listen to speech and listeners' comprehension.

Authors in [13, 14] pointed out that a lot of filled pauses and segment prolongations are followed by silence of specific length and the distribution and duration of filled pauses influences transcription of spontaneous oral presentations.

Deme and Markó [6] found that the filled pauses are more frequent in turn-initial positions in a sentence than prolongations, but they are preceded by silent pauses of average duration. Similarly, duration of silent pauses following prolongations in turn-final positions is longer than those following filled pauses in the similar places.

Prylipko et al. [22] investigated a wide range of hesitation fillers and showed their predictive potential in spontaneous speech. They concluded that all hesitation filler is a good predictor of the following word and omission of fillers from the context significantly increase the perplexity of LM.

Ohta et al. [16] studied the effect of modeling of filled pauses and silences for responses of the conversational agent in a spoken dialogue system. Experimental results showed that the filled and silent pauses positioned at the sentence boundaries can enhance the user comprehension and improve the naturalness of the responses of a spoken dialogue system. The authors pointed out that silent and filled pauses in spontaneous speech are produced by two main factors. The first is the psychological factor such as breathing, where positions are irrelevant to the linguistic breaks and the second, where pauses are placed at the boundaries of information units.

Based on the Switchboard corpus study, Duchateau et al. [7] note that the average amount of filled pauses and other hesitation fillers in conversational speech is about 6 events per 100 words, as it was confirmed in our previous work [32]. Several other research studies also indicate that filled pauses and other hesitation fillers occurring in spontaneous speech are language independent [39, 40].

It should be noted that the age and gender of a speaker play an important role in the modeling of hesitation fillers in spontaneous speech. Men are characterized by a higher degree of spontaneity and imperfection of a speech than the women. Also, older speakers bring more interruptions into the speech than younger speakers. The decisive factor is the complexity and length of the sentences [5].

All these results indicate the importance of the filled pauses and other hesitation fillers in design and development of a spontaneous speech recognition system. Moreover, representation of filled pauses in the language model training data helps to decrease model perplexity as well as errors in the speech recognition [32].

8.3   Source Text Data

The source text data (summarized in the Table 8.1), represented by manually annotated speech transcripts, that were used for analysis and classification of hesitation fillers in the Slovak language, come from the following datasets [24, 26, 33, 37]:

- **Short statements**—TV series "100 názorov" brings a diversity of views through short statements of the famous people from various regions of Slovakia.[1] The collection of speech recordings were realized between years 2011 and 2012;
- **Judicial readings**—gender-balanced acoustic database includes about 236 h of judicial readings from the court recorded in studio conditions and conference rooms using close-talk microphones, realized between years 2011 and 2012;
- **Parliament plenary sessions**—include about 125 h of 90% male and 10% female semi-spontaneous speech recorded in the main conference hall of the National Council of the Slovak Republic using gooseneck condenser microphones.[2] The speech recordings were realized between the years 2007 and 2013;
- **TEDx talks**—a collection of 220 Slovak TEDx talks and Jump Slovakia lectures realized between 2010 and 2016 years was gathered from the official YouTube channels.[3] Manually annotated part of the speech database covers 50 randomly selected talks and lectures in total duration of 11.25 h;
- **Court TV shows**—acoustic database consists of about 80 h of speech recordings obtained from TV series The Court. The speech recordings were realized between years 2008 and 2010 and are characterized by a relatively high degree of spontaneity created by professional and amateur actors;
- **TV discussions**—include about 13.5 h of speech recordings obtained from the two popular Slovak late-night TV talk shows "Večera s Havranom[4]" and "Pod Lampou[5]", recorded in 2012 and 2015 years respectively;

---

[1] http://www.100nazorov.sk/.

[2] http://tv.nrsr.sk/archiv/schodza.

[3] https://www.youtube.com/user/TEDxTalks, https://www.youtube.com/user/jumpslovensko.

[4] https://www.rtvs.sk/televizia/archiv/11690.

[5] https://www.tyzden.sk/podlampou/.

Table 8.1 Source speech databases

Speech database	Short statements	Judicial readings	Plenary sessions	TEDx talks	TV shows	TV discus.	TV news	radio news
Duration	32:08:32	236:02:45	125:14:05	11:16:46	80:17:20	13:32:09	296:37:05	15:58:45
Male	26:57:33	118:32:10	99:41:13	07:58:05	34:55:15	12:22:51	160:52:10	08:21:15
Female	04:48:53	114:34:33	22:32:20	03:02:28	27:46:37	00:51:51	132:49:33	07:25:35
Speakers	176	271	441	52	482	36	15 987	561
Male	142	134	343	36	264	30	10 322	391
Female	34	137	98	16	218	6	5 665	170
Speech rate	112.80	128.16	130.62	136.84	126.58	127.53	158.23	159.76

Table 8.2 Statistics of reflexes, filled pauses, speaker noise and prolongations on speech databases

Sspeech database	Short statements	Judicial readings	Plenary sessions	TEDx talks	TV shows	TV discus.	TV news	Radio news
Tokens	315 514	2 119 359	1 201 231	135 438	627 566	143 959	2 867 422	141 466
Regular words (%)	86.82	88.50	89.71	86.36	89.39	89.91	94.67	95.00
Reflexes (%)	7.09	8.08	7.09	7.09	6.09	5.07	4.04	4.06
Filled pauses (%)	3.39	0.35	1.42	2.27	1.53	2.81	0.70	0.28
Speaker noise (%)	1.65	2.94	1.24	2.51	1.55	1.50	0.26	0.16
Prolongations (%)	0.71	0.11	0.39	0.69	0.91	0.14	0.21	0.42
Speech rate	112.80	128.16	130.62	136.84	126.58	127.53	158.23	159.76

- **TV news**—the mixture of TV broadcast news databases consists of about 296.5 h of speech recordings acquired between years 2010 and 2016 from the main and morning TV news from various Slovak TV stations (Jednotka, Dvojka, Markiza, TV JOJ and TA3);.
- **Radio news**—include about 16 h of speech recordings acquired from the popular radio news from various Slovak radio stations (Rádio Slovensko, Rádio Lumen, Rádio Expres and Rádio VIVA) in 2010.

Each of the speech databases was manually annotated, double-checked and corrected by out team of trained human annotators with using the Transcriber tool [4].

The statistics about the percentage of reflexes, filled pauses, speaker noise and prolonged words in the mentioned corpora are summarized in the Table 8.2. Manually annotated speech transcripts were used for automatic extraction of selected hesitation fillers for the purpose of modeling of filled pauses and prolongations to improve automatic transcription of spontaneous speech. Moreover, the speech transcripts contained a large amount of different labels for the same hesitation fillers that were corrected or removed. We have created a set of tools in the PERL programming language for this purpose.

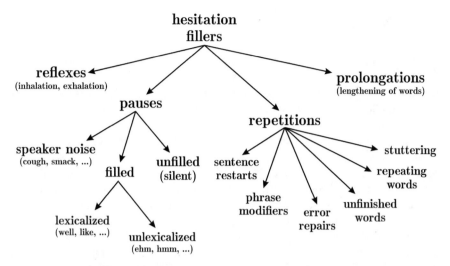

Fig. 8.1 Classification of hesitation fillers

8.4 Analysis of Hesitation Fillers in Spontaneous Speech

Hesitation fillers represent the breaks in phonation, which are usually filled with certain sounds. Such fillers are semantic and show that speaker needs an additional time to formulate the next piece of utterance [10]. Different types of hesitation fillers can be classified into 4 categories according to their duration, frequency of occurrence, and their character (see Fig. 8.1). Inspired by Rose [23], and his extensive research[6,7] we distinguish between reflexes, pauses, repetitions and prolongations. Each category will be further explained in the following sections.

8.4.1 Reflexes

Reflexes are a common part of speech. Into the group of reflexes we can include *inhalation* [i] and *exhalation* [ex]. They belong to the hesitation fillers with activity that repeats periodically. Their duration is the shortest among all non-speech events, but the frequency of reflexes is the highest. It depends on the present mental state of the speaker and his health. The frequency of breathing is indirectly proportional to the volume of the lungs and directly proportional to the speakers' speaking rate. Our analysis of the manually annotated speech transcripts showed that reflexes have rate between 6 and 7% on average, which is the highest value among all hesitation fillers.

---

[6]http://old.filledpause.com.

[7]http://filledpause.com.

8.4.2 Pauses

Hesitation fillers can be divided according to the way of expression into *filled* (i.e. vocalized fillers) and *unfilled* (also called *silent pauses*). The pause filler announces moment when the speaker pauses and thinks what she or he is going to say next. This time ends with a short speech activity or silence. Therefore, we distinguish between the silent and filled pauses. A short pause during hesitation is called a silent pause. When the speaker fills hesitation with some sound, such as *schwa* [ə] or *non-verbal sounds* like [hm], [ehm], or [uh] we speak about filled pauses. Speaker noise is a special type of pause in spontaneous speech . The speaker noise can be represented by *cough* [txh], *smack* [bb], or other activity produced by the speaker. In general, pauses alone do not have informational value, but they only fill the time until the speaker is prepared to continue. They can be used as a marker of the sentence boundaries or change of the subject of speech. As Pakhomov and Savova reported, the filled pauses have a certain distribution and well-defined function in discourse [19]. The frequency of pauses is comparable with common words in a language and it depends on the nature and spontaneity of the speech. Our analysis was based on the observation of selected types of filled pauses in manually annotated speech transcripts. The summary of our analysis is given in Table 8.3.

8.4.3 Repetitions

Repetitions significantly alter the speech content. They can be divided into several basic categories:

(a) *sentence restarts* that mark quick end of the current thought and begin of another;
(b) *phrase modifiers* that show that the speaker corrected himself;
(c) *error repairs* that arise at the moment when the speaker mistakes and usually happen at the ends of words;
(d) *unfinished words* in the form of prefixes, roots or letter combinations;
(e) *repeating words* that usually appear at the beginning of the utterance.

Initial vowel sounds during the *stuttering* is a special case of the repetitions and an accompanying phenomenon of nervousness, anxiety, fear, aggression, congenital or pathological speech disorders.

8.4.4 Prolongations

Prolongation is intentional advance of the word length in the utterance. It happens when a speaker suddenly realized something and wants to change the topic or give weight to his statements. The speech decoder often incorrectly recognizes the prolongation as a long word ending with a vowel. The other possible error is that the

Table 8.3 Statistics on filled pauses on particular subset of the speech database

Short statements		Judicial readings		Plenary sessions		TEDx talks		TV shows		TV discus.		TV news		Radio news	
ə	7 669	ə	4 442	ə	15 276	ə	2 550	ə	7 243	ə	2 954	ə	5 219	ə	303
m	545	s	211	s	00191	mm	0007	m	505	s	153	eh	4 098	s	10
ah	349	m	163	m	165	m	6	mhm	311	m	115	ehm	1 288	ah	6
s	248	mhm	100	oh	60	s	2	hm	188	ah	41	ee	1 027	h	5
oh	113	oh	61	h	54	ah	2	s	53	oh	22	eeh	819	m	5
hm	76	ah	36	mhm	39	oh	2	h	35	h	11	ehh	573	hm	5
uh	24	uh	35	eh	24	ach	1	mh	20	hm	9	hmm	410	uh	3
eh	22	hm	27	uh	20	au	1	uhm	16	uh	8	ah	153	eh	2
mm	14	h	21	š	20	mhm	1	uh	14	eh	5	aah	121	mhm	2
mhm	4	eh	16	hm	19	uh	1	hmm	5	mhm	5	mm	113	hh	1

Table 8.4 Statistics on prolongations on particular speech subset of the database

Short statements		Judicial readings		Plenary sessions		TEDx talks		TV shows		TV discus.		TV news		Radio news	
a:	786	a:	175	a:	962	a:	483	a:	1 461	a:	46	a:	1 004	a:	99
že:	254	na:	94	že:	580	že:	82	že:	598	že:	18	že:	410	že:	64
je:	74	o:	61	o:	288	je:	25	no:	271	je:	7	na:	257	na:	47
na:	60	podľa:	42	na:	260	ale:	21	ale:	158	sa:	6	je:	166	sa:	31
ale:	55	že:	37	sa:	117	to:	20	ja:	152	ako:	5	to:	150	o:	19
ako:	38	sa:	35	je:	94	na:	15	to:	136	to:	5	sa:	132	to:	17
sa:	36	do:	31	ale:	87	o:	12	sa:	124	ale:	5	o:	90	je:	17
to:	34	za:	21	za:	75	sa:	9	na:	91	no:	4	ale:	80	aby:	14
ta:k	30	pre:	20	aby:	61	ta:k	6	je:	70	pre:	3	aby:	61	ale:	11
aby:	19	je:	18	do:	61	za:	6	so:m	66	na:	3	sme:	56	ako:	9

prolongation is recognized as a pair phonemes between two words and non-existent ending word is inserted. A representative sample of words that are often prolonged in Slovak is summarized in the Table 8.4. The table shows that most of the prolonged words are monosyllabic prepositions, conjunctions, pronouns or particles.

8.5 Handling of Filled Pauses and Prolongations

Hesitation fillers cause considerable problems for any continuous speech recognition system. The filled pauses or prolongations are often recognized as short (monosyllabic) words with similar acoustic-phonetic properties. In general, hesitation fillers in the discourse are usually formed by filled and silent pauses, repeatings and prolonged words and combination of these phenomena.

We focused our research only on the filled pauses and prolongations, and modeling of their distribution because these two categories are the two most frequent hesitation fillers in spontaneous speech. In addition, filled pauses and prolongations play similar role and are considered to be complementary [16].

Several authors present various method for modeling of filled pauses and prolongations. One major group is based on stochastic processes and hidden-event language modeling [12, 17, 20, 22, 28, 34]. Another group uses random insertion of filled pauses and word lengthenings in the language model training data [19].

In this research, we propose a hybrid method that combines hidden-event filled pause modeling with random distribution of word lengthenings for robust statistical language modeling and spontaneous speech recognition in Slovak.

8.5.1 Statistical Modeling of Filled Pauses and Prolongations

It is clear from previously published works that only real occurrences of the selected hesitation phenomena and the surrounding word context can create a robust language model. The central question of statistical modeling is the optimal prediction of filled pauses and prolongations and its treatment in the language model history. Inspired by research works mentioned in the Sect. 8.2, we have proposed methods for implicit modeling and representation of filled pauses and prolongations in statistical models and recognition dictionary.

The complete process of statistical modeling and distribution of filled pauses and prolongations over a corpus of Slovak written texts is depicted in the Fig. 8.2 and described in the following sections.

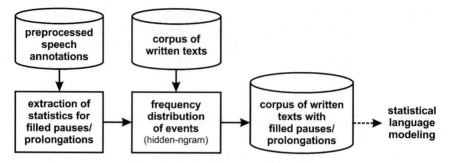

Fig. 8.2 The block diagram of statistical modeling and distribution of hesitation fillers and prolongations over a corpus of Slovak written texts

8.5.1.1   Source Data Processing

In the first step, we have processed the text data obtained from manually annotated speech transcripts into a form suitable for automatic analysis, classification, statistical modeling and distribution of selected types of hesitation fillers. We have created a set of scripts written in the PERL language for this purpose. A variety of hesitation fillers that were observed in the speech transcripts were unified and classified into four categories (according to the categories described in the Sect. 8.4).

8.5.1.2   Counting Frequency and Implicit Statistical Modeling of Filled Pauses

In the second step, we created language models with inserted filled pauses from manually annotated speech transcripts. The hidden-event language modeling was used because the exact locations of filled pauses in the sentences would be difficult to predict with a high accuracy. The hidden-event language models were created using the `ngram-count`[8] tool from the SRILM Toolkit [36], as it was reported in [15]. Frequencies and locations of filled pauses were estimated from annotations and used for their uniform distribution over a corpus of Slovak written texts.

8.5.1.3   Distribution of Hesitation Fillers over a Corpus of Written Texts

We have used the `hidden-ngram`[9] tool from the SRILM Toolkit [36] for frequency distribution of filled pauses as hidden events over a corpus of written texts. The tool uses a simple hidden Markov model-based classifier and the hidden-event n-gram model with filled pauses trained on the corpus of manually annotated speech tran-

---

[8]http://www.speech.sri.com/projects/srilm/manpages/ngram-count.1.html.

[9]http://www.speech.sri.com/projects/srilm/manpages/hiddenngram.1.html.

scripts [15, 35]. We have created customized tools in the PERL language that use random insertion based on the a-posteriori probabilities of prolongations in the given word context over a corpus of written texts, as it was published in [20]. The estimation of the a-posteriori probabilities of prolongations in the given word context was obtained from the manually annotated speech transcripts. A representative sample of the most frequent hesitation fillers in Slovak (see Tables 8.5 and 8.6) was uniformly distributed over a corpus of written texts.

8.5.1.4   Training of the Language Models on the Prepared Text Corpora

The last step is identical with statistical estimation of n-gram probabilities and training language models on large text corpora. The training data include the distribution of a representative sample of filled pauses and prolonged words that emerged from the statistical analysis of manually annotated speech transcripts obtained from the Slovak speech databases.

8.5.2   Acoustic Modeling of Filled Pauses and Prolongations

There are three different approaches of implicit acoustic modeling of filled pauses:

1. all filled pauses are modeled with one context-independent AM;
2. each filled pause has its unique context-independent AM;
3. each filled pause is modeled with context-dependent AMs designed for regular words according to its acoustic-phonetic properties [40].

As can be seen in [39, 40], experimental results performed on the Slovenian BNSI broadcast news transcription task show that filled pauses modeled with context-dependent AMs designed for regular words achieved the best speech recognition results. Therefore, we decided to use this kind of implicit acoustic modeling of filled pauses for automatic transcription of spontaneous Slovak speech. The advantage of such modeling is that it assures enough training data for the most common (vocalized) filled pauses (see Table 8.5) and there is no need for additional acoustic models. A problem may occur only if (unvocalized) unfilled pauses or speaker noises in spontaneous speech appear.

Filled pauses are modeled and represented in the speech recognition dictionary by a small set of phonetic classes, in similar way as it was reported in [40]. In this case, filled pauses with similar acoustic-phonetic properties were grouped together in one of the eight phonetic classes, summarized in the Table 8.5.

A similar principle of implicit acoustic modeling has been used for the most frequent prolongations, summarized in the Table 8.6.

Table 8.5 A representative sample of filled pauses in a small set of phonetic classes in the recognition dictionary with their phonetic transcription

Phonetic classes in the speech recognition dictionary

[ah]		[ehm]		[hm]		[m]		[oh]		[s]		[uh]		schwa	
AH	aah	EH	eeh	HM	hmhm	M	mm	O	oo	S	ss	U	uu	[@]	O=
	aa:h		ee:h		hmh		m		oo:		šš		uu:		O=:
	ah		eheh		hmm	MH	mhm		o:				u:		
	auh		eh		hm		mh	OH	ohh			UH	uh		
	ax		ex	HH	hh	MN	mn		ohm				u:h		
	a:h		e:h		h	MO	moh		oh				ux		
	a:x		e:x				mo		ooh				u:x		
AA	aa	EHM	eehm			ME	meh		oo:h			UHM	uhmm		
	aa:		ehmm				me		ox				uhm		
AHM	aahm		ehm				me:		o:h			UM	um		
	aa:hm		e:hm						o:x				u:m		
	ahmm	EE	ee					OU	ouh				umm		
	ahm		ee:						ou						
	a:hm	EM	emm												
AM	am		em												
	aum		e:m												
		EW	ev												
			eu												

Table 8.6 A representative sample of the 50 most frequent prolongations in Slovak

Prolonged words									
a:	alebo:	či:	ja:	lebo:	po:	pri:	so:m	teda:	u:ž
a:j	bola:	čiže:	je:	na:	podľa:	proste:	ste:	tie:	vlastne:
aby:	bolo:	čo:	ju:	nie:	pre:	sa:	ta:k	to:	za:
ako:	bude:	do:	kde:	no:	preto:	si:	ta:m	toho:	zo:
ale:	by:	ešte:	keby:	o:	pretože:	sme:	takže:	tu:	že:

8.5.3 Representing Filled Pauses and Prolongations in the Speech Recognition Dictionary

As Kipyatkova et al. concluded [10], the extension of the speech recognition dictionary with statistical models for each type of the hesitation fillers allows the recognition system to avoid false recognition of hesitation fillers as short words.

Considering the fact that the filled pauses are represented by a small set of phonetic classes, it is necessary to determine the conditional probability that filled pause belongs to a class. The log probabilities are estimated from the manually prepared annotations of the Slovak speech databases. Moreover, all filled pauses appear in output recognition hypotheses as transparent words { }.

The prolongations are modeled as regular words (with multiple pronunciation variants), but they can be distinguished from them by the underscore symbol at the place where word is usually prolonged (for example: "a_j", "aby_", etc.).

An example of representing selected filled pauses and prolongations in the recognition dictionary is depicted in the Fig. 8.3 and Fig. 8.4, respectively.

8.6 Speech Recognition Setup

8.6.1 Language Modeling

The trigram model of the Slovak language was generated using the SRILM Toolkit [36], smoothed with the Witten-Bell discounting algorithm and restricted to the vocabulary size of 406 k unique words. The reference model has been trained on the web-based corpus of Slovak written texts[10] of more than 282 million of tokens in 15.8 million of sentences [30].

---

[10]http://nlp.web.tuke.sk.

[ah]	@-0.325164	AH	{}	a a: h sp
[ah]	@-0.325164	AH	{}	a h sp
[ah]	@-0.325164	AH	{}	a u h sp
[ah]	@-0.325164	AH	{}	a x sp
[ah]	@-0.325164	AH	{}	a: h sp
[ah]	@-0.325164	AH	{}	a: x sp
[ah]	@-1.091080	AHM	{}	a a h m sp
[ah]	@-1.091080	AHM	{}	a a: h m sp
[ah]	@-1.091080	AHM	{}	a h m m sp
[ah]	@-1.091080	AHM	{}	a h m sp
[ah]	@-1.091080	AHM	{}	a: h m sp
[ah]	@-2.170262	AM	{}	a m sp
[ah]	@-2.170262	AM	{}	a u m sp
[ehm]	@-0.150974	EH	{}	e e h sp
[ehm]	@-0.150974	EH	{}	e e: h sp
[ehm]	@-0.150974	EH	{}	e h e h sp
[ehm]	@-0.150974	EH	{}	e h sp
[ehm]	@-0.150974	EH	{}	e x sp
[ehm]	@-0.150974	EH	{}	e: h sp
[ehm]	@-0.150974	EH	{}	e: x sp
[ehm]	@-0.766267	EHM	{}	e e h m sp
[ehm]	@-0.766267	EHM	{}	e h m m sp
[ehm]	@-0.766267	EHM	{}	e h m sp
[ehm]	@-0.766267	EHM	{}	e: h m sp
[ehm]	@-2.263196	EM	{}	e m m sp
[ehm]	@-2.263196	EM	{}	e m sp
[ehm]	@-2.263196	EM	{}	e: m sp
[ehm]	@-2.661136	EW	{}	e u sp
[ehm]	@-2.661136	EW	{}	e v sp
[hm]	@-0.254494	HM	{}	h m m sp

Fig. 8.3 Representing selected filled pauses in the recognition dictionary

8.6.2 Acoustic Modeling

In the process of speech transcription we have used gender-dependent acoustic models for both genders. Both AMs are based on triphone context-dependent 3-state hidden Markov models (HMMs) with 32 Gaussian mixtures on a state and feature vectors containing 39 mel-frequency cepstral (MFC) coefficients, along with zeros, delta and acceleration coefficients and cepstral mean normalization removal (MFCC_0DAZ) [31]. The training process was modified when tree-based state tying

a_	[a]	a a: sp
a_	[a]	a: sp
a_j	[aj]	a a i_ sp
a_j	[aj]	a a j sp
a_j	[aj]	a: i_ sp
a_j	[aj]	a: j sp
aby_	[aby]	a b i i: sp
aby_	[aby]	a b i: sp
ako_	[ako]	a k o o: sp
ako_	[ako]	a k o: sp
ale_	[ale]	a l e e: sp
ale_	[ale]	a l e: sp
alebo_	[alebo]	a l e b o o: sp
alebo_	[alebo]	a l e b o: sp
bola_	[bola]	b o l a a: sp
bola_	[bola]	b o l a: sp
bolo_	[bolo]	b o l o o: sp
bolo_	[bolo]	b o l o: sp
bude_	[bude]	b u J_ e e: sp
bude_	[bude]	b u J_ e: sp
by_	[by]	b i i: sp
by_	[by]	b i: sp
či_	[či]	tS i i: sp
či_	[či]	tS i: sp
čiže_	[čiže]	tS i Z e e: sp
čiže_	[čiže]	tS i Z e: sp
čo_	[čo]	tS o o: sp
čo_	[čo]	tS o: sp
do_	[do]	d o o: sp
do_	[do]	d o: sp

Fig. 8.4 Representing selected prolongations in the recognition dictionary

for HMMs was replaced by the triphone mapping algorithm [24]. Acoustic models have been trained on manually annotated speech recordings from the database of judicial readings, broadcast news, real parliament proceedings, and the Court TV shows (see Table 8.1) in total duration of more than 600 h.

8.6.3 Decoding

We configured a speech recognition setup based on open source recognition engine Julius [11] for automatic transcription of speech and evaluation of the quality of the proposed LMs. Julius uses two-pass recognition strategy, where the input speech utterances are decoded in the first pass with bigram LM, and the second pass uses reverse trigram model to narrow the search space of the results from the first pass.

8.6.4 Evaluation

The evaluation data sets (summarized in the Table 8.7 and Table 8.8) contain randomly selected speech utterances from the database of the Slovak TEDx talks by ten representative speakers, for both phenomena (filled pauses and prolongations) in spontaneous speech respectively. These speech segments were removed from acoustic model training and contain a total of 12 082 regular words in 1 082 short speech segments (phrases or sentences). The selected speech utterances for each speaker contain at least one filled pause or prolongation in each speech segment. It should be noted that the chosen speakers have the highest percentage of filled pauses or prolongations in each speech segment among all speakers in the entire database of the Slovak TEDx talks.

Word error rate and perplexity measures were used to evaluate a contribution of the proposed LMs with filled pauses and prolongations into the process of automatic transcription of the Slovak TEDx talks using speech recognition system.

Table 8.7 Set of representative speakers (*Eval*01) containing the highest percentage of filled pauses in each speech segment among all speakers in the entire database of the Slovak TEDx talks

*Eval*01 speaker ID	Speech segments	Regular words	Reflexes (%)	Filled pauses (%)	Speaker noise (%)	Prolonged words (%)	Speech rate (wpm)
pna14ke	138	1 231	8.33	6.87	1.10	2.11	138.65
jbe13ba	53	482	6.44	6.05	3.60	0.90	145.76
str12ba	65	743	6.26	4.99	4.31	0.18	145.39
skl14ba	106	1 385	5.32	4.65	3.44	0.17	156.13
mpe14ba	66	498	8.99	4.49	1.97	0.69	157.40
jbe14tn	91	1 174	5.56	4.42	3.56	0.66	145.76
mci14ba	57	546	8.79	3.80	2.71	1.52	128.60
mgo13ke	81	1 033	5.46	3.78	4.18	0.40	146.20
mji14ba	84	986	7.16	3.4	3.31	1.93	157.07
jma13ke	46	618	4.20	3.22	6.13	1.09	136.68

Table 8.8 Set of representative speakers (*Eval02*) containing the highest percentage of prolongations in each speech segment among all speakers in the entire database of the Slovak TEDx talks

Eval02 speaker ID	Speech segments	Regular words	Reflexes (%)	Filled pauses (%)	Speaker noise (%)	Prolonged words (%)	Speech rate (wpm)
pna14ke	43	470	8.33	6.87	1.10	2.11	138.65
jma14tn	21	221	7.41	2.29	1.44	1.94	136.22
mji14ba	58	622	7.16	3.43	3.31	1.93	157.07
jho13ba	22	260	6.40	1.17	2.00	1.82	165.75
sca14ke	32	379	6.00	1.48	2.62	1.69	133.71
mbe12ba	18	258	6.43	2.14	4.29	1.61	151.80
mci14ba	25	258	8.79	3.80	2.71	1.52	128.60
ssz14ke	30	320	9.75	1.68	3.19	1.47	145.22
jdo12ba	22	183	9.44	3.41	0.46	1.15	160.07
zkr14ke	24	415	5.63	1.81	1.81	1.09	170.73

Word error rate (*WER*) is computed by comparing reference manually annotated speech transcription against the recognized result as follows:

$$WER = \frac{N_{SUB} + N_{DEL} + N_{INS}}{N_{REF}} \times 100 \ [\%], \tag{8.1}$$

where N_{SUB} refers to the number of substituted words, N_{DEL} is related to words, which are missed out, N_{SUB} indicates the number of words incorrectly added by the recognizer and N_{REF} is the total number of words in the reference.

Perplexity (*PPL*) is defined as a reciprocal value of the (geometric) average probability assigned by the language model to each word in the evaluation data and is related to the cross-entropy $H(W)$ by the equation:

$$PPL = 2^{H(W)} = \frac{1}{\sqrt[n]{P(W)}} = \frac{1}{\sqrt[n]{P(w_1 w_2 \ldots w_n)}}, \tag{8.2}$$

where $P(w_1 w_2 \ldots w_n)$ is the probability of sequence of n words in history.

8.7 Experiments

The evaluation of the language models was based on calculating the model perplexity (*PPL*) and the performance of the automatic transcription system using a word error rate (*WER*) for the ten individual speakers obtained from the database of the Slovak TEDx talks (for more details see Sect. 8.6.4). The experiments were performed with

Table 8.9 Perplexity (*PPL*) and word error rate (*WER*) values evaluated on speech segments with filled pauses for each speaker in the *Eval*01 data set

Eval01 speaker ID	Speech segments	Regular words	Reference lang. model		Filled pauses and prolong. modeling	
			PPL_B	WER_B (%)	PPL_{HF}	WER_{HF} (%)
pna14ke	138	1 231	702.56	53.53	695.56	46.63
jbe13ba	53	482	760.47	36.72	763.16	32.99
str12ba	65	743	1 265.34	44.69	1 252.34	41.28
skl14ba	106	1 385	1 299.71	52.27	1 301.16	49.03
mpe14ba	66	498	723.34	43.57	712.59	38.35
jbe14tn	91	1 174	769.36	35.18	756.65	29.47
mci14ba	57	546	785.00	36.08	786.79	35.35
mgo13ke	81	1 033	1 071.98	60.89	1 070.28	59.54
mji14ba	84	986	619.59	48.38	606.59	45.33
jma13ke	46	618	620.89	57.77	608.53	54.05
Averaged values			861.87	46.91	855.37	43.20

trigram LMs containing the most frequent filled pauses and prolongations (LMs with HFs). All models were smoothed with the Witten-Bell discounting algorithm and restricted to the same vocabulary as the reference LM.

Tables 8.9 and 8.10 summarize the experimental results of *PPL* and *WER* values for ten selected speakers for the reference LM and LMs containing filled pauses and prolongations. The model perplexity PPL_B of the reference LM varies between 729.29 and 861.87, depending on the evaluation data. The word error rate WER_B of the reference model reaches 46.31% on average for evaluation data containing filled pauses in each speech segment and 37.25% on average for evaluation data containing prolongations.

As we can see from obtained results summarized in the Table 8.9, filled pauses in the language model training data helps to decrease perplexity PPL_{HF} about 0.75% (relative) on average. In the case of adding of the most frequent prolongations into the LM training data, there was a slight increase in PPL_{HF} because of new words in form of word lengthenings were added into the LM vocabulary. On the contrary, we have seen significant improvement in recognition performance when compared to the reference modeling. We observed a relative reduction in WER_{HF} from 0.69 to 9.32% (4.56% on average) for evaluation data set with at least one word prolongation in each speech segment and from 2.02 to 16.23% (7.90% on average) for evaluation data with filled pauses, depending on the particular speaker. Note that all experiments were performed with the same language model in which filled pauses and prolongations were modeled together.

The second experiment was focused on the examination of the impact of the filled pauses and prolongations modeling on the selected speech segments that did not contain any hesitation fillers. We examined whether the effect of modeling of

Table 8.10 Perplexity (*PPL*) and word error rate (*WER*) values evaluated on speech segments with prolongations for each speaker in the *Eval02* data set

Eval01 speaker ID	Speech segments	Regular words	Reference lang. model		Filled pauses and prolong. modeling	
			PPL_B	WER_B (%)	PPL_{HF}	WER_{HF} (%)
pna14ke	43	470	437.64	41.49	439.15	39.79
jma14tn	21	221	695.59	30.77	697.52	30.32
mji14ba	58	622	493.04	46.46	474.29	46.14
jho13ba	22	260	1 018.16	30.38	1 019.10	28.85
sca14ke	32	379	732.22	42.48	733.35	38.52
mbe12ba	18	258	638.96	46.51	641.24	44.57
mci14ba	25	258	635.14	28.29	636.36	27.13
ssz14ke	30	320	1 164.84	37.19	1 174.06	34.38
jdo12ba	22	183	834.66	34.43	839.17	32.79
zkr14ke	24	415	642.65	34.46	640.59	33.01
Averaged values			729.29	37.25	729.48	35.55

Table 8.11 Overall speech recognition results evaluated on speech segments without and with filled pauses and prolongations (hesitation fillers) and combination of both data sets together

Evaluation data set	Segments without hesitation fillers		Segments with hesitation fillers		Overall performance	
	WER_B (%)	WER_{HF} (%)	WER_B (%)	WER_{HF} (%)	WER_B (%)	WER_{HF} (%)
extended Eval01	42.59	42.30	46.91	43.20	44.64	42.73
extended Eval02	38.48	37.14	37.25	35.55	38.11	37.14

filled pauses and prolongations does not cause any degradation in recognition of "clean" speech segments or decrease of the overall recognition performance. We selected only "clean" speech segments without filled pauses and prolongations from the database of the Slovak TEDx talks for all ten speakers in both *Eval01* and *Eval02* data sets. In this way, we have created new extended versions of both evaluation data sets, *extended Eval01* and *Eval02*.

In the case of the "clean" speech segments, we observed a relative reduction in WER_{HF} in comparison to the reference WER_B up to 0.68% on average for the *Eval01* and 2.55% on average for the *Eval02* evaluation data (see Table 8.11). The overall word error rate decreased from 44.64 to 42.73% on average by merging the "clean" speech segments without filled pauses and those with filled pauses for speakers in *Eval01* data set. We observed overall reduction in the word error rate for the *Eval02* data from 38.11 to 37.14% on average.

The experimental results show the importance of hesitation fillers and prolongations modeling for improving recognition performance in the design of an automatic transcription system for spontaneous speech.

8.8 Conclusion

This chapter addresses the issue of the use of hesitation fillers and prolongations in the context of modeling the Slovak language and their influence on the recognition performance in the task of automatic transcription and subtitling of the Slovak TEDx talks. From the performed experiments we can conclude that the usage of the frequency of occurrence of the most frequent hesitation fillers and prolongations (from the manually annotated speech transcripts) and hidden-event filled pauses modeling can contribute significantly to the robustness and performance of the language models. These results were enhanced by implicit acoustic modeling of filled pauses and prolongations and representing filled pauses using phonetic classes with similar acoustic-phonetic properties in the speech recognition dictionary.

We have also examined the influence of such hesitation phenomena as repeating words on the speech recognition performance in our previous work [32]. As shown by the results obtained, modeling of selected repetitions did not produce a significant improvement in the recognition of spontaneous speech due to the fact that the tested speech recordings contain only a small amount of this kind of phenomena.

In our further research, we want to focus on statistical modeling of such hesitation fillers as *error repairs* and *repeating, unfinished* or *stuttered words* are. Also, more sophisticated decoding such as Kaldi [21] for eliminating common recognition errors is necessary.

Acknowledgements The research in this chapter was partially supported by the Ministry of Education, Science, Research and Sport of the Slovak Republic under the research projects KEGA 055-TUKE-4/2016: *"Transfer of Substantial Results of Research in Speech Technology into Education"* and VEGA 1/0511/17: *"Personalized Acoustic and Language Modeling"*, and by the Slovak Research and Development Agency under the research project APVV-15-0517: *"Automatic Subtitling of Audiovisual Content for Hearing Impaired"*.

References

1. Adell J, Bonafonte A, Escudero D (2010) Modelling filled pauses prosody to synthese disfluent speech. In: Proceedings of speech prosody, Chicago, IL, paper 624
2. Baranyi P, Csapó A (2012) Definition and synergies of cognitive infocommunications. Acta Polytech Hung 9(1):67–83
3. Baranyi P, Csapó A, Sallai GY (2015) Cognitive infocommunications. Springer International Publishing Switzerland
4. Barras C, Geoffrois E, Wu Z, Liberman M (2001) Transcriber: development and use of a tool for assisting speech corpora production. Speech Commun 33(1–2):5–22
5. Beňuš Š, Enos F, Hirschberg J, Shriberg E (2006) Pauses in deceptive speech. In Proceedings of speech prosody, Dresden, Germany, paper 212
6. Deme A, Markó A (2012) Lengthenings and filled pauses in the spontaneous speech of Hungarian adults and children. In Proceedings of workshop of fluent speech: combining cognitive and educational approaches, Ultrecht, Netherlands

7. Duchatea J, Laureys T, Wambacq P (2004) Adding robustness to language models for spontaneous speech recognition. In Proceedings of COST278 and ISCA tutorial and research workshop on robustness issues in conversational interaction, Norwich, UK, paper 11

8. Gavuvain JL, Adda G, Lamel L, Adda-Decker M (1997) Transcribing broadcast news: the LIMSI Nov96 Hub4 System. In Proceedings of 1997 darpa speech recognition workshop, Chantilly, Virginia

9. Hládek, D, Ondáš, S, Staš, J (2014) Online natural language processing of the Slovak language. In Proceedings of 5th IEEE international conference on cognitive InfoCommunications, CogInfoCom 2014, Vietri sul Mare, Italy, pp 315–316

10. Kipyatkova I, Karpov A, Verkhodanova V, Železmý M (2013) Modeling of pronunciation, language and nonverbal units at conversational Russian speech recognition. Int J Comput Sci Appl 10(1):11–30

11. Lee A, Kawahara T (2009) Recent development of open-source speech recognition engine Julius. In: Proceedings of 2009 Asia-Pacific signal and information processing association: annual summit and conference, APSIPA ASC 2009, Sapporo, Japan, pp 131–137

12. Liu Y, Shriberg E, Stolcke A (2003) Automatic disfluency identification in conversational speech using multiple knowledge sources. In: Proceedings of EUROSPEECH, Geneva, Switzerland, pp 957–960

13. Moniz H, Mata AI, Viana MC (2007) On filled pauses and prolongations in European Portuguese. In: Proceedings of INTERSPEECH Antwerp, Belgium, pp 2645–2648

14. Moniz H, Trancoso I, Mata AI (2009) Classification of disfluent phenomena as fluent communicative devices in specific prosodic contexts. In: Proceedings of INTERSPEECH Brighton, UK, pp 1719–1722

15. Nunes R, Neves L (2008) Filled pauses modeling. L^2F—Spoken language system laboratory, INESC-ID Lisboa, Technical Report, Lisboa, Portugal, p 9

16. Ohta K, Kitaoka N, Nakagawa S (2014) Modeling filled pauses and silences for responses of a spoken dialogue system. Int J Comput 1998–4308(8):136–143

17. Ohtsuki K, Furui S, Sakurai N, Iwasaki A, Zhang Z-P (1999) Recent advances in Japanese broadcast news transcription. In: Proceedings of EUROSPEECH, Budapest, Hungary, pp 671–674

18. Oviatt S (1995) Predicting spoken disfluencies during human-computer interaction. Comput Speech Lang 9(1):19–35

19. Pakhomov SA, Savova G (2000) Filled pause distribution and modeling in quasi-spontaneous speech. In: Proceedings of the 14th international congress of phonetic sciences, ICSP 1999, San Francisco, CA, pp 31–34

20. Peters J (2003) LM studies on filled pauses in spontaneous medical dictation. In: Proceedings of HLT-NAACL 2003, Edmonton, Canada, pp 82–84

21. Povey D, Ghoshal A, Boulianne G, Burget L, Glembek O, Goel N, Hannemann M, Motlicek P, Qian Y, Schwarz P, Silovsky J, Stemmer G, Vesely K (2011) The Kaldi speech recognition toolkit. In: Proceedings of IEEE 2011 Workshop on Automatic Speech Recognition and Understanding, ASRU 2011, Waikoloa, HI, US, pp 1–4

22. Prylipko D, Vasilenko B, Stolcke A, Wendemuth A (2012) Language modeling of nonverbal vocalizations in spontaneous speech. In: Sojka P et al (eds.), Text, Speech and Dialogue, LNAI 7499, Springer International Publishing Switzerland, pp 488–495

23. Rose RL (2013) Crosslingual corpus of hesitation phenomena: a corpus for investigating forst and second language speech performance. In: Proceedings of Interspeech, Lyon, France, pp 992–996

24. Rusko M, Juhár J, Trnka M, Staš J, Darjaa S, Hládek D, Sabo R, Pleva M, Ritomský M, Ondáš S (2016) Advances in the Slovak judicial domain dictation system. In: Vetulani, Z et al (Eds.), Human Language Technology Challenges for Computer Science and Linguistics, LNAI 9561, Springer International Publishing Switzerland, pp 16–27

25. Siu M-H, Ostendorf M (1996) Modeling disfluencies in conversational speech. In: Proceedings of ICSLP, Philadelphia, PA, pp 386–389

26. Sabo R (2008) Anotovaná rečová databáza parlamentných nahrávok (Annotated speech database of parliament proceedings. In: Rusko M et al (eds) Akustika a spracovanie reči. Slovakia, Bratislava, pp 131–135 (in Slovak)

27. Shriberg EE (1994) Preliminaries to a theory of speech disfluencies. Ph.D. thesis, University of California, Berkeley, p 406

28. Schramm H, Aubert XL, Meyer C, Peters J (2003) Filled-pause modeling for medical transcription. In: Proceedings of IEEE workshop an spontaneous speech processing and recognition, SSPR 2003, Tokyo, Japan, paper TM06

29. Somiya M, Kobayashi K, Nishiyaki H, Sekiguchi Y (2007) The effect of filled pauses in a lecture speech on impressive evaluation of listeners. In: Proceedings of INTERSPEECH 2007, Antwerp, Belgium, pp 2645–2648

30. Staš J, Juhár J (2015) Modeling of Slovak language for broadcast news transcription. J Electric Electron Eng 8(2):39–42

31. Staš J, Viszlay P, Lojka M, Koctúr T, Hládek D, Kiktová E, Pleva M, Juhár J (2015) Automatic subtitling system for transcription, archiving and indexing of Slovak audiovisual recordings. In: Proceedings of the 7th Language & Technology Conference, LTC 2015, Poznań, Poland, pp 186–191

32. Staš J, Hládek D, Juhár J (2016) Adding filled pauses and disfluent events into language models for speech recognition. In Proc. of the 7th IEEE international conference on cognitive InfoCommunications, CogInfoCom 2016, Wroclaw, Poland, pp 133–137

33. Staš J, Koctúr T, Viszlay P (2016) Automatická anotácia a tvorba rečového korpusu prednášok TEDxSK a JumpSK (Automatic annotation and building of a speech corpus of TEDxSK and JumpSK talks). In: Proceedings of the 11th workshop on intelligent and knowledge oriented technologies and 35th conference on data and knowledge, WIKT & DaZ 2016, Smolenice, Slovakia, pp 127–132 (in Slovak)

34. Stolcke A, Shriberg E (1996) Statistical language modeling for speech disfluencies. In: Proceedings of ICASSP, Atlanta, GA pp 405–408

35. Stolcke A, Shriberg E, Hakkani-Tur D, Tur G (1999) Modeling the prosody of hidden events for improved word recognition. In: Proceedings of EUROSPEECH, Budapest, Hungary, pp 311–314

36. Stolcke A (2002) SRILM—An extensible language modeling toolkit. In Proc. of ICSLP, Denver, CO, pp 901–904

37. Viszlay P, Staš J, Koctúr T, Lojka M, Juhár J (2016) An extension of the Slovak broadcast news corpus based on semi-automatic annotation. In: Proceedings of LREC, Portorož, Slovenia, pp 4684–4687

38. Watanabe M (2009) Features and roles of filled pauses in speech communication: a corpus-bases study of spontaneous speech. Hituji Syobo Publishing, p 147

39. Žgank A, Rotovnik T, Maučec MS (2008) Slovenian spontaneous speech recognition and acoustic modeling of filled pauses and onomatopoeas. WSEAS Trans Signal Process 4(7):388–397

40. Žgank A, Maučec MS (2010) Modeling of filled pauses and onomatopoeas for spontaneous speech recognition. In: Shabtai NR (ed) Advances in speech recognition. Croatia, Sciyo, pp 67–82

Chapter 9
Enhancing Air Traffic Management Security by Means of Conformance Monitoring and Speech Analysis

Milan Rusko, Marián Trnka, Sakhia Darjaa, Jakub Rajčáni, Michael Finke and Tim Stelkens-Kobsch

Abstract This document describes the concept of an air traffic management security system and current validation activities. This system uses speech analysis techniques to verify the speaker authorization and to measure the stress level within the air-ground voice communication between pilots and air traffic controllers on one hand, and on the other hand it monitors the current air traffic situation. The purpose of this system is to close an existing security gap by using this multi-modal approach. First validation results are discussed at the end of this article.

9.1 Introduction

Modern Air Traffic Management (ATM) is a very complex process which is essential for the steadily growing air traffic. ATM can be seen as a set of services involving different kinds of specialized and commercial-off-the-shelf systems. In the last years, the awareness for ATM security in general as well as its security gaps more and more evolved.

M. Rusko (✉) · M. Trnka · S. Darjaa
Institute of Informatics, Slovak Academy of Sciences (SAV), Bratislava, Slovakia
e-mail: milan.rusko@savba.sk

M. Trnka
e-mail: trnka@savba.sk

S. Darjaa
e-mail: utrrsach@savba.sk

J. Rajčáni
Faculty of Arts, Department of Psychology, Comenius University, Bratislava, Slovakia
e-mail: jakub.rajcani@uniba.sk

M. Finke · T. Stelkens-Kobsch
Institute of Flight Guidance, German Aerospace Center (DLR), Braunschweig, Germany
e-mail: Michael.Finke@dlr.de

T. Stelkens-Kobsch
e-mail: Tim.Stelkens-Kobsch@dlr.de

© Springer International Publishing AG, part of Springer Nature 2019
R. Klempous et al. (eds.), *Cognitive Infocommunications, Theory and Applications*,
Topics in Intelligent Engineering and Informatics 13,
https://doi.org/10.1007/978-3-319-95996-2_9

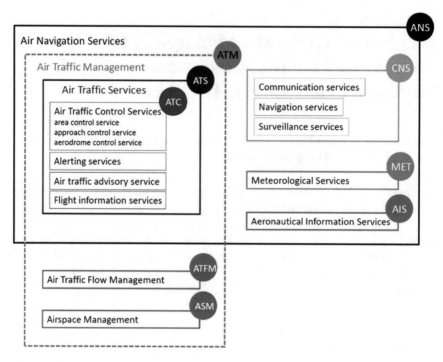

Fig. 9.1 Components of ANS [24]

To discuss the approach taken in this article a comprehensive overview needs to be established. Looking at Fig. 9.1, the Air Traffic Management may be seen as the core of the Air Navigation Services (ANS) together with Communication, Navigation and Surveillance (CNS). Associated to ATM are the Air Traffic Services (ATS) including Air Traffic Control (ATC), the Air Traffic Flow Management (ATFM) and the Airspace Management (ASM).

ATM consists of lots of complex processes and control systems and assures that [4]:

- aircraft are guided safely through the sky and on the ground and
- airspace is managed to accommodate the changing needs of air traffic over time.

Safety related issues or incidents inevitably arise with growing air traffic, but intensive research of recent years has produced a colorful bouquet of countermeasures, mitigation means and procedures to prevent, avoid or mitigate the potential hazard. These management methods are well established and used [16].

When looking at security this is rarely the case. Attackers have managed to exploit different vulnerabilities of ATM recurrently and have shown off the existing capability gaps with hazardous effects. Although the consciousness about vulnerabilities of ATM and its sub-parts rises, the ATM notices a lack to answer efficiently and consistently, which also holds true for the capability to anticipate future attacks. Potential attackers are continuously adapting to new security measures and are always up to date regarding cutting-edge technologies. Indeed the increasing implementation of Commercial Off The Shelves (COTS) hard- and software allows new exploitation possibilities never imagined before, as a lot of these systems typically never were designed for secure applications or critical infrastructures.

One of the main on-going activities relating to ATM security is led by the Single European Sky ATM Research (SESAR). SESAR is the technological pillar of the Single European Sky (SES). Its role is to define, develop and deploy what is needed to increase ATM performance and to build Europe's intelligent air transport system.

The projects and programs invented by SESAR and intended to enhance security need to examine ATM security in two complementary dimensions:

1. Assuring self-protection/resilience of the ATM System.
2. Providing collaborative support to the wider field of aviation security.

Initial work in the field of ATM security has been initiated after September 11th, 2001, and after major critical infrastructure incidents in 2003. The FP7 project GAMMA is the first project aiming to build a holistic solution for ATM Security as both an initial methodological approach and as a concept. GAMMA furthermore developed security prototypes and conducted validations of the proposed solution and the addicted prototypes. One of the envisaged prototypes shall add a security layer to ATC.

As a starting point, Air Traffic Control services consist of the following sub-services [17]:

- Area control service: separation and control of IFR flights in high altitudes,
- Approach control service: separation and control of IFR flights during their arrival and departure phase,
- Aerodrome control service: control of flights on the maneuvering area or in the vicinity of an aerodrome.

One of the most important means for providing ATC service is the issuance of ATC clearances, which are standardized worldwide to a very high degree. According to [20] an ATC clearance is defined as an authorization for an aircraft to proceed under conditions specified by an air traffic control unit. Usually, standardized phraseology is used to submit ATC clearances via voice communication, which normally consists of [21]:

- The aircraft call-sign to which the ATC clearance is addressed (e.g., "Lufthansa one two three"),
- A specific keyword characterizing the ATC clearance (e.g., "descend")
- One or more parameters for this specific clearance (e.g., "flight level seven zero").

Depending on the type of the ATC service different clearances are commonly used [20]:

- Area and Approach Control Service: en-route/approach clearances, level instructions, re-routings, radar vectors, speed instructions, instructions to change to another radio frequency.

- Aerodrome Control Service (Tower): take-off, landing, taxi, departure, start-up and push back clearances and clearances related to flights operating under visual flight rules (VFR).

These clearances are used to maintain a safe and orderly flow of traffic by maintaining prescribed separation values (e.g., radar separation) and/or establish an approach or departure sequence. The main purpose is to avert any danger of collisions [17].

Although the use of data link technology is more and more introduced especially in the Area Control Service, analogue and unsecured radio telephony is still the most common and most important way to transmit ATC clearances. Taking part in air-ground voice communication requires a formal authorization, but this authorization is currently not verified during the conversation. This creates a significant security risk, because physically an unauthorized person can easily intrude the air-ground voice communication, imitate the air traffic controller and insert false ATC clearances. Examples for recent events are [27, 41].

The work presented herein was achieved within the Global ATM Security Management Project (GAMMA Project) by a collaboration of the German Aerospace Center (DLR) and the Slovak Academy of Science (SAV).

This article describes the multi-modal approach chosen when designing a security system which is capable of identifying unauthorized and/or malicious ATC clearances within air-ground voice communication.

9.2 Multi-modal Approach to Close the Security Gap in Air-Ground Voice Communication

The work presented in this article proposes a security system which consists of a set of independent modalities and which is called Secure ATC Communication (SACom). Due to safety reasons an invasive manipulation of the existing air-ground voice communication (as it would be in case of encryption techniques, adding watermarks or when blocking content) was excluded by the basic prototype requirements defined in the GAMMA project [11].

The modalities composed in the SACom system are the following:

- Speaker verification (vocal),
- Stress detection (vocal),
- Conformance monitoring (non-vocal),
- Conflict detection (non-vocal).

All four modalities are implemented in dedicated modules, which are at first completely independent from each other, but which are later correlated by an additional correlation module to produce an overall score. This overall score shall represent the current security status of ATC voice communication. Note that stress detection is not related to linguistic stress (i.e. focus or emphasis in the intonation) but it concerns the emotional and physiological state of being under stress.

The SACom system is intended to be used at controller working positions (CWPs), but using it as a pilot assistance tool is also thinkable and possible, provided that access to required data can be ensured. The functions of the SACom need the following data as input:

- Air-ground voice communication audio stream,
- Database of authorized speakers including audio examples of their voice,
- Air traffic surveillance data (radar data),
- Reliable data about valid given ATC clearances.

Apart from its purpose as a security-related controller assistance system, the SACom is also able to automatically alert an entity at a higher level of security management, which may be a security operations center or a similar installation at local, National or European level.

At present time, a prototype for the SACom system is already available and has been validated as a standalone system as well as in an integrated configuration together with other GAMMA prototypes and one or more SMPs.

9.3 Vocal Modalities

9.3.1 Speaker Authorization

This modality is implemented in a dedicated SACom module and continuously listens to the audio stream of the air-ground voice communication. Further, the voices of all speakers taking part in this communication are compared with a database of authorized speakers.

According to the radio communication standards, the administration and supervision of radio communication regulations is to be done by a designated national authority [22]. This includes also the authorization of persons to take part in aeronautical telecommunications. For example, in Germany this national authority responsible for administration and supervision is the Federal Ministry of Economic Affairs and Energy while the Federal Network Agency is entitled to authorize the participation in aeronautical radio communication by issuing appropriate radio telephony certificates [9], which could be supplemented by a database of authorized speakers. As a conclusion, a general administration and authorization process is already in place and it seems to be compatible with the general requirements of speaker authorization technology.

The boundary conditions and challenges when introducing speaker authorization to air-ground voice communication are multifaceted:

- Speech rate: depending on the traffic situation and the ATC environment the air-ground voice communication may be very busy, involving a large number of transmissions following rapidly after each other.

- Short transmissions: the speech may contain utterances of only a few seconds in length, may be cropped by physical effects or user errors, may be (partly) interfered by other unintentional transmissions or by transmissions on a very adjacent frequency.
- Signal quality: The audio quality of aeronautical radio voice communication directly depends on the distance between the transmitter and the receiver; atmospheric effects, weather effects or terrain shading effects play a significant role and may considerably decrease the audio quality to a very poor signal-to-noise ratio.
- Variety in voice communication equipment: On the side of air traffic control as well as on board of an aircraft, a large number of different radio communication sets from different manufacturers all over the world are certified for the use in air-ground voice communication, therefore a speaker authorization function has to be compatible with all of these sets and must be internationally standardized.
- Background noise: In an ATC operations room as well as in an aircraft cockpit, some background sounds can be present which will be transmitted in addition to the pure message (e.g., background conversations, other controllers working at nearby stations, incoming transmissions on other frequencies/radios, different warning sounds, air conditioning etc.).
- Large number of possible authorized speakers: In principle every pilot, every controller, ground personnel moving vehicles on the maneuvering area of an aerodrome, persons employed in search and rescue or in aerodrome flight information service are to be considered as authorized speakers, which are most likely several hundred thousands of people all over the world. A more tactical authorization management process is therefore needed.
- International participants: In aviation it is very likely that speakers from different nations speaking with different accents are possible participants.

The scenario of use of the speaker verification in the GAMMA project is difficult from several points of view. The authorization task requires open-set speaker recognition, which is also known as Multi-target detection [40] or Open-set, text-independent speaker identification (OSTI-SI) [10]. The actual incoming speech utterance is compared to the models of all the authorized speakers. If the maximum score of all these comparisons is lower than a pre-defined threshold, the speaker is declared to be non-authorized. There are dozens of speakers, that are authorized to communicate in certain flight sector in one moment and the number of potential violators is practically unlimited. This makes the task really difficult.

In the GAMMA project we did not aim at designing a perfect speaker verification system, but rather to prove a concept of using speech verification technology to improve the robustness of the ATM voice communication against intruders. Because of that a standard speaker verification system was developed using Kaldi research toolkit [44]; the i-vector approach [2] was used with probabilistic linear discriminant analysis (PLDA) scoring [23]; and LibriSpeech corpus [45] was used for Universal Background Model (UBM) training (2500 speakers, 3 min of speech per each).

As it was known in advance, that the number of participants (ATCOs and pseudo-pilots) in the validation experiments will be limited to several persons, the evaluation

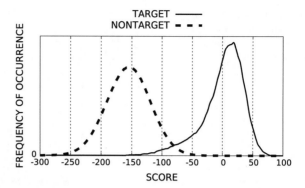

Fig. 9.2 Score distributions for target and non-target speakers with the VoxForge test set

had to be done on a third-party speech database. A part of the VoxForge database [45] was chosen as a test set to determine the classification properties of the speaker verification function of the system. 400 speakers served as target (authorized) speakers. The enrollment recordings consisted of several utterances giving approximately one minute of speech per speaker. The utterances were taken from different recording sessions when available. Other 1500 speakers were chosen as non-target (impostors, non-authorized). The total number of 80,000 test utterances was used with the length of 2–10 s each.

At the beginning the speaker verification function was tested. Each of the authorized speakers was tested against his own utterances (target) and against utterances of each of all the other speakers (non-target). The similarity between the two compared speech samples is expressed in a form of score. The results are shown on the score distribution graph in Fig. 9.2. It can be seen that the verification works well on the VoxForge data, as the score distributions of target and non-target speakers overlap only minimally and the Equal Error Rate (EER) was only 3, 2%. The reliability of speaker verification on a given test data in a form of the Detection Error Trade-off (DET) Curve [30] is presented in the Fig. 9.3.

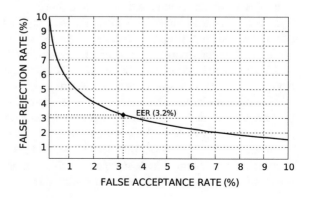

Fig. 9.3 The graph of false rejection rate versus false acceptance rate (DET curve) of the SV system tested on VoxForge speech database

Fig. 9.4 The impact of the number of authorized persons on the authorization error

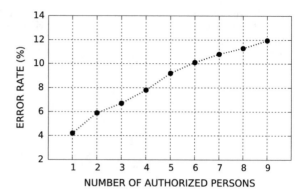

First tests of the authorization task were done on a very small database of 10 speakers and 410 test samples, recorded in the environment of the DLR's Air Traffic Management and Operations Simulator (ATMOS).

The results presented on Fig. 9.4 indicate that an increasing number of authorized persons has an effect of increasing the authorization error.

This is an expected consequence of the multiple speaker verification. While the error was as low as 4% for a simple verification of one speaker, it reached 12% when the number of authorized persons increased to 9. The authorization procedure is very sensitive to the number of authorized persons, which therefore should be kept as small as possible.

9.3.2 Stress Detection

Similar to the "Speaker Authorization" module, also the "Stress Detection" module continuously monitors the radio voice communication. It searches for known voice patterns that are typical for speech under stress. This function estimates the stress level of each utterance and provides a stress score.

It is expected that unusual situations like security (but also safety) events lead to stressful situations at the pilot's and controller's side. The measured stress score can contribute to a correlation process and therefore helps to identify such situations.

As this technology is also based on voice communication analysis, the same boundary conditions as for speaker authorization apply.

The following understanding of stress is determined by purposes of this study and may not be applicable in other contexts. As a general term, stress can be defined as a state in which physiological or psychological integrity of an individual is threatened via external or internal means [31]. The notion of threat to individual well-being is common, possibly central in defining stress. External or internal adverse forces that serve as threatening stimuli we refer to as stressors.

When using the term "speech under stress", we assume that the speaker is in a state of stress, therefore some form of pressure applied to the speaker results in perturbations of the speech production and consequently the acoustic signal [13].

Although stress was originally considered a non-specific response, changes in speech that are a result of both involuntary or autonomic bodily changes and voluntary effort, are dependent on a particular stressor. Hansen et al. [14] proposed a taxonomy of stressors, based on the mechanisms in which they influence speech process. For example, stressors with direct physical impact on speech (e.g., acceleration, vibration) are considered "zero order". Following are chemical and biological stressors—"first order", perceptual stressors (e.g., Noise, Lombard effect)—"second order", and psychological and emotional stressors—"third order". All of these types of stressors should be taken into consideration in ATM security. Moreover, due to conscious and unconscious interpretation of the stressful stimuli [14, 26] all lower degree stressors may be accompanied by a third level emotional effect.

Currently, a number of research works examined speech under stress [25, 28, 36], while other studies examined emotions in speech [28, 42]. Although, these two lines of research have been done separately, there are many common points in both approaches. In psychological theory, concepts of emotions and stress are largely interconnected [26]. Based on dimensional models of affect [35], stress may be associated with high arousal (physiological activation) and low emotional valence (unpleasantness).

In order to simplify the complex problem of stress, changes in arousal in speech were examined, using actors who read security warning messages on three levels of arousal. Although this design has its limitations, using acted stress or emotion is a standard procedure in study of emotional speech [38].

A novel speaker-independent acute-stress detector is based on emotional-arousal evaluation from speech. It was was designed using Kaldi [43]; the models use 1024 Gaussian mixtures [32].

The models of speech with different levels of arousal were created by adapting the universal model. The adaptation database of acted speech with various levels of arousal was recorded following the methodology used in [34] (15 speakers, 7 min of speech per arousal level).

Three levels of arousal neutral, increased, and high (levels 1, 2, 3) were considered in the first version. The classification accuracy is expressed in terms of recall and precision (see Table 9.1). Recall is defined as the number of true positive decisions (stress level correctly identified) over the number of true positives plus the number of false negatives (stress incorrectly rejected). Precision is defined as the number of

Table 9.1 Identification accuracy test results

	Stress level 1	Stress level 2	Stress level 3
Precision (%)	72.97	64.46	80.85
Recall (%)	92.77	51.18	75.59

true positive decisions over the number of all positive decisions. The identification accuracy of three stress levels (from 7 speakers) was tested on 553 utterances per level and speaker. The results are shown in Table 9.1.

Note, that Arousal Level 2 is lower both in recall and precision than Arousal Levels 1 and 3. This is due to the fact, that the range (the strength) of the acted arousal is very speaker-dependent. While the first and third levels created relatively distinct clusters, the Arousal Level 2 utterances from one speaker were often uttered similarly to Level 1 or Level 3 of other speakers. It may be needed in future to consider speaker-dependent stress-detection, having a neutral-speech model available for each authorized speaker.

For more details on the stress detector see [37].

9.4 Correlation with Non-vocal Modalities

As a next step, the speaker verification and the stress detection function are combined with aircraft conformance monitoring and conflict detection functions. Both functions are based on one hand on surveillance data and on the other hand on given ATC clearances.

Surveillance data can for example be delivered by:

- Primary or secondary radar systems.
- Multi-Lateration systems.
- Automatic Dependent Surveillance—Broadcast (ADS-B) [46].
- Any other system capable of providing live data about aircraft state vectors with an update rate comparable to modern radar systems.

Data regarding valid ATC clearances can for example be delivered by:

- Electronic Flight Strips (EFS) [47–49].
- Radar systems with mouse input [29].
- Speech recognition [15].

9.4.1 Conformance Monitoring

The basic principle of conformance monitoring is a continuous cross-check of gathered surveillance data (which contains all current aircraft state vectors) with given ATC clearances (which represent the expected behavior of all aircraft and the expected development of the traffic situation). In case a deviation between both is detected the conformance monitoring will issue a warning to notify the air traffic controller (or the user of the SACom System) about the particular aircraft and its behavior.

Conformance monitoring is already a well-defined functionality of ATC systems [5], but it has been used for the first time in the GAMMA project in the frame of security management.

These functions are originally designed for safety purposes and include also trajectory prediction and updating as well as reminder functions to the controller if routine actions need to be performed. Although they were not designed and used for ATM security purposes, these functions can nevertheless also be helpful in the context of ATM security management. If a pilot executes a successfully injected fake ATC clearance a conformance monitoring function would immediately provide an indication because this fake ATC clearance is not contained in the set of valid ATC clearances as none of the above listed systems captures this clearance.

As mentioned previously, precise and latest data about aircraft state vectors of all aircraft under control have to be available. Alternatively, operational systems already use the Downlink of Airborne Parameters (DAP) capability of transponder mode S to use the selected altitude as input for monitoring tools [3]. This however requires the considered aircraft to be equipped with an appropriate transponder.

The second required input is precise and complete data about the latest given ATC clearances. An overview of common ATC clearances for each type of ATC unit is given in the introduction section.

The work presented in this article is focused on approach and area control service and related procedures. The following classes of non-conformance can therefore be defined for these services:

- Vertical non-conformance: a deviation from the assigned level.
- Lateral non-conformance: a deviation from the assigned route or heading.
- Horizontal speed non-conformance: a deviation from the assigned speed.
- Vertical speed non-conformance: a deviation from the assigned rate of climb/descend.
- Trajectory non-conformance: a lateral, vertical or timely deviation from an assigned 4D-trajectory.

Further, several types of non-conformance were defined [33]:

1. The aircraft still maintains the previous ATC clearance and does not execute the new one,
2. The aircraft executes the new ATC clearance in an incorrect way,
3. The aircraft deviates from a valid ATC clearance without having received a new clearance.

After an ATC clearance has been issued the pilot may need some time to acknowledge the clearance and to operate the airplane controls (Reaction time). In addition to that, it may take some more time until the airplane fully complies with the new ATC clearance (e.g. until a turn is complete). It is further important to define tolerance areas. Depending on regulations or environmental conditions such as the weather, small deviations from the assigned level are natural and are already considered in appropriate ATC directives and navigational performance standards [18, 20]. In case

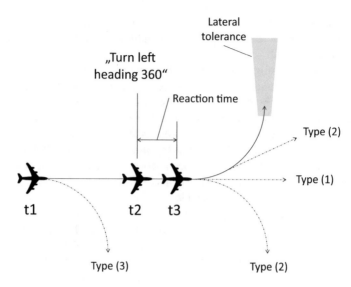

Fig. 9.5 Lateral non-conformance as used in GAMMA (t1, t2 and t3 represent consecutive steps in time)

of radar vectoring guidance, wind drift effects have to be taken into account. Regarding horizontal speed conformance, the difference between Indicated Airspeed (IAS) and Ground speed—which depends on the pressure of the outside air and therefore also on the current level of the aircraft—has to be considered carefully.

Figure 9.5 illustrates the three defined types of Lateral non-conformance as used in GAMMA in case of an instructed left turn.

9.4.2 Conflict Detection

A conflict detection function continuously monitors all aircraft in a defined area and predicts conflicts between two or more aircraft. For traffic flying according to instrument flight rules (IFR), a conflict is given when the applicable minimum vertical or lateral separation is not ensured and a risk of a collision exists [20].

For the purpose of conflict detection, it is not sufficient to consider just the current aircraft state vectors; also the given ATC clearances have to be taken into account. To illustrate this, Fig. 9.6 shows an example for a situation where vertical separation is ensured, but a risk of false alarms exists if conflict detection algorithms are only based on the present aircraft state vectors. Therefore, modern conflict detection algorithms should be (and are usually) based on predicted aircraft trajectories.

Conflict detection and prediction is already a standard functionality of modern ATC systems. According to EUROCONTROL, two types of conflict detection and prediction are distinguished: medium-term conflict detection (MTCD) [6] and short-

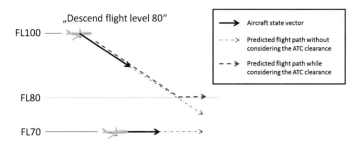

Fig. 9.6 Relevance of the ATC clearance for conflict detection algorithms

term conflict alert (STCA) [7]. MTCD is based on trajectory calculations up to 20 min in advance. In contrast to this, STCA uses specific aircraft state vectors with a pre-warning time of up to 2 min.

Conflict detection functions can therefore be enhanced to predict conflicts which would be caused by unsuitable or fake ATC clearances before they are executed.

For validation purposes, the conflict detection functions of the SACom are modeling STCA functions at its present state. These are based on both, aircraft state vector and predicted trajectories, taking given ATC clearances and the context situation ("this aircraft intends to approach and land at this airport") into account.

9.4.3 Overall Threat Indicator

Within the GAMMA concept, the SACom prototype collects individual vocal and non-vocal indicators and correlates them to an overall threat indicator score as a basis for automatic reporting. This score shall represent the current security status of ATC voice communication and can either be used to support situational awareness of the air traffic controller or can serve as a basis for automatic reporting to an entity on a higher security management level.

The following data is used as input for the correlation process:

- Speaker identity and stress score per utterance transmitted via air-ground voice communication.
- Aircraft identification and class of non-conformance as long as the deviation is still present.
- Aircraft IDs in case of a conflict between two aircraft is detected.

The used correlation process includes the following steps:

1. Definition of a time period for correlation.
2. Define which speaker verification scores/stress detection scores are considered as relevant event.
3. Count all relevant events within the defined time frame.

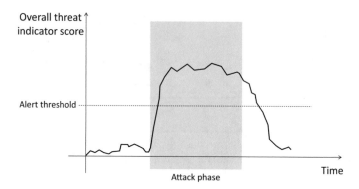

Fig. 9.7 Expected profile of the overall threat indicator during an unauthorized insertion of a fake ATC clearance

4. Apply weighting factors specific to the individual modality.
5. Sum up to an overall threat indicator score.
6. Define an alert threshold as trigger for alerting functions or automatic reporting functions.

During normal operations the overall threat indicator score should be constantly low or zero. As soon as the attack starts, it is expected that security relevant events trigger more than one modality at once, causing a rapidly increasing overall threat indicator score. An alert threshold can then be defined for issuing an alert (see Fig. 9.7).

9.5 Validation Methodology

The project GAMMA strives for a holistic approach to develop security prototypes and integrating them into an enhanced ATM security concept. In order to achieve this, the European Operational Concept Validation Methodology E-OCVM) [8] was taken as a guideline. E-OCVM is widely accepted as standard for validations in ATM. The project therefore developed an operational concept, which applies the elaborated management capabilities regarding security situations as an additional layer for air navigation services. In this context a set of key performance indicators (KPI) have been postulated, which reflect the security requirements and objectives. Within the validation activities also these KPI have been evaluated and proven to be adequate.

During the lifetime of the project a validation plan has been developed [12] which is based on E-OCVM and follows other guidelines provided by SESAR [39]. This led to a list of validation objectives describing the feasibility of the measures and the entire concept within the development phase prior to industrialization. However, the validation work needed to remain focused on a level of sub-solutions which reflect partially integrations of the holistic security system concept. In other words the full spectrum could not be validated in one attempt, which is a result of the ATM/CNS system complexity [12].

Nevertheless while setting up the validations, a baseline for the different scenarios needed to be defined. This was achieved by taking as baseline the existing non-interoperable operational concepts and system functionalities concerning security. This baseline further on served to demonstrate and validate the GAMMA benefits.

The KPI postulated provide a benchmark to rate the benefit of the developed additional controls during the validations. Especially the security KPI help to weigh the performance of the innovated solutions for improving the security of the ATM system. The KPI used within the validations were selected as they are based on the most feared events, which were identified before. The KPI are also selected to reflect the initially stated security objectives in an ideal way.

The selected and proven KPI may be reused as assessment criteria for new developed security controls in ATM, when it comes to validation. Moreover these KPI may also be utilized to measure the effectiveness of security prototypes and the integration of security prototypes into security management concepts. Taking the security KPI into account the benefit of new developments can be compared to the well-defined baseline.

The validation campaign of the SACom prototype is divided in two iterations. The first iteration took place from April to October 2016 and was focused on the SACom as a standalone system. The second iteration took place in the first half of 2017 and examined the SACom working together with other GAMMA prototypes in an integrated platform. Each iteration involved several exercise runs containing a series of human in the loop simulations conducted in real-time. The simulations took place in the Air Traffic Management and Operations Simulator (ATMOS) located in the DLR(German Aerospace Center) premises in Braunschweig (see Fig. 9.8). For simulating the air-ground voice communication an open source VoIP radio communication simulation system was used. Trained pseudo pilots communicate with the air traffic controller and steer the simulated aircraft.

In these exercise runs the Approach control unit of Düsseldorf Airport with moderate realistic traffic load was simulated. The indications of the SACom prototype were made available to the participating air traffic controller in a subset of simulations, depending on the distinct exercise step.

The following steps are included in one exercise run:

1. Introduction/Briefing: The controller is briefed about the trial campaign and the simulation environment.
2. Voice Enrollment: The controller's voice is recorded and registered in the database of authorized speakers.
3. ATMOS Training: A short simulation with low traffic load to make the controller familiar with the used equipment and local procedures.
4. Short Simulations: a set of 20 simulations of 3–6 min containing prepared traffic situations with scripted safety or security events; the SACom indications are not visible to the air traffic controller (baseline situation).

Fig. 9.8 Air traffic management and operations simulator (ATMOS) at the DLR's institute of flight guidance, Braunschweig

5. SACom Training: A short simulation with low traffic load to make the controller familiar with the SACom indications.
6. Attack Scenario: a single simulation of 45 min and average traffic load. This simulation contains a phase of normal operations lasting 15 min; then an attack phase of 20 min where an unauthorized person injects false ATC instructions into the air-ground voice communication; and finally again a phase of 10 min with normal operations.
7. De-Briefing: The controller fills out standardized questionnaires related to human factors and subjective feedback about the simulations and the prototype is collected.

Active air traffic controllers from the German air navigation service provider DFS were invited to take part in the exercise runs.

9.6 First Results and Discussion

As already mentioned, the standalone validation trials took place in 2016. The following sections give an overview of the results and lessons learned so far.

Relative Quantity

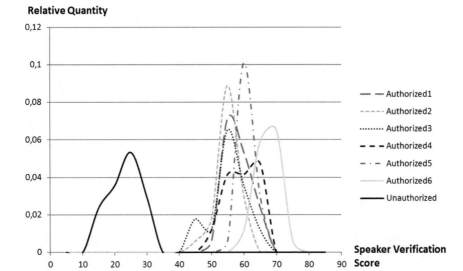

Fig. 9.9 Results of the speaker verification function for a separate trial using DLR's ATMOS with 6 authorized and one non-authorized speaker

9.6.1 Speaker Verification

As one outcome, the speaker verification can basically be considered as usable in air-ground voice communication. In the used simulation setup with VoIP audio quality, the speaker verification function could easily be integrated in existing working procedures and equipment while high speaker verification scores were obtained throughout (see Fig. 9.9). The utterances of the unauthorized speaker could clearly be distinguished from authorized utterances.

The relative quantity shown on the y-axis of Fig. 9.9 is given with decimal values as it describes the amount of occurrences relative to the whole population which is normalized to 1.

On the x-axis the speaker verificationscore is shown. This is an internal relative measure of the speaker verification system representing the similarity of the voice of the incoming utterance to the model created during the enrollment of the authorized speaker. A low Speaker Verification Score indicates a mismatch of the voice of an utterance with the database of authorized speakers.

Especially in a busy traffic environment, clearly splitting the audio into individual utterances before the speaker verification is done can be a real challenge. During the validation trials, best results were achieved when periods of silence longer than 300 ms were considered as break between two utterances.

In addition to that, the audio processing needed some time, which led to a delayed display of the speaker verification result to the air traffic controller. As a consequence the controller sometimes had difficulties to link the displayed result with a recent utterance.

The feedback of the air traffic controllers who took part in the single validations showed that they prefer to keep the current number of continuous indications on their screens at a minimum. Nevertheless, there was one controller (of six) who rated the continuous speaker verification indication as very helpful in handling the attack scenario (exercise step 6). This controller stated that the indicator raised his attention every time when it indicated unauthorized intervention. In this moment he was explicitly expecting the consequences, which prepared him to react immediately.

9.6.2   Stress Detection

As simulations are artificial situations no severe safety consequences need to be feared as it would be in reality, therefore inducing a higher level of stress was another big challenge. Controllers are also trained and used not to show their stress in the voice communication.

It has to be mentioned that the stress detection function sometimes indicated higher stress scores following a clear trend (for 2 of 6 controllers). Unfortunately, this trend could not be successfully correlated with other (subjective and unprecise) stress assessments applied for the exercise runs. Figure 9.10 shows the stress detection results during exercise step 6 for one of these controllers to give an example (measured stress score with regard to the consecutive stress measurement, red lines mark the period of the attack).

The results presented on Fig. 9.10 show that the output of the stress detector is still very noisy. This results from the infancy of stress-detection. Unclear definition of

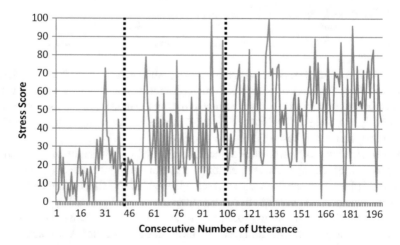

Fig. 9.10 Results of the stress detection function for exercise step 6 of a selected exercise run

stress, short utterances, small training database, need for better speaker normalization and channel compensation, unavailability of recordings representing real stress, and unclear levels of stress induced during the experiments are just a selection of reasons, why the on-line stress-detection can not work more reliably at the moment. Nevertheless a clear trend is visible (ascending signal average from approximately the beginning of the attack). This really encourages further research with regard to stress detection.

9.6.3 Conformance Monitoring

The conformance monitoring clearly showed the ability to identify deviations from the cleared flight path much faster than an air traffic controller. Conformance monitoring algorithms also showed a higher detection rate regarding aircraft deviations. Both indicators show the potential of monitoring tools for both, safety and security purposes. In the following two tables (Tables 9.2 and 9.3) we compare the detection rate and average time until detection of the SACom and the air traffic controller.

Table 9.2 Comparison of determined detection rates of the air traffic controller and the conformance monitoring function of the SACom prototype

Exercise run ID	ATCO detection rate (%)	SACom detection rate (%)	Difference (%)
VL-C1	92.0	88.0	−4
GP-C2	76.7	93.3	+16.6
AL1-C3	91.7	95.8	+4.1
AL2-C4	80.0	96.7	+16.7
AL3-C5	84.6	88.5	+3.9
AL4-C6	85.0	85.0	0

Table 9.3 Comparison of determined average times until detection of the air traffic controller and the conformance monitoring function of the SACom prototype

Exercise run ID	ATCO time until detection (s)	SACom time until detection (s)	Difference (s)
VL-C1	41.6	16.5	−25.1
GP-C2	39.4	11.8	−27.6
AL1-C3	43.1	15.8	−27.3
AL2-C4	38.7	14.5	−24.2
AL3-C5	38.9	13.9	−25.0
AL4-C6	34.7	14.1	−20.6

Obviously several factors directly influenced the performance of the air traffic controller, e.g.:

- The surprise effect and controller's expectations: especially in the first simulations (first short scenarios of exercise step 4) the controller expected a more or less normal traffic flow and was sometimes not aware of (or even surprised by) aircraft deviations.
- The focus on the traffic flow: the traffic scenario required more attention during the last phase of approaches before starting the final descent; therefore aircraft deviations occurring outside of the focus were detected later.
- The effect on safety which a possible deviation would have: if a deviation counteracts the latest preplanning of the air traffic controller it is rather discovered than a deviation which makes no difference.

Specific to the attack scenario, the conformance monitoring function was basically rated by the controllers as useful for this purpose, but indications have to be redesigned to better meet human factors. Additionally reliability in terms of false alarms has to be improved.

9.6.4 Conflict Detection

Controllers are highly trained to predict and avoid potential conflicts between two flights, therefore only very few situations occurred where the prescribed minimum separation was not ensured and a conflict was present. The performance of the prototype was comparable to the performance of the air traffic controllers, but the conflict detection function was nevertheless rated as most useful in general and also specific to the simulated attack scenario.

9.6.5 Correlation

In principle, the correlated score showed the expected behavior. The main difference is that, due to unavoidable false alarms which are also fed into the correlation process, there is always a 'basic' score level even if there are no real events in the simulation. The correlation function has to be calibrated to this basic score level. Provided that the alert threshold is correctly set, this correlation process can be used to give a security alert and also an "all-clear" signal as soon as the attack is over.

Especially the "all-clear" signal could be very beneficial as one of six controllers, who was fully aware of the security incident in exercise step 6 and who applied several countermeasures (higher separation, safety-focused guidance accepting capacity loss, informing pilots with broadcasting warnings on the radio frequency) did not notice when the attack was over and continued to apply these countermeasures without any reason.

9.7 Conclusion

The concept of using the information on human identity and behavior in combination with information communication technology to create a more secure and robust system for air traffic control exactly fits into the concept of CogInfoCom [1].

In the performed exercise runs, especially the speaker verification function and the conformance monitoring function are suitable to identify fake ATC clearances or their effects when executed by a pilot.

Clearly distinguishing a security threat from a safety event is still challenging as the effects may be very similar, e.g.:

- Emergency situations or unexperienced pilots may also be under stress.
- Technical failures or mistakes can also lead to deviations from the ATC clearance.
- ATC errors or ATC training situations may also lead to conflict situations.

Although the multi-modal approach and setup tested in the work described herein was already very successful, more independent indicators than those described in this document are needed to clearly distinguish deliberate attacks from safety events. Another point which needs to be improved is the reliability of the collected indicators. As this investigation shows, the research in the field of voice analysis of ATC operations regarding stress is still in its infancy and has a long way to go until an operational deployment can commence.

At this point of research it has to be noted that a human operator will be indispensable for the near and medium-term future as he (as a human) is able to analyze the situation very rapidly and to verify the indication as soon as an alert is raised. When it comes to tactical decisions and treatment again the human assessment and evaluation is of paramount importance. Therefore, beside the development of the system itself also a comprehensive training is needed before the SACom is brought into operation.

Acknowledgements The basic ideas of this article were briefly presented at CogInfoCom 2017. Rusko and Finke [33] The research leading to the results presented in this article has received funding from the European Union FP7 under grant agreement no 312382. More information can be found under http://www.gamma-project.eu.

References

1. Baranyi P, Csapo A, Sallai Gy (2015) Cognitive Infocommunications (CogInfoCom). Springer International. ISBN 978-3-319-19607-7
2. Dehak N, Kenny PJ, Dehak R, Dumouchel P, Ouellet P (2011) Front-end factor analysis for speaker verification. IEEE Trans Audio Speech Lang Process 19(4):788–798
3. Edmunds A (2010) Mode S—Helping to Reduce Risk. Hindsight 10 Magazine, Eurocontrol, Winter
4. EUROCONTROL, Air traffic management (ATM) explained. http://www.eurocontrol.int/articles/air-traffic-management-atm-explained

5. EUROCONTROL: Specifications for Monitoring Aids. Edition 1.0, 15th July 2010. ISBN 978-2-87497-033-7
6. EUROCONTROL: Specification for Medium-Term Conflict Detection. Edition 1.0, 15th July 2010. ISBN 978-2-87497-034-4
7. EUROCONTROL: Specification for Short Term Conflict Alert. Edition 1.0, 22th Nov 2007
8. EUROCONTROL (2010), European Operational Concept Validation Methodology, Version 3.0. https://www.eurocontrol.int/publications/european-operational-concept-validation-methodology-eocvm
9. Federal Ministry of Economic Affairs and Energy: Verordnung über Flugfunkzeugnisse (Flug-FunkV). Issued 20th Aug 2008, updated 7th Feb 2012
10. Fortuna J et al (2004) Relative effectiveness of score normalisation methods in open-set speaker identification. In: Proceedings of the speaker and language recognition workshop (Odyssey), pp 369–376
11. GAMMA Consortium, Deliverable 6.2: Prototype Requirements. Oct 2015
12. GAMMA consortium, 2015, D5.1 Validation Exercise Plan
13. Hansen JH, Patil S (2007) Speech under stress: analysis, modeling and recognition. In: Speaker Classification I, pp 108–137
14. Hansen JHL et al (2000) The impact of speech under 'stress' on military speech technology. NATO Proj 4 Rep
15. Helmke H, Rataj J, Muehlhausen T, Ohneiser O, Ehr H, Kleinert M, Oualil Y, Schuldner M, Klakow D (2015) Assistant-based speech recognition for atm applications. In: 11th FAA/Eurocontrol ATM seminar. Lissabon/Portugal, June 2015
16. ICAO International Civil Aviation Organization: Universal Safety Oversight Audit Programme—Continuous Monitoring Manual. doc. 9735, 2011
17. ICAO International Civil Aviation Organization: Air Traffic Services. Annex 11, 13th Edition, 2001
18. ICAO International Civil Aviation Organization: Performance-based Navigation (PBN) Manual. Doc. 9613, 3rd Edition, 2008
19. ICAO International Civil Aviation Organization: Manual on Global Performance of the Air Navigation System. Part I & II. Doc. 9883, February 2008
20. ICAO International Civil Aviation Organization: Procedures for Air Navigation Services—Air Traffic Management. Doc. 4444, 15th Edition, 2007
21. ICAO International Civil Aviation Organization: Manual of Radiotelephony. Doc. 9432, 4th Edition, 2007
22. ICAO International Civil Aviation Organization: Aeronautical Telecommunications. Annex 10, Volume II, 6th Edition, 2001
23. Kanagasundaram A, Vogt R, Dean D, Sridharan S (2012) PLDA based speaker recognition on short utterances. In: Proceedings of oddysey speaker and language recognition workshop. June 2012
24. Kreuz M (2015) Modellierung von Flugsicherungsprozessen auf Basis von System Dynamics. Forschungsbericht/DLR, Deutsches Zentrum für Luft- und Raumfahrt, p 33
25. Kurniawan H, Maslov AV, Pechenizkiy M (2013) Stress detection from speech and galvanic skin response signals. In: Computer-based medical systems, pp 209–214
26. Lazarus RS (1993) From psychological stress to the emotions: a history of changing outlooks. personality: critical concepts. Psychology 4:179
27. LiveATC: Fake ATC in Action (LTBA-Istanbul). 25th May 2011. http://www.liveatc.net/forums/atcaviation-audio-clips/25-may-2011-fake-atc-in-action-(ltba-istanbul)
28. Macková L, Čižmár A, Juhár J (2015) A study of acoustic features for emotional speaker recognition in I-vector representation. Acta Electrotechnica et Informatica 15(2):15–20
29. Maij A, van Dijk H (2015) Air traffic controller working position—a comparison between a single large display and multiple display set-up. Natl Aerosp Lab
30. Martin AF et al (1997) The DET curve in assessment of detection task performance. In: Proceedings of Eurospeech '9, Rhodes, Greece, Sept 1997, vol 4, pp 1899–1903

31. Mc Ewen B, Lupien, S.: Stress: hormonal and neural aspects. In: Ramachandran et al (2002) Encyclopedia of human brain, vol 4. Academic Press, pp 129–139
32. Reynolds DA (1995) Robust text-independent speaker identification using gaussian mixture speaker models. IEEE Trans Speech Audio Process 3(1):72–83
33. Rusko M, Finke M (2016) Using speech analysis in voice communication : a new approach to improve air traffic management security. In: IEEE proceedings of 7th international conference on cognitive infocommunications CogInfoCom 2016. Wroclaw, Poland, pp. 181-186. ISBN 978-1-5090-2643-2
34. Rusko M, Darjaa S, Trnka M, Ritomský M, Sabo R (2014) Alert! ... calm down, there is nothing to worry about, warning and soothing speech synthesis. In: LREC, pp 1182–1187
35. Russell JA (1980) A circumplex model of affect. J Pers Soc Psychol 39(6):1161–1178
36. Ruzanski E, Hansen JH et al (2005) Effects of Phoneme Characteristics on TEO Feature-Based Automatic Stress Detection in Speech. In: ICASSP, vol 1. pp 357–360
37. Sabo R, Rusko M, Ridzik A, Rajčáni J (2016) Stress, arousal, and stress detector trained on acted speech database. In: 18th International conference on speech and computer SPECOM 2016, Budapest, Aug 2016, pp 675–682
38. Scherer KR (2003) Vocal communication of emotion: a review of research paradigms. Speech Commun 40:227–256
39. SESAR Joint Undertaking, Project 16. 02. 04, 2013, Security Validation Process, 16. 02. 04-D03
40. Singer E, Reynolds D (2004) Analysis of multi-target detection for speaker and language recognition. In: Proceedings of the speaker and language recognition workshop (Odyssey), pp 301–308
41. The Age: Lone-Wolf Radio Hoaxer Hacks Melbourne air Traffic Control. Nov 2016. http://www.theage.com.au/victoria/lonewolf-radio-hoaxer-hacks-melbourne-air-traffic-control-afp-20161107-gsk12o.html
42. Womack BD, Hansen JH (1996) Classification of speech under stress using target driven features. Speech Commun 20(1):131–150
43. January 2016. http://kaldi-asr.org/
44. January 2016. http://www.openslr.org/12/
45. Accessed on the 22nd of Jan 2016. http://www.voxforge.org/home
46. July 2016. http://www.multilateration.com/surveillance.html
47. NAVCANatm, NAVCANstrips page, July 2016. http://www.navcanatm.ca/en/navcansuite/navcanstrips.aspx?gclid=CJPCy7el7c0CFfEV0wod9IALcQ
48. July 2016. http://saab.com/security/air-traffic-management/air-traffic-management/e-strip/
49. July 2016. http://www.frequentis.com/en/sp/solutions-portfolio/air-traffic-management/products-and-solutions/air-traffic-control-and-automation/electronic-flight-strip-handling-system/

Chapter 10
Compassion Cluster Expression Features in Affective Robotics from a Cross-Cultural Perspective

Barbara Lewandowska-Tomaszczyk and Paul A. Wilson

Abstract The present paper focuses on a comparison between expressive features of principal members of the British English emotion cluster of COMPASSION (*empathy, sympathy* and *compassion*) and their prototypical dictionary equivalents in Polish, *empatia* and *współczucie*. A cross-cultural asymmetry between the English and Polish clusters is argued to increase in the case of the Polish emotion *sympatia*, which presents a more peripheral correspondence pattern than the more polysemous concept of English *sympathy* and is shown to belong to a different Emotion Cluster (LOVE/HAPPINESS). The expression features of empathy, sympathy and compassion in British English and Polish need to be tuned accordingly in socially interactive robots to enable them to operate successfully in these cultures. The results showed that British English *compassion* is characterized by more positive VALENCE and more of a *desire to act* than Polish *współczucie*. Polish *empatia*, as juxtaposed to British English *empathy*, which has a wider range of application, presents a less consistent pattern of correspondences. The results further showed that although the processes of emotion recognition and expression in robotics must be tuned to culture-specific emotion models, the more explicit patterns of responsiveness (British English for the compassion model in our case) is also recommended for the transfer to make cognitive and sensory infocommunication more readily interpretable by interacting agents.

B. Lewandowska-Tomaszczyk
State University of Applied Sciences in Konin, Konin, Poland
e-mail: blt@konin.edu.pl

P. A. Wilson (✉)
University of Łódź, Łódź, Poland
e-mail: p.wilson@psychology.bbk.ac.uk

© Springer International Publishing AG, part of Springer Nature 2019
R. Klempous et al. (eds.), *Cognitive Infocommunications, Theory and Applications*,
Topics in Intelligent Engineering and Informatics 13,
https://doi.org/10.1007/978-3-319-95996-2_10

10.1 Introduction

At the outset it is important to underscore the main tenet of the present investigation, namely that emotions are conceptually organized in clusters of similar emotions that have a prototype structure, rather than being discrete entities that function independently of other emotions. A key feature is the proximity between emotion clusters, which exerts an influence on the internal structure of individual emotion clusters. Emotion clusters are characterized by fuzzy boundaries and are susceptible to a number of cultural influences such as religion, face, honor, and individualism-collectivism. Each of these cultural factors has an influence on ways emotions cluster, with the latter identified as one of the most significant in this respect. For example, although HAPPINESS and LOVE are two separate clusters in British English, they form one schema in Polish [24]. Our demonstration that emotion clustering is at the heart of the meaning of emotions in different cultures dictates the need for Affective Robotics modeling to incorporate such conceptual architecture if it is to allow social robots to successfully navigate the challenge of interacting with humans at an emotional level in different cultures.

The expression features that are relevant to the modeling of social robots in terms of their affective competence are derived from properties pertaining to the body, behavior and language, which are central to cross-cultural differences in the prototypes of emotions. In the current chapter we identify expression and recognition elements of empathy, sympathy and compassion in Polish and British English within the context of social robotic intelligent engineering. Such an attempt at modeling is relevant to the SID (Social Importance Dynamics Model) advanced by [27], which proposes that social importance (SI), a term that is synonymous with status, is a function of our willingness to act in favor of another social being and consider their needs and well-being to be more important than one's own. The main factors that are identified in determining how much SI is ascribed to an individual are interpersonal relations, group membership and role, task interdependence, personal attributes, and conformity to standards. [27] demonstrate how the SID can be used to model behavior on the basis of cultural dimensions. Specifically, they model a scenario in which the amount of SI attributed to an elder in a particular culture is a function of power distance of that culture, which is characterized by the degree of acceptance of unequal power distribution by less powerful society members [17]. Our work extends this research by additionally focusing on the need for cross-cultural differences in the conceptual representation of emotions to be modeled in social robots. For example, the results presented in [43] show the necessity for emotion property profiles to be adjusted in robotic systems to facilitate their emotional competence across cultures. In order to identify the key properties for empathy, sympathy and compassion, one needs to gain further insight into the nature of these emotions. We aim to support this argument, especially in terms of recognition features pertaining to emotions, but we additionally assert the necessity in social robotic modeling to emphasize an error type that is based more on false positive features (i.e. the offer of help when it is unnecessary) than false negative features (i.e. necessary help is not forthcoming), which is more consistent with cultures in which patterns of responsiveness are more explicit.

10.1.1 Empathy, Sympathy and Compassion

Contrasting empathy, sympathy and compassion allows a deeper assessment of how these emotions function in contexts that foreground the plight of others. Although these emotions often occur concomitantly, it has been proposed that there is a specific temporal order, with empathy preceding sympathy and compassion, which might be followed by prosocial behavior, depending on the circumstances [37].

10.1.1.1 Empathy

A key feature of empathy is the acknowledgment and comprehension of the negative emotions that another individual is experiencing [15]. However, it is important to further stress that, unlike sympathy and compassion, empathy involves sharing the feelings of another individual. As [37] explain, whereas the affective sharing element of empathy is characterized by *feeling with* another individual, in sympathy and compassion such sharing may be conjectured to be weaker as they are rather characterized by feelings *towards* another individual. However, historically all of these items in both languages are derived from older forms possessing the element of *sharing with*. Therefore, an individual who has an empathetic response to a person who is feeling worried will also experience worry.

10.1.1.2 Sympathy

Sympathy is an emotional response that arises when one becomes aware that someone is affected by a negative event [15]. As this awareness represents what [37] refer to as feeling *for* another individual as opposed to feeling *with* them, it is consistent with the above discussion proposing that the sharing of emotions is the domain of empathy rather than sympathy.

Research on the differences between compassion and sympathy has produced somewhat mixed results, which is consistent with the historical development of the meanings of all the lexical forms investigated here. [3] showed that for American participants compassion and sympathy loaded highly on the empathy factor in four out of five studies. Although this suggests a close relationship between compassion and sympathy, in the other study sympathy loaded on the distress factor whereas compassion loaded on the empathy factor. Similarly, in the results of the emotions sorting study performed by [35] on American participants (see Sect. 10.2.3 for more details on this methodology), compassion was a member of the LOVE cluster and sympathy was included in the SADNESS cluster.

10.1.1.3 Compassion

Although compassion is similar to sympathy in that it is evoked in response to the plight of others, it is associated with relatively more negative states as well more of an active response that is characterized by a desire to help; however, this might not be manifested behaviorally [15].

There is an apparent paradox between the positive and negative VALENCE of emotions (e.g. warmth vs. sorrow, respectively) that can both be associated with compassion. As [11] note, whereas the negative VALENCE inherent in compassion is based on the reaction to the plight of others, the positive VALENCE that additionally characterizes this emotion derives from the possible interpersonal bond that one might develop with the suffering individual and the possible help offered. [16] similarly view compassion in terms of the feelings of sorrow and concern about the suffering of others, as well as a desire to ease their suffering. Consistent with this view of compassion, [6] showed that compassion can feel pleasant or unpleasant depending on the VALENCE of events in the audio clips that were presented as mood inductions. Specifically, in comparison with unpleasant feelings of compassion, which were induced by showing the plight of others, a more pleasant version of this emotion was elicited by more neutral clips. Further evidence highlighting the ambivalence of the VALENCE of compassion comes from a study by [36] in which functional magnetic resonance (fMRI) recordings were taken as subjects were presented with images designed to induce either compassion or pride. The results suggested that images associated with compassion, such as those showing poverty and vulnerable infants, activated both feelings of personal distress as well as caregiving tendencies. The clear message emerging from these studies is that compassion is an emotion that can be conceptualized as either positive or negative.

The ambivalent VALENCE of compassion is a likely candidate for the source of cross-cultural variation in the meaning of this emotion, especially within the context of individualism-collectivism. Although a detailed discussion of individualism-collectivism is not the main focus of the current study, it is necessary to consider the main issue that is relevant to the present discussion, namely, self versus other focus of orientation that pertain to individualism versus collectivism, respectively. Whereas self-focus in individualism is characterized by personal autonomy, personal goals, personal attitudes, and individual responsibility for actions, other-focus in collectivism is characterized by the focus on interpersonal ties, shared aims and the preservation of social harmony [5, 17, 26, 41, 42]. The greater focus on interpersonal relationships in collectivistic cultures would probably engender an outward focus on the suffering person and hence the more salient meaning of compassion is likely to be sorrow, sadness or distress. Individualists, however, being more independent, autonomous entities, are more likely to focus on themselves when confronted with an individual who is suffering, which makes self-control and, thus, the possible help that they will provide in that situation more conceptually salient to them.[1] In the light of the focus of the present chapter, it is important to determine the relevance of individualism and collectivism to British English versus Polish compassion. Polish culture is usually deemed to be collectivistic in nature (e.g. [39]). Nevertheless, although it scores 60 on the individualistic versus collectivistic dimension, which demonstrates that it is relatively more collectivistic than the individualistic culture of the United Kingdom (89 on the individualistic vs. collectivistic dimension), it is clear that Poland is not as collectivistic as countries such as China (20 on the individualistic vs. collectivistic dimension) [18]. In terms of individualistic versus collectivistic influences on the conceptualization of compassion, it can thus be concluded that although positive VALENCE is probably more salient in the more individualistic, British English version of *compassion*, the Polish equivalent, *współczucie*, would appear to be characterized more by negative VALENCE.

To conclude, although empathy, sympathy and compassion are similar emotions that are often elicited in concert when one is confronted with the plight of others, our discussion highlights important distinguishing features. A major distinction is that whereas empathy is characterized by sharing the emotion(s) that another individual is experiencing (i.e. feeling *with*), sympathy and compassion do not involve such sharing and can be thought of as emotions in which one feels *for* another person.[2] Although sympathy and compassion are similar in this respect, empirical research has not produced a clear pattern regarding the relationship between these two emotions. However, it does appear that a desire to help might be more salient in compassion, although it might not be manifested behaviorally. This is consistent with the precedence of empathy over compassion with regard to temporal order. Another feature that appears to be relatively more salient in compassion is its capacity to engender both positive and negative feelings, with the Polish variant possibly having a relatively more negative meaning due to its more collectivistic status than the more individualistic British English form.

10.1.2   Background to the Current Study

The chapter attempts to make a contribution to our wider project in which emotions have the potential to determine possible responses in intelligent engineering systems. It is the monitoring and regulatory mechanisms of emotions that function on the basis of behavior modeling parameters and the moral system that combines the cognitive and evaluative properties that form the basis upon which emotion-sensitive socially interacting robots might perform behavior selection to achieve stability

---

[1] Apart from individualism-collectivism, Hofstede has developed other dimensions pertaining to culture, namely power distance, uncertainty avoidance, masculinity, indulgence, and long term orientation [18]. However, these are beyond the scope of the present focus.

[2] In Polish this shift in meaning is accompanied by a historical syntactic change: from the older form *współczuć z kimś* 'to co-feel with somebody' to the contemporary structure *współczuć komuś* lit. 'to co-feel [to] somebody' (Dative case).

(homeostasis) and viability (survival). When robots and humans engage in social interaction, emotions dictate a behavioral response in terms of an orientation towards (approach strategy) or away from (avoidance strategy) a goal, or the display of still more complex behavior [23].

Our focus in the present paper is on a comparison between the expressive features of principal members of the English COMPASSION cluster, which contains the emotions of *empathy, sympathy* and *compassion*, and their prototypical dictionary equivalents of Polish *empatia* and *współczucie*. An asymmetry identified between the English and Polish clusters is shown to increase in the case of the Polish emotion *sympatia*. The latter presents different cluster membership (LOVE/JOY) and a correspondence pattern with English *sympathy*, mainly with reference to their common properties expressing warmth, understanding, harmony and compatibility.

Although empathy, sympathy and compassion are all characterized by a response to the plight of others, it is the differences between these emotions both within and across cultures that determine potential action selection. From the above discussion of these emotions, it would appear that the salient features in this respect are VALENCE and the degree of engagement in a helping response. An important feature of the current focus is to determine and compare the central features of British English versus Polish empathy, sympathy and compassion. A major influence on these core features in both cultures are the associations that these emotions have both with each other and similar emotions within the SYMPATHY/EMPATHY/COMPASSION cluster, and with other emotions in different emotion clusters. This viewpoint is consistent with our proposal in [23] that both the recognition and production of emotions are determined by clusters of emotions as opposed to solitary emotions.

As argued in the present chapter, the properties of the empathy, sympathy and compassion in the British variant of English only asymmetrically correspond to the accepted lexicographic Polish equivalents. Such differences need to be taken into consideration in the intelligent engineering of emotions in social robots.

10.1.2.1   Aims

Rather than considering emotions as distinct, unified representations, the current study underscores the conceptual point of view that emphasizes the cluster structure of emotions. It is argued that this clustering structure needs to be a central element of intelligent engineering emotions modeling in social robotics. It is additionally important that social robots should be equipped with the awareness of differences both within and across cultures with respect to empathy, sympathy and compassion. Moreover, given the intimate, personal nature of the situations in which these emotions are present, social robots also should be taught the capacity to imitate humans. In this respect, an important social skill is the ability to learn and update the opinions and values etc. of the significant humans they have close contact with.

By employing the GRID, online emotions sorting, and corpus methodologies the main aim is to compare the different types of empathy, sympathy and compassion that a social robot would need to be able to competently encode and decode in social interactions in British English versus Polish culture. In this sense it is an extension of our earlier work [25]. The comparison of the important features of these emotions allows an assessment of their relative presence in people's lives. A central comparative element pertains to the importance of VALENCE and the degree of engagement in a helping response.

10.2 Materials and Methods

The complex nature of emotions that are conceptually structured in clusters rather than single emotions underscores the need for empirical investigation to employ diverse methodologies in a cross-disciplinary approach. To meet these demands, the present study employed the GRID, online emotions sorting, and language corpus methodologies (see below). Dimensions and components are an integral part of the GRID instrument in its investigation of emotion prototype structure. Support for the dimensional structure of emotion concepts has come from a number of studies [11, 12, 34]. In addition to enabling analyses based on these emotion dimensions (of which VALENCE is particularly relevant to our analyses on compassion—see below), the wide range of emotion features in the instrument allows relevant components to be isolated and analyzed (see analyses on *desire to act* vs. *desire not to act* features of compassion below). A main advantage of the online emotions sorting methodology is the enriched meaning of emotion clusters that it provides, which extends horizontally, providing information on the relationships between clusters. Using this methodology we can not only establish that the British English version of *compassion* is more positive than its more negative Polish counterpart, *współczucie*, but additionally discover, by determining its prototypical location within clusters and how it relates to emotions in other clusters, the possible reasons for this cross-cultural difference. The corpus methodology provides more detailed materials on the specificity of particular emotion terms, which also includes information on how they function in various contexts. The context is recovered by examining the distribution (i.e. occurrence) of a particular term in various structures and identifying the frequency of their co-occurrence with neighboring words. The type of meaning they represent and the structure in which they are used make it possible to identify both the VALENCE and intensity with which the emotion terms are used in discourses.

10.2.1 GRID

The GRID instrument was developed at The Swiss Centre for Affective Sciences (University of Geneva) in collaboration with Ghent University and investigates emotion patterns in 23 languages in 27 countries. The GRID methodology comprises an online set of questions whereby 144 emotion features are rated on the basis of 24 prototypical emotion terms. Each of the six major components of emotion are represented: appraisals of events (31 features); psycho-physiological changes (18 features); facial, vocal or gestural expressions (26 features); action tendencies (40 features); subjective experiences (22 features) and emotion regulation (4 features). There are an extra three features that are associated with other qualities, including the social acceptability and frequency of the emotion. The likelihood of presence of each of these features is assessed on the basis of each of the emotion terms. The instrument provides a detailed analysis as it allows emotion concepts to be compared cross-culturally in terms of the six categories of emotions that have been determined [8, 28, 34].

Apart from its focus on components, the GRID instrument is additionally dimensional in its approach. This involves the identification of a small number of dimensions that represent the emotion domain. [9] argues that the assessment of experiences pertaining to emotions and affect is a central aspect of dimensional approaches. [11] highlight the consistency that is present in their data between dimensions and components. The optimal number of dimensions has been the subject of recent discussion, with the earlier proposal of [30] suggesting three (EVALUATION, POTENCY and ACTIVATION), and later approaches settling on AROUSAL and VALENCE (e.g. [44]). The complete set of data from all the GRID languages has produced a structure comprising four dimensions (VALENCE, POWER, AROUSAL and NOVELTY) [13]. Further results have demonstrated the stability of this structure and that it corresponds with the componential data.

As noted above, VALENCE is central to the emotions of empathy, sympathy and compassion and its presence in the GRID analyses requires an understanding of its nature. Personal pleasure and the attainment of goals are salient features of the VALENCE dimension. It is also characterized by avoidance and approach action potentials, and pleasant versus unpleasant emotions. Features included in this factor are, for example, "felt good", "consequences positive for person", "smiled", "felt negative", "frowned", and "wanted to destroy whatever was close".[3]

10.2.1.1 Procedure

Subjects completed the GRID in a controlled Web study [32], in which each participant was presented with four emotion terms randomly chosen from the set of 24 and asked to rate each in terms of the 144 emotion features. They rated the likelihood that each of the 144 emotion features can be inferred when a person from their cultural group uses the emotion term to describe an emotional experience on a 9-point scale ranging from *extremely unlikely* (1) to *extremely likely* (9), with *neither unlikely, nor likely* appearing in the center (5). Participants rated all four emotions terms on the

---

[3]Descriptions of the POWER, AROUSAL and NOVELTY dimensions are not provided as these dimensions are not included in the GRID analyses in the present study.

basis of a certain emotion feature sequentially on separate screens, followed by the next emotion feature and so on.

10.2.1.2   Participants

33 British English-speaking subjects (all born and raised in Britain and spent most of adult lives in Britain; 21 females; mean age 23.2 years) rated *compassion* and 29 Polish-speaking subjects (all born and raised in Poland and spent most of adult lives in Poland; 26 females; mean age 25.6 years) rated *współczucie*.[4]

10.2.2   Language Corpus Methodology

English and Polish language corpus data in the form of concordances and the frequencies of words co-occurring minimally 5 times (collocations), whose significance was computed by the TTEST association score, were used for analysis to identify the salience of particular linguistic meanings. The language materials were drawn from the British National Corpus (100 million words) and the National Corpus of Polish (www.nkjp.pl) (300 million units of balanced data, normalized to 100 million units). Although the BNC and NKJP are comparable in structure, the narrow contexts of examined emotions may diverge in English and Polish. The data underwent a cognitive linguistic analysis with regard to the type of construal of the examined emotion event scenarios [21] in terms of the structural properties of the language used [23].

10.2.3   Online Emotions Sorting Methodology

In the emotions sorting methodology participants are required to categorize emotion terms. These terms are categorized on the computer desktop in the online variant. This task has been employed in a comparison of Indonesian versus Dutch emotion concepts [10] and an analysis of the concept of pleasure [7].

NodeXL [38] was employed to compare the conceptual propinquity between empathy, sympathy and compassion and between these emotions and those in the LOVE, HAPPINESS and SADNESS clusters in Polish versus British English. This application allows such comparisons by the creation of emotion co-occurrence matrices in the two languages. These *co-occurrences* are also termed *interconnections* and pertain to how many times participants place emotion terms in the same categories.

---

[4]Each of the participants was randomly presented with 4 of the 24 prototypical emotion terms to rate and each of these terms was rated separately on each of the 144 emotion features.

10.2.3.1   Selection of Terms

The selection of emotion terms was carried out on the basis of egalitarian principles.[5] To ensure the independence of the choice of emotion terms in each language, the selection of the British English and Polish emotion terms was conducted independently. This meant that neither language was dominant over the other and that neither was used as a point of reference to the other, ensuring a selection of British English emotion terms that are independent of the Polish emotion terms.

Various references and Internet sources were initially consulted to compile a list of 569 British English emotion terms and 322 Polish emotion terms. Then each of these emotion terms was placed in one of the following categories: happiness/joy, love, anger, hate, fear, sadness, surprise, compassion, contentment, guilt, shame and pride. The authors then had consultations to decide on the emotion terms that were the most prototypical of the original lists of emotion terms in the respective languages. Care was taken to ensure that each of the above emotion categories was represented as equally as possible. There were 200 emotion terms in the list of the most prototypical British emotions and 199 emotion terms in the list of the most prototypical Polish emotions.

In the next phase, the final list of 135 emotions was determined on the basis of participant ratings of prototypicality. There were 29 Polish participants (mean age = 33.8 years, 15 females) and 22 British English participants (mean age = 51.4 years, 11 females). The emotion prototypicality ratings task was the same for the British English and Polish participants, who were required to rate each of the emotion terms (200 for the British English participants and 199 for the Polish participants) on sheets of paper on the basis of the extent to which they deemed them to be emotions on a 9-point scale. The 135 British English emotion terms and the 135 Polish emotion terms that had the highest mean prototypicality scores were selected as the emotion terms to be used in the main experiment.

10.2.3.2   Procedure

The task required participants to group 135 emotions that were presented on a computer desktop into categories according to how they judged the similarity of the emotions. There were no restrictions with respect to the amount of time they could take to do the task and the number of categories that they could form. Participants were recruited either by Internet forum advertisements or by responding to direct contact. The subjects, who were given instructions in their own mother tongue, conducted the experiment on their own and were free to choose the time and location. Initial instructions informed the participants that the experiment was designed to determine which emotions are deemed to belong to the same group. They were further specifically informed that they would be required to categorize 135 emotions

---

[5]Due to space restrictions what is presented in this section is a summary of the procedure followed to select the emotion terms.

into groups on the computer screen. Participants were also told that they needed to think carefully about the categorization task and to have as many groups of emotions that they wanted, which were not restricted in terms of the number of emotions. The participants were then shown a video that demonstrated the procedure, which was followed by a practice session with food items before the proper experiment began.

10.2.3.3   Participants

There were 58 British English-speakers (all raised and born in Britain and spent most of adult lives in Britain; mean age 42.7 years, 27 females) and 58 Polish-speakers (all born and raised in Poland and spent most of adult lives in Poland; mean age 35.8 years; 27 females). These participants were different to those in the GRID study.

10.3   Results

10.3.1   GRID

Principle components analysis (PCA) with varimax rotation was conducted on the complete set of data from the Polish and British English participants. There were 124 Polish subjects (mean age 23.2 years; 95 females) and 201 British English subjects (mean age 21.5 years; 124 females).[6] The four-dimensional structure produced was the same as demonstrated by [13] and represented 81.8% of the overall variance (VALENCE—52.9% of the variance, POWER—15.5%, AROUSAL—8.3%, and NOVELTY—5.1%). If a feature had a 0.6 loading on a certain dimension it was included in that dimension as long as it did not load higher on a different dimension. The analysis made a comparison between Polish *współczucie* and British English *compassion* on the basis of the VALENCE dimension features. The higher and lower means for this dimension were identified for each subject. The 2 × 2 Anova that was conducted on these means had one within-subjects variable (VALENCE: (positive VALENCE GRID features vs. negative VALENCE GRID features)). There was also a between-subjects variable (*language group*: Polish *współczucie* vs. British English *compassion*).

There was a significant interaction between VALENCE and *language group*, F (1, 58) = 34.88, $p < 0.001$. This interaction was analyzed further by contrasts. It can be seen in Fig. 10.1 that *compassion* was characterized by more positive VALENCE than *współczucie* (F (1, 58) = 44.44, $p < 0.001$). It can also be seen in Fig. 10.1 that *współczucie* was characterized by more negative VALENCE than *compassion* (F (1, 58) = 4.41, $p < 0.05$). Figure 10.1 additionally shows that *compassion* is associated with more positive VALENCE than negative VALENCE (F (1, 58) = 31.17, $p < 0.001$).

---

[6]As each participant rated 4 of the 24 prototypical emotion terms on the 144 emotion features, the overall amount of subjects in the British English and Polish datasets is larger than the numbers of participants that rated the specific emotion terms in each language.

Fig. 10.1 *Compassion* and *współczucie* means on the VALENCE dimension

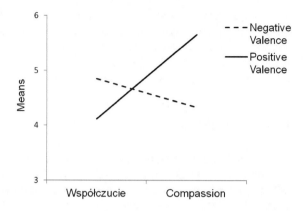

It can also be seen in Fig. 10.1 that *współczucie* is associated with more negative VALENCE than positive VALENCE (F (1, 58) = 8.20, $p < 0.01$).

The analysis additionally made a comparison between Polish *współczucie* and British English *compassion* on the basis of GRID features pertaining to *desire to act* (*wanted to act, wanted to tackle situation, felt urge to be attentive, felt urge to be active, wanted to undo what was happening, wanted to go on with what doing, required immediate response*) versus *desire not to act* (*lacked motivation to pay attention to what was going on, wanted to do nothing, lacked motivation to do anything, wanted to submit to situation, felt urge to stop what doing*). The means for these dimensions were computed for each participant. The 2 × 2 Anova that was conducted on these means had one between-subjects variable (*language group*: Polish *współczucie* vs. British English *compassion*) and one within-subjects variable (*action*: *desire to act* vs. *desire not to act*).

There was a significant interaction between *action* and *language group*, $F(1, 58) =$ 10.09, $p < 0.01$. This interaction was analyzed by contrasts. There was a significant difference between *desire to act* and *desire not to act* for *compassion*,

Fig. 10.2 *Compassion* and *współczucie* means on *desire to act* versus *desire not to act*

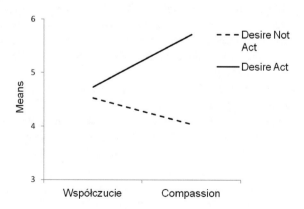

Table 10.1 Independent t-test results of GRID expression features for *compassion* versus *współczucie*

GRID expression features	Compassion means	Współczucie means	T	p
Heartbeat getting faster	5.53	3.68	2.9	0.01
Sweat	3.94	2.79	2.06	0.05
Felt hot	4.25	3.0	2.2	0.05
Blushed	4.84	2.68	3.65	0.01
Smiled	6.25	3.14	5.34	0.01
Showed tears	5.53	3.68	2.9	0.01
Decreased volume of voice	6.03	7.96	−3.9	0.01
Had a trembling voice	4.66	6.64	−3.35	0.01
Has an assertive voice	5.16	3.64	2.59	0.05
Changed melody of speech	6.84	7.57	−2.1	0.05
Wanted to sing and dance	4.84	1.64	6.18	0.01

$F(1, 58) = 664.64$, $p < 0.001$. Figure 10.2 shows that *compassion* was characterized by a higher *desire to act* than *desire not to act*. Additionally, it can be seen in Fig. 10.2 that *compassion* was characterized by more of a *desire to act* than *współczucie* ($F(1, 58) = 13.68$, $p < 0.01$). However, *desire to act* was not significantly different to *desire not to act* for *współczucie* and *compassion* was not significantly different to *współczucie* on *desire not to act*.

Emotion expression features were further analyzed in the GRID data. Table 10.1 shows the significant results of Independent t-tests conducted on *współczucie* versus *compassion* features pertaining to five varieties of perceptual features of emotions: linguistic (e.g. *short utterance*), paralinguistic (e.g. *trembling voice*), facial (e.g. *frowned*), body movement (e.g. *abrupt bodily movements*), and physiological (e.g. *quicker breathing*).

It can be seen in Table 10.1 that the results of the VALENCE features are consistent with the relatively more positive VALENCE of *compassion* compared with *współczucie* (*smiled, had a trembling voice* and *wanted to sing and dance*). The greater *desire to act* in *compassion* is supported by the higher rating of *had an assertive voice* in *compassion* compared with *współczucie*. The elevated values of *compassion* in comparison with *współczucie* on psychophysiological features (*heartbeat getting faster, sweat*, and *felt hot*) could be considered as support for the greater *desire to act* of the British English version as these features are usually associated with greater activity.

10.3.2 Language Corpus Data

10.3.2.1 Empathy

A contrastive analysis of the English and Polish collocational profiles indicates that whereas English *empathy* involves both individuals and, significantly more often, facts and events that are a matter for concern and need attention (examples 1–3), Polish *empatia* directly addresses individuals.

(1) His face betraying mournful empathy with Gentle's bruising.
(2) They tend to have more empathy towards anything foreign.
(3) To introduce... a widening empathy with environmental concerns.

Polish *empatia* is mostly used with negatively colored causes affecting interactants and is, similar to English *empathy*, a synonym of understanding, particularly *rozumienie czyjegoś bólu czy cierpienia* 'understanding of someone's pain or suffering', i.e. suffering of disadvantaged individuals, as in:

(4) pełnych empatii i zrozumienia dla losu ludzi bezdomnych 'full of empathy and understanding for the fate of homeless people'.

In its extended sense, *empatia* covers both understanding of the interactant's emotional state as well as the acceptance and tolerance of a person's possible fault(s).

In both Polish and English empathy implies openness about one's feelings in terms of language and behavior.

10.3.2.2 Sympathy

Whereas, as indicated above, *compassion* implies sharing the sorrows and worries of other people, and a desire to help, *sympathy* is a broader, polysemous term. It signifies a general feeling of kinship with others, both in positive and negative contexts. These senses that are used in terms of the one-word form *sympathy* in English, correspond to two distinct words in Polish: one implying negative sources—*współczucie*; and the other—*sympatia*, considered rather a translational false friend to the English term *sympathy*. *Sympatia* is defined in Polish as "a feeling involving liking somebody or something, favorably disposed towards somebody or something; friendly; attraction to somebody" [40]. This Polish word therefore only partly overlaps with English *sympathy* in terms of co-feeling with the positive feelings of others (see Table 10.2 for Polish Verbal collocates of *sympatia*).

10.3.2.3 Compassion

Whereas the feeling of *empathy* primarily involves the mental state of understanding, both Polish and English *compassion* arise as a result of experiencing a sense of

Table 10.2 *Współczucie* 'compassion' general collocates

N^o	Collocate	Collocations	Freq	Chi^2
1	głęboki 'deep'	głębokie-współczucie 17, współczucie-głębokie 1	18	28,899.38
2	budzić 'wake'	budzi-współczucie 9, budzić-współczucie 6, budzą-współczucie 4, budzącego-współczucie 3, budzącą-współczucie 3, budzące-współczucie 3, budził-współczucie 2, budziła-współczucie 2, budzącymi-współczucie 1, budziły-współczucie 1, budzący-współczucie 1, budzisz-współczucie 1, współczucie-budzi 1	37	20,500.46
3	litość 'pity'	litość-współczucie 10, współczucie-litość 3, litości-współczucie 1	14	15,740.49
4	empatia 'empathy'	współczucie-empatia 5, empatię-współczucie 1	6	14,787.31
5	szczery 'sincere'	szczere-współczucie 7, współczucie-szczere 1	8	12,305.52
6	wyrażać 'express'	wyrażali-współczucie 4, wyraża-współczucie 3, wyrażał-współczucie 3, wyrażamy-współczucie 2, wyrażając-współczucie 2, wyrażać-współczucie 2, współczucie-wyraża 1, wyrażano-współczucie 1, wyrażającą-współczucie 1, wyrażałem-współczucie 1, wyrażam-współczucie 1	21	10,926.5
7	życzliwość 'kindness, friendliness'	współczucie-życzliwość 3, życzliwość-współczucie 3	6	10,705.39

co-feeling with another individual's misfortune. This is indicated in the lists of collocates in both languages such as *love* 'miłość', *heart* 'serce', *help* 'pomoc', etc. A detailed analysis reveals a somewhat less positive character of lower Polish nominal collocates, e.g. *upomnienie* 'reminder, admonishment', *krew* 'blood', *troska* 'worry (also implying care)/problem', and, surfacing more frequently than in English, *łzy* 'tears'.

It is evident in the collocation lists that Polish *współczucie*, apart from its correlation with positive emotions such as *odwaga* 'courage', *nadzieja* 'hope', *empatia* 'empathy', *czułość* 'tenderness', and *poszanowanie* 'respect', is also highly correlated with more negative emotions such as *żałość* 'sorrow, worry', *żal* 'sadness, sorrow, longing', *współwina* 'mutual guilt' or *pustka* 'emptiness'.

10.3.2.4   Collocations

Clustering of the analyzed concepts with other emotion terms is observed in their collocational combinations. Collocations are patterns in which two words co-occur

Table 10.3 *Compassion* adjectival collocates

N^o	Collocate	POS[a]	A[b]	TTEST[c]	M13[d]
1	genuine	adj	4.0	1.79	7.25
2	overwhelming	adj	3.0	1.63	7.30
3	Christian	adj	3.0	1.27	5.10
4	deep	adj	3.0	1.16	4.79
5	human	adj	4.0	0.84	4.78
6	greater	adj	3.0	0.69	3.91

[a]Part of Speech
[b]Raw frequency
[c]t-test
[d]Mutual Information (MI) test

more frequently than by chance as, for example, *współczucie* with the Verb *wzbudzić współczucie* lit. 'wake up compassion', the Adjective *głębokie współczucie* 'deep compassion' or the Noun *litość i współczucie* 'pity and compassion'. Such frequent combinations reveal the semantic coloring of one word sense by another characteristic of a particular culture. *Współczucie* is strongly associated with *litość* 'pity' in Polish on the one hand, and on the other, although less often, with empathy, friendliness and kindness. *Litość* is also connected with *politowanie*, which possesses elements of negative pity (*litość* is marked as linked to *politowanie* in the online emotions sorting results—see Fig. 10.4). *Politowanie* is expressed by an experiencer who has a sense of superiority or even contempt toward the person s/he pities, thus also showing elements of negative pride (Polish *pycha*). In the consulted Polish corpora *politowanie* has the strongest co-occurrence links with the negative emotions of *oburzenie* 'indignation', *pogarda* 'contempt', *smutek* 'sadness' and it forms a fixed phrase with *śmiech* 'laughter' in *wzbudzać śmiech i politowanie* 'to cause laughter and negatively evaluated pity'. There are fewer frequently marked elements of pity and traces of positive compassion in *politowanie* as evident as in the corpus example:

(5) Oświadczam wszystkim, że kończę z paleniem. W pracy niedowierzająco kiwają głowami i patrzą z politowaniem 'I say I give up smoking. At work, people nod incredulously and look with *politowanie* (pity, compassion, disbelief)'.

As identified in the English collocation patterns (Table 10.3), all Adjectival collocates of *compassion* are positive.

The Polish Adjectival collocates of *współczucie* on the other hand demonstrate the presence of more negative clustering (see Table 10.4).

10.3.2.5   Parallel Corpus Data

Parallel corpus materials, that is published translation texts (http://paralela.clarin-pl.eu/) [31], confirm these displaced (i.e. asymmetrical) equivalence patterns [22]

Table 10.4 *Współczucie* 'compassion' adjectival collocates

N°	Collocate	POS	A	TTEST	M13
1	głęboki	adj	87.0	9.12	18.41 'deep'
2	pełny	adj	84.0	8.46	16.48 'full'
3	szczery	adj	55.0	7.33	18.13 'sincere'
4	serdeczny	adj	38.0	6.10	17.09 'cordial'
5	godny	adj	19.0	4.19	13.22 'worth/dignified'
6	ludzki	adj	23.0	3.96	11.56 'human'
7	zwykły	adj	13.0	2.96	9.89 'ordinary'
8	falszywy	adj	10.0	2.86	10.03 'fake, false'
9	biedny	adj	10.0	2.82	9.86 'poor'
10	prawdziwy	adj	13.0	2.21	8.78 'true'
11	buddyjski	adj	5.0	2.19	10.27 'Buddhist'
12	bezinteresowny	adj	5.0	2.17	9.95 'disinterested, selfless'
13	godni	adj	4.0	1.98	11.22 'worth'
14	bolesny	adj	5.0	1.94	7.59 'painful'
15	przejęty	adj	4.0	1.82	7.51 'concerned'
16	zdolny	adj	5.0	1.76	6.88 'able'
17	gorący	adj	5.0	1.73	6.80 'hot'
18	bezmierny	adj	3.0	1.72	10.62 'enormous'
19	zdawkowy	adj	3.0	1.70	9.10 'desultory'

between the English SYMPATHY, COMPASSION, PITY and SORROW clusters and Polish *współczucie* (examples 6–9).

(6) Pol. i tym razem zgromadzilo je nie tyle *współczucie*, ile ciekawość widzenia okazałego pogrzebu
Eng. and even now their faces betrayed more curiosity to see a grand funeral than any *sympathy*.

(7) Pol. To *współczucie* nie powstrzyma mnie od niczego
Eng. I may pity the man; but *compassion* will not stand in my way.

(8) Eng. face to see the divine force of faith in him and his great *pity* for all that was poor, suffering, and oppressed in this world
Pol. by wyczytać w nich boską moc wiary i wielkie *współczucie* dla wszystkich, którzy żyją w nędzy, cierpią i są prześladowani.

(9) Eng. I remember *feeling sorry* for the family
Pol. Pamiętam *współczucie* dla rodziny.

The corpus data show that English *compassion* has a higher likelihood of positive evaluation than *współczucie*. The higher correlation pattern between *compassion* and positive emotions and between Pol. *współczucie* and negative emotions is consistent with this finding. The corpus results show that *compassion* is associated with love

Fig. 10.3 British English
EMPATHY/SYMPATHY/
COMPASSION cluster

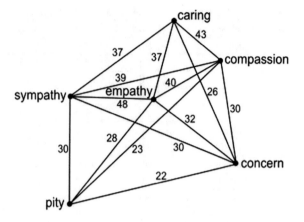

relatively more frequently and that *współczucie* displays negative aspects of pity (*politowanie*).

10.3.3　Online Emotions Sorting Results

Interconnections were analyzed within the Polish and within the British English EMPATHY/SYMPATHY/COMPASSION clusters (see Figs. 10.3 and 10.4), and also between the two language variants of empathy, sympathy and compassion and emotions chosen from the SADNESS cluster (despair, despondence, depression, and sad-

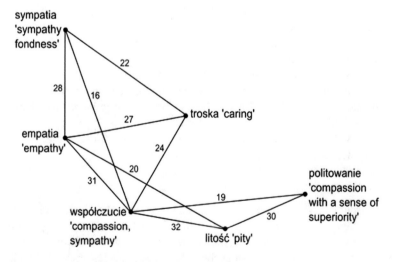

Fig. 10.4 Polish EMPATHY/SYMPATHY/COMPASSION cluster

ness), the HAPPINESS cluster (joy and happiness), and the LOVE cluster (adoration, tenderness, fondness and love)—see Table 10.5.

10.3.3.1 Polish Versus British English
EMPATHY/SYMPATHY/COMPASSION Clusters

Figure 10.3 shows that the British English EMPATHY/SYMPATHY/COMPASSION cluster is generally more cohesive than the corresponding cluster in Polish (see Fig. 10.4). The three emotions under scrutiny in the present paper have particularly close interconnections in the British English cluster (*sympathy–empathy* (48), *sympathy–compassion* (39), and *empathy–compassion* (40)) in comparison with the Polish cluster (*sympatia* 'sympathy, fondness'*–empatia* 'empathy' (28), *sympatia* 'sympathy, fondness'*–współczucie* 'compassion, sympathy' (16), and *empatia* 'empathy'*–współczucie* 'compassion, sympathy' (31)). As discussed below, the possible reasons for this are the meaning of liking that is salient in *sympatia*, and the negativity that characterizes *współczucie*, possibly as a consequence of its propinquity with *politowanie* 'compassion with a sense of superiority'.

EMPATHY

The differences between British English *empathy* and Polish *empatia* are ambivalent with respect to VALENCE. The data from the language corpora underscore the negativity of *empatia* in relation to *empathy* and this corresponds with the higher co-occurrences between *empathy* and its respective LOVE cluster emotions (e.g. *empathy–caring* (37)) than between *empatia* and its respective LOVE cluster emotions (*empatia–troska* 'caring' (27)). This also demonstrates that *empathy* is associated with a focus that is relatively more interpersonal in nature. However, the possible more positive VALENCE of *empatia* is supported by the relatively greater co-occurrences that *empatia* has with HAPPINESS cluster emotions.

SYMPATHY

The friendly attraction and liking that characterizes *sympatia* is clearly evident in its interconnections with the LOVE cluster emotions (*miłość* 'love, affection' (32) and *lubienie* 'liking, fondness' (43)), which are higher than the corresponding interconnection values in British English (*sympathy–love* (21) and *sympathy–fondness* (27))—see Table 10.5. The relatively more positive VALENCE that is also likely to accompany such attraction and liking is visible in Table 10.3 in the relatively greater co-occurrences between *sympatia* and *szczęście* 'happiness' (28) and between *sympatia* and *radość* 'joy, glee, delight' (23) than the corresponding British English interconnections (*sympathy–happiness* (8) and *sympathy–joy* (7)).

COMPASSION

Table 10.5 corresponds with the GRID results in showing that *compassion* is more positive than *współczucie*. This is demonstrated most directly in the SADNESS cluster emotions. Despite the HAPPINESS cluster emotions having similar co-occurrences

Table 10.5 Interconnections between British English versus Polish compassion, empathy and sympathy

British English Emotions	C(Br)[a]	E(Br)[b]	S(Br)[c]	Polish Emotions	W(Pl)[d]	E(Pl)[e]	S(Pl)[f]
love	25	22	21	miłość 'love, affection'	6	17	32
fondness	32	29	27	lubienie 'liking, fondness'	14	28	43
tenderness	32	29	28	lubienie, czułość 'tenderness'	12	22	31
adoration	24	25	22	podziw 'adoration, admiration'	7	10	19
				uwielbienie 'adoration, admiration'	9	15	25
happiness	10	11	8	szczęście 'happiness'	7	14	28
joy	8	8	7	radość 'joy, glee, delight'	7	18	23
sadness	3	1	3	smutek 'sadness, sorrow'	9	3	1
depression	0	1	1	depresja 'depression'	8	2	0
despondence	2	2	2	zniechęcenie 'despon-dence'	6	2	0
				przygnębienie 'depression, despondence, upset'	9	3	0
despair	1	2	1	rozpacz 'despair'	8	1	0

[a] *Compassion*
[b] *Empathy*
[c] *Sympathy*
[d] *Współczucie*
[e] *Empatia*
[f] *Sympatia*

between *compassion* and *współczucie*, the higher interconnections between the LOVE cluster emotions and *compassion* in comparison with *współczucie* is consistent with the more positivity of *compassion* (e.g. *compassion–tenderness* (32), *współczucie–*

lubienie, czułość 'tenderness' (12)). This propinquity between *compassion* and LOVE cluster emotions also indicates the interpersonal features of *compassion*.

A reason for the relatively more negative VALENCE of *współczucie* is possibly based on its conceptual proximity to *politowanie* 'compassion with a sense of superiority' (see the corpus data and the co-occurrence value of 19 in Fig. 10.4, above).

10.4 Discussion and Conclusions

The intricate relationships between empathy, sympathy and compassion demonstrate the need to assess within and between cluster relationships rather than the discrete emotion properties in each instance. The complementary results from the GRID, online emotions sorting, and language corpus methodologies allow an in-depth comparison of the British English versus Polish variants of these emotions and further highlight the advantage of using a composite approach.

10.4.1 Empathy

The comparison between *empathy* and *empatia* appears to show the most ambivalent results. As demonstrated by the data from the language corpora, *empatia* is associated with more negative VALENCE than *empathy*. Nevertheless, the results from the online emotions sorting study also underscore some positive elements in *empatia*. The corpus results also highlight that *empathy* has a wider scope in terms of its application and is employed in a number of events and situations that have a sympathetic tone.

10.4.2 Sympathy

In comparison with the uniquely positive contexts that characterize Polish *sympatia*, which corresponds to English *liking* and *friendship*, British English *sympathy* incorporates features of negativity, and is elicited, similar to *compassion*, by unfavorable events and circumstances. This can additionally be seen when one compares the most frequent Adjectival collocate corpus items, which show a great deal of consistency: *compassion* (*great, overwhelming, poor, human, deep*)—t-test average 2.35; *sympathy* (*considerable, deepest, deep, public, great*)—t-test average 2.35. The co-occurrence values from the online emotions sorting results are also consistent with this greater proximity between *sympathy* and *compassion* (39) than between *sympatia* 'sympathy, fondness' and *współczucie* 'compassion, sympathy' (16), which should rather be considered members of different, albeit related, emotion clusters; the former—part of the JOY/HAPPINESS/LOVE cluster(s), while the latter—part of the COMPASSION cluster.

10.4.3 Compassion

Consistent with predictions, what is clearly demonstrated in the results is that *compassion* has a more positive VALENCE than *współczucie*. This can be seen in the more positive VALENCE of *compassion* in comparison with *współczucie* in the GRID results. It is also evident in the relatively closer proximity between *compassion* and the LOVE cluster emotions and between *współczucie* and the SADNESS cluster emotions in the online emotions sorting data. The corpus data is consistent with this in showing that *compassion* has a greater positivity than *współczucie*.

There are three possible reasons for the VALENCE effects shown for *compassion* and *współczucie*. First, the greater positivity of *compassion* is consistent with predictions on the basis of this individualistic British English variant engendering a more inward, personal focus that might make the help the individual can offer more salient, which can be compared with the possibly more outward oriented, more collectivistic *współczucie* that is more likely to highlight the suffering that is being experienced. The second possible facet of the explanation concerns the meaning of *sympatia*, which possesses uniquely positive VALENCE in Polish and is located in the MIŁOŚĆ (love) cluster rather than the WSPÓŁCZUCIE (compassion) cluster. In comparison, although English *sympathy* does include the positive elements of liking, when used in the negative sense it typically refers to Emotion Events expressing an irreversible loss (*grief*), similar to Pol. *współczucie*. The third possible reason for more negative VALENCE effects for Polish *współczucie* centers on the linguistic link between *współczucie* and the unambiguously negative *politowanie* (pity combined with a sense of contempt and superiority).

The greater *desire to act* that is shown for *compassion* in the GRID data would appear to be consistent with the relatively more positive VALENCE that characterizes this emotion. First, if one has a greater orientation towards the help one can offer a person in distress, as individualism possibly suggests, then it is clear that this has the probability of being manifested behaviorally. Second, the negativity associated with *współczucie* might preclude action if, for example, the level of stress is overbearing. Third, the propinquity between *współczucie* and *politowanie* would rather elicit contempt than the offer of help.

10.4.4 Affective Robotics Modeling

The three methodologies used in the current study show, converging, consistent evidence in support of our argument that the competence of social robots depends on the tuning of profiles of expression features [43]. Some relevant studies in this respect include the development of an algorithm that makes an association between emotion values and multi-sensor input [4], the need for multimodality in emotion perception [19], and the relationship between emotions and dialog acts [29]. The conclusions that we present have a particular relevance to modeling in connection with emotion *recognition* features in our analyses, as well as instances of ambiguity and lexical polysemy; for example, English *sympathy*, which can be construed either negatively or positively depending on the context. The modeling of such different scenarios needs to involve a variety of cluster features that can be decoded by social robots. Alternatively, the emotional event scenarios that need to be acquired by such robots for the *encoding* and *decoding* of such emotions need to be tuned to the more expressive variant of this emotion event. We argue that in the case of *empatia/współczucie* and *sympatia* this is more associated with British as opposed to Polish prototypes. In the English concept active involvement and *desire to act* on the part of the emotion experiencer are relatively more salient. What is more, the amount of expression with respect to *empathy/sympathy/compassion*-related quality of voice, volume and gesticulation, including tactile features such as *hugs*, might additionally add to the heightened efficacy of the expressivity of the emotions analyzed, especially concerning *compassion* and *współczucie*.

Affective robots need to deal with challenges in the real world; for example, assisting individuals who need support, such as children, individuals who have special needs or senior citizens. In these contexts, robots need to be able to mimic the gestures, actions and linguistic expressions that are expressed more actively in comparison with situations in which they interact with people who are not in need of such assistance. In this respect, the role of robots would be to provide a simulation of caretakers in recognizing even minute signs of, especially negative, situations, and to have a clear expression of the offer to assist and be active in caretaking actions that are at the heart of *empathy/compassion/sympathy* contexts, conforming to basic assumptions of affective and therapeutic robotics (cf. EmotiRob in [33] on emotion synthesis). However, rather than simplifying all expression properties, we propose to model those bundles of emotion expression features which are more explicitly articulated in human behavior, gesture and language. The SID [27] offers a potential basis to such modeling, which can be extended to include the cluster properties of empathy, sympathy and compassion in Polish versus British English. Within this modeling context, future studies are necessary to map the profiles of features employed in the present study on representations of image-schematic prototypes [20] and structures of Emotion Events [23]. The detection and simulation of empathy, sympathy and compassion in such social robots means that they can be viewed as artificial cognitive systems within the field of affective computing [2]. The present study can make a contribution to applications pertaining to engineering whereby a robot is modeled to have the ability to infocommunicate with the human cognitive architecture through a transfer of information based on *sensor-bridging* [2]. Through the engagement of different types of communication via *representation-bridging* such a type of transfer either facilitates the flow of information from an artificial agent to a human, or vice-versa. Broadening the scope of application, it is clear that our results are relevant to other cognitive infocommunication systems apart from emotion-sensitive socially interacting robots, such as frameworks of home assistance to meet the needs of the elderly and handicapped [14], and the development of automatic computational systems to detect empathy in spoken, call centre conversations [1].

References

1. Alam F, Danieli M, Riccardi G (2016) Can we detect speakers' empathy? A real-life case study. In: 7th IEEE international conference on cognitive infocommunications (CogInfoCom), pp 59–64. https://doi.org/10.1109/CogInfoCom.2016.7804525
2. Baranyi P, Csapo A (2012) Definition and synergies of cognitive infocommunications. Acta Polytechnica Hungarcia 9(1):67–83
3. Batson CD, Fultz J, Schoenrade PA (1987) Distress and empathy: two qualitatively distinct vicarious emotions with different motivational consequences. J Personal 55(1):19–39
4. Berthelon F, Sander P (2013) Emotion ontology for context awareness. In: 4th IEEE international conference on cognitive infocommunications (CogInfoCom), pp 59–64. https://doi.org/10.1109/CogInfoCom.2013.6719313
5. Choi I, Nisbett RE, Norenzayan A (1999) Causal attribution across cultures: variation and universality. Psychol Bull 125:47–63
6. Condon P, Barrett LF (2013) Conceptualizing and experiencing compassion. Emotion 13(5):817–821
7. Dube L, Le Bel J (2003) The content and structure of laypeople's concept of pleasure. Cognit Emot 17(2):263–295
8. Ellsworth PC, Scherer KR (2003) Appraisal processes in emotion. In: Davidson RJ, Scherer KR, Goldsmith H (eds) Handbook of affective sciences. Oxford University Press, New York, pp 572–595
9. Fontaine JJR (2013) Dimensional, basic emotion, and componential approaches to meaning in psychological emotion research. In: Fontaine JJR, Scherer KR, Soriano C (eds) Components of emotional meaning: a sourcebook. Oxford University Press, Oxford, pp 31–45
10. Fontaine JJR, Poortinga YH, Setiadi B (2002) Cognitive structure of emotion terms in Indonesia and The Netherlands. Cognit Emot 16(1):61–86
11. Fontaine JJR, Scherer KR (2013) The global meaning structure of the emotion domain: Investigating the complementarity of multiple perspectives on meaning. In: Fontaine JJR, Scherer KR, Soriano C (eds) Components of emotional meaning: a sourcebook. Oxford University Press, Oxford, pp 106–125
12. Fontaine JJR, Scherer KR, Roesch EB, Ellsworth PC (2007) World of emotions is not two-dimensional. Psychol Sci 18(12):1050–1057
13. Fontaine JJR, Scherer KR, Soriano C (eds) (2013) Components of emotional meaning: a sourcebook. Oxford University Press, Oxford
14. Fülöp I.M., Csapó A, Baranyi P (2013) Construction of a CogInfoCom ontology. In: 4th IEEE international conference on cognitive infocommunications (CogInfoCom), pp 811–816. https://doi.org/10.1109/CogInfoCom.2013.6719210
15. Gladkova A (2010) Sympathy, and in English and Russian: a linguistic and cultural analysis. Culture Psychol 16(2):267–285
16. Goetz JL, Dacher K, Simon-Thomas E (2010) Compassion: an evolutionary analysis and empirical review. Psychol Bull 136(3):351–374
17. Hofstede G (1980) Cultures consequences: international differences in work-related values. Sage, Beverly Hills
18. Hofstede G. Country comparison. https://www.geert-hofstede.com/countries.html. Accessed 12 Feb 2016
19. Hunyadi L (2015) On multimodality in the perception of emotions from materials of the HuComTech corpus. In: 6th IEEE international conference on cognitive infocommunications (CogInfoCom), pp 489–492. https://doi.org/10.1109/CogInfoCom.2015.7390642
20. Lakoff G (1987) Women, fire, and dangerous things. What categories reveal about the mind. The University of Chicago Press, Chicago
21. Langacker RW (1987/1991) Foundations of cognitive grammar, vols 1 and 2. Stanford University Press, Stanford
22. Lewandowska-Tomaszczyk B (1996) Depth of negation—a cognitive semantic study. Łódź University Press, Łódź

23. Lewandowska-Tomaszczyk B, Wilson PA (2013) English fear and Polish strach in contrast: GRID approach and cognitive corpus linguistic methodology. In: Fontaine JJR, Scherer KR, Soriano C (eds) Components of emotional meaning: a sourcebook. Oxford University Press, Oxford, pp 425–436
24. Lewandowska-Tomaszczyk B, Wilson PA (2015) It's a date: love and romance in time and space. In: Paper at international workshop: love and time, University of Haifa, 8–10 March 2015
25. Lewandowska-Tomaszczyk B, Wilson PA (2016) Compassion, empathy and sympathy expression features in affective robotics. In: 7th IEEE international conference on cognitive infocommunications (CogInfoCom), pp 65–70. https://doi.org/10.1109/CogInfoCom.2016.7804526
26. Markus HR, Kitayame S (1991) Culture and the self: implications for cognition, emotion, and motivation. Psychol Rev 98(2):224–253
27. Mascarenhas S, Prada R, Paiva A, Hofstede G (2013) Social importance dynamics: a model for culturally-adaptive agents. In: Aylett R, Krenn B, Pelachaud C, Shimodaira H (eds) Intelligent virtual agents, vol 8108. Lecture notes in computer science. Springer, Berlin, Heidelberg, pp 325–338
28. Niedenthal PM, Krauth-Gruber S, Ric F (2006) The psychology of emotion: interpersonal, experiential, and cognitive approaches. Psychology Press, New York
29. Ondas S, Mackova L, Hladek D (2016) Emotion analysis in DiaCoSk dialog corpus. In: 7th IEEE international conference on cognitive infocommunications (CogInfoCom), pp 151–156. https://doi.org/10.1109/CogInfoCom.2016.7804541
30. Osgood CE, May WH, Miron MS (1975) Cross-cultural universals of affective meaning. University of Illinois Press, Chicago
31. Pęzik P (2016) Exploring phraseological equivalence with paralela. In: Gruszczyńska E, Leńko-Szymańska A (eds) Polish-language parallel Corpora Warsaw. Instytut Lingwistyki Stosowanej UW, pp 67–81
32. Reips U-D (2002) Standards for internet-based experimenting. Exp Psychol 49:243–256
33. Saint-Aime S, Le Pevedic B, Duhaut D (2008) EmotiRob: an emotional interaction model. In: The 17th IEEE international symposium on robot and human interactive communication, RO-MAN, pp 89–94
34. Scherer KR (2005) What are emotions? And how can they be measured? Soc Sci Inf 44:693–727
35. Shaver P, Schwartz J, Kirston D, O'Connor C (1987) Emotion knowledge: further exploration of a prototype approach. J Pers Soc Psychol 52:1061–1086
36. Simon-Thomas ER, Godzik J, Castle J, Antonenko O, Ponz A, Kogan A, Keltner D (2012) An fMRI study of caring vs. self-focus during induced compassion and pride. Soc Cognit Affect Neurosci 7:635–648
37. Singer T, Lamm C (2009) The social neuroscience of empathy. Ann New York Acad Sci 1156:81–96
38. Smith M, Ceni A, Milic-Frayling N, Shneiderman B, Mendes Rodrigues E, Leskovec J, Dunne C (2010) NodeXL: a free and open network overview, discovery and exploration add-in for Excel 2007/2010/2013/2016. http://nodexl.codeplex.com/ from the Social Media Research Foundation, http://www.smrfoundation.org
39. Szarota P, Cantarero K, Matsumoto D (2015) Emotional frankness and friendship in Polish culture. Pol Psychol Bull 46(2):181–185
40. Szymczak M (1981) Słownik Języka Polskiego. PWN, Warszawa
41. Triandis HC (1995) Individualism and collectivism. Westview Press, Boulder, CO
42. Triandis HC (2001) Individualism-collectivism and personality. J Pers 69:907–924
43. Wilson PA, Lewandowska-Tomaszczyk B (2014) Affective robotics: modelling and testing cultural prototypes. Cognit Comput 6(4):814–840
44. Yik MSM, Russell JA, Barrett LF (1999) Structure of self-reported current affect: integration and beyond. J Pers Soc Psychol 77:600–619

Chapter 11
Understanding Human Sleep Behaviour by Machine Learning

Antonino Crivello, Filippo Palumbo, Paolo Barsocchi, Davide La Rosa, Franco Scarselli and Monica Bianchini

Abstract Long-term sleep quality assessment is essential to diagnose sleep disorders and to continuously monitor the health status. However, traditional polysomnography techniques are not suitable for long-term monitoring, whereas, methods able to continuously monitor the sleep pattern in an unobtrusive way are needed. In this paper, we present a general purpose sleep monitoring system that can be used for the pressure ulcer risk assessment, to monitor bed exits, and to observe the influence of medication on the sleep behavior. Moreover, we compare several supervised learning algorithms in order to determine the most suitable in this context. Experimental results obtained by comparing the selected supervised algorithms show that we can accurately infer sleep duration, sleep positions, and routines with a completely unobtrusive approach.

A. Crivello (✉) · F. Palumbo · P. Barsocchi · D. La Rosa
Information Science and Technologies Institute, National Research Council of Italy, Pisa, Italy
e-mail: antonino.crivello@diism.unisi.it; antonino.crivello@isti.cnr.it

F. Palumbo
e-mail: filippo.palumbo@isti.cnr.it

P. Barsocchi
e-mail: paolo.barsocchi@isti.cnr.it

D. La Rosa
e-mail: davide.larosa@isti.cnr.it

A. Crivello · F. Scarselli · M. Bianchini
Department of Information Engineering and Mathematics, University of Siena, Siena, Italy
e-mail: franco@diism.unisi.it

M. Bianchini
e-mail: monica@diism.unisi.it

11.1 Introduction

Recently, sleep assessment and related evaluation research has grown steadily [28]. In this field, a hard task to reach is related to obtaining sleep dataset and, in the last years, many researchers have tried to address this problem [16]. These works demonstrate how to gather sleep features through accurate sleep session logs, in particularly developing systems able to capture sleeping behavior in terms of regularity, timing of bed time, number of night awakenings, sleep onset and sleep disorders [25].

These disorders are summarized in two different, secondary and primary, grades. The main primary sleep problems identified are related to breathing disorder (SDB), Rapid eye movement sleep behavior disorder (RBD), Restless Leg Syndrome (RLS) and Periodic Limb Movement in Sleep (PLMS). Instead, secondary sleep disorders are due to discomfort and protracted pains, dyspnoea, and medical treatment which can interfere with sleep quality [19]. Several subjects might suffer from coexisting sleep disorders. This aspect can lead to situations of psychiatric disorders, especially in case of persistent insomnia [21]. Furthermore, human sleep behavior can change as a consequence of life-style modifications (i.e., bereavement, retirement, environmental changes) [22].

ICT solutions can help to better manage patients with chronic disease and to overcome these different challenges. In particular, in [5, 6] authors show the impact of "Humans and ICT interaction", demonstrating that these technologies can efficiently support researchers in this field, leading the development of artificial devices able to understand cognitive processes and consequently improving human well-being and human-machine interactions. Our propose represents an example of how human can co-evolve with ICT solutions and more than interaction. It also shows that the scientific investigation in this field has much higher level of discussion than an interaction or an interface between human and measurement.

In this work, we show a system able to identify patient movements and bed posture during sleep sessions using a completely unobtrusive system. In fact, our propose is composed of forty-eight Force Sensing Resistor (FSR) placed in a rectangular grid pattern above the slats and below the mattress. FSRs are connected to a single-board Raspberry responsible for gathering and sending data collected to a central unit using a middleware layer. Our proposal overcomes classical problems of actigraphy-based solutions. In fact, these technologies use wrist and/or wearable devices, particularly complicated to use in a real test bed scenario. Instead, our work is based on cheap technology and does not require active interaction between users and system and, furthermore, is able to detect different postures. This information is particularly useful, for example, in order to assure pressure ulcer prevention. Regarding this illness, especially elderly people deal with the inability of repositioning or to reach desirable positions, promoting blood circulation problems and, indeed, ulcers. This condition, namely bedsores [8], may be early identified and addressed from nursing care and caregivers through an efficient and continuous monitoring, preventing worsening of these symptoms. A correct posture detection benefit is two-fold. On one hand, self-movements can be inferred and consequently prognostications can be made. On the other hand, caregivers can adopt right care program designed to meet accurately elderly needs and avoiding bedsores [9].

In this paper several machine learning approaches are investigated and discussed in order to demonstrate our results in terms of sleep posture detection. It is worth to notice how this system does not require any a priori information about the users.

The rest of the paper is organized as follows. In Sect. 11.3 the state-of-the-art on long-term sleep monitoring systems is presented. In Sect. 11.4 is described our proposal. Section 11.4.2 describes the methodology adopted to recognize sleep postures and in Sect. 11.5 are presented the experimental results in a real test bed with different classification algorithms. Finally, Sect. 11.6 presents results discussion and draw conclusions.

11.2 Related Work

In 1995, the Standards of Practice Committee of the American Sleep Disorders Association (ASDA) commissioned a task force to evaluate the role of actigraphy in sleep medicine. The term actigraphy refers to methods using wristband-like devices able to gather data through users' movements. In general, these system use piezoelectric transducers considering three axes and counting numbers of movements into a predefined time window.

Furthermore, ASDA focused on paper revisions on this topic [24], trying to draw guidelines [4]. As consequence, ASDA considered actigraphy-based tools acceptance from clinicians and researchers as an important milestone. Nowadays, these devices are continuously growing in sleep research, and the increasing number of works in this field is a clear demonstration [3, 27]. Actigraphy-based tools have several strengths: easy sleep behavior monitoring and ability to infer about sleep patters in a time window. In spite of these points, we have to consider also important weaknesses.

In [1] authors show that about a third of adolescents and children sleep recordings were not enough in order to perform sleep understanding. Indeed, an accurate patient reporting activity is needed in order to score records gathered because actigraphy-based devices (sleep and drowsiness can be easily mistaken) or, in general, wearable devices, are not reliable due to misplacement, or users have poor understanding of the treatment protocol. Consequently, a report log have to contain several information about users' behavior before and just after a sleep session, including unforeseen events. Otherwise, information gathered from wearable devices and report would be disjointed. The situation continues to worsen when people who suffer of mental illness are considered. In fact, wearable devices may be not allowed for security reason or they can be damaged from the users on purpose. In a real scenario these problems have to be considered. In [8, 20] authors propose wearable devices in order to infer about users' posture during a sleep session but it is not apparent how the problems can be overcome. In [26] authors shows an unobtrusive method able to detect users' posture and respiratory pattern. This interesting work is based on expensive devices, used as sensors, called Kinotex. They are made by the Canadian

Space Agency for haptic purpose. In [23] a textile approach is shown. Authors use their textile system just over the mattress. The main problem is represented by an accurate calibration task of the system related to each users. Finally, our system is able to merge the inexpensive feature of [26] and the unobtrusive feature of [23] placing, under the mattress, a grid of force sensor resistors (FSRs) and gathering force pressure information in order to infer about position and movements.

11.3   The Sleep Monitoring System

In this section, we describe the developed platform in terms of needed hardware and software artifacts. The proposed system has been designed in order to provide an effective solution both from a cost and a deployment point of view. It allows to unobtrusively provide data to the application layer and to be easily integrated in different pervasive computing scenarios, exploiting the presence of an open source middleware infrastructure.

11.3.1   Hardware Components

The platform for the data collection is based on the Raspberry Pi computing board (Fig. 11.1a) which features an ARMv6 CPU, 512 MB of RAM and various input and output interfaces including USB ports, display ports, digital I/O and an I2C bus. The operating system running on the platform is Raspbian, a dedicated Linux distribution specifically tuned for the Raspberry Pi board. The hardware platform exposes a 26-pin expansion header upon which additional shields can be stacked.

To gather the body force distribution acting on the bed, special sensors called FSR (Force Sensing Resistor) are employed. These sensors are composed of several

(a)   **(b)**

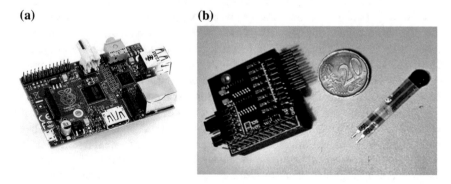

Fig. 11.1 The Raspberry Pi platform (**a**) and a shield for the analog-to-digital conversion together with an FSR sensor (**b**)

layers including a conductive polymer which variates its resistance depending on the force acting on the surface. The FSR sensors feature an extremely low thickness (below 0.5 mm), they are inexpensive and resist well to shocks. To acquire the pressure samples, it is necessary to convert the sensors resistance value in analog form, to a digital one. For this reason, the analog-to-digital converter shields are used (Fig. 11.1b). These shields, stacked on the Raspberry Pi board, take in input the voltage drop generated by the FSRs by means of a partition resistor and make available the digital value to the application.

Each shield is equipped with two Microchip MCP3424 A/D converter units and, in turn, each of these units feature a 18-bit, four channels delta-sigma ADC with differential inputs and self offset and gain calibration. Consequently, the shields could handle up to 8 sensors each. These units embed also a PGA (Programmable Gain Amplifier) with a gain factor up to $8\times$ in order to be able to amplify weak signals before conversion. The ADC units communicate with the data collecting application through the Raspberry Pi I2C bus, using addresses that have been manually set via hardware jumpers. Due to the addressing space, each Raspberry Pi can handle at most 4 ADC shields, resulting in a maximum of 32 deployable sensors per board.

11.3.2   Software Architecture

The proposed architecture has been designed to achieve a high level of flexibility and scalability. On the software side, to allow a distributed data collection and dispatch, a middleware layer has been employed. This is particularly needed when deploying several monitoring systems in different sites while a main server is in charge of receiving and processing the data. The middleware introduces an abstraction layer upon which hardware devices and software entities can seamlessly interoperate by adopting a common model for communications and data representation [7].

This interoperability layer is composed of two stacked sub-layers: on the top the middleware core that exposes the common API and below, the communication connector which implements the publish/subscribe paradigm. A client module, exploiting the middleware functionalities, can publish new sensors or query the sensors already announced in a given environment and obtain those which are available together with the functionalities they export. The communication layer is based on the Message Queue Telemetry Transport (MQTT), a lightweight protocol for machine-to-machine connectivity especially developed for high throughput and small footprint overhead. The middleware is formed by two logical channels, called buses, that are:

- **the context bus**, which is used to distribute the data generated by the communicating entities (e.g. sensing devices and/or pure software modules)
- **the service bus**, which is used to announce the availability of the data-source entities together with their characterizing functionalities and data format.

The buses are based on the MQTT concept of topic, a hierarchical organized string used to identify the channels to which clients can both subscribe for updates or publish data. The middleware API hides all the complexity of the data subscription and distribution process, relieving the clients from the low-level aspects by providing them with an efficient and straightforward interface.

11.4 The Proposed Solution

The purpose of the proposed method is two-fold: (i) inferring the different stages of the user's sleep session and (ii) detecting the different postures. Our approach is particularly useful in reaching the user's satisfaction, because it does not require any wearable devices (unobtrusiveness). As shown in Fig. 11.2, the proposed system uses forty-eight Force Sensing Resistor (FSR) sensors placed as a grid below the mattress and above the slats of the bed. These sensors are placed considering where the pressure mainly occur: knees, back, and chest.

The proposed algorithm evaluates when force values, due to the pressure, change in their amplitude. In fact, this observation lead us to detect when the user is still in a position or when a motion occurs. The algorithm considers two different stages: sleep stages identification (when a sleep session starts/ends and when movements and/or limited muscle activity occur) and detection of the user's postures (left lateral, right lateral, prone, supine).

11.4.1 Stage Detection

In order to better explain how the stage detection algorithm works, we show in Fig. 11.3 a typical behavior of six different FSR time series, together with the ground-truth data of the sleep stages, that have been collected by a video camera inside the room.

When the user gets in the bed, the status of the pressed sensors drastically changes, then it stabilizes at a new high pressure value, and finally, when the user changes his/her position in the bed after a period of time, the pressure stabilizes itself at the original value.

A straightforward sum of all the values from each FSR sensor is not enough, since it can lead to false positives and false negatives, as shown in Fig. 11.4. In order to overcome this issue, a stage detection algorithm must take into account only the variation of the most stressed FSR sensors. As shown in the Fig. 11.4, only if the red zone changes, the algorithm must detect a movement. Moreover, the algorithm can consider as movements also external events (for example someone who makes the bed), therefore the presence of a detection filter is needed to avoid possible misclassification.

Relying on these considerations we propose the stage detection algorithm described as follows:

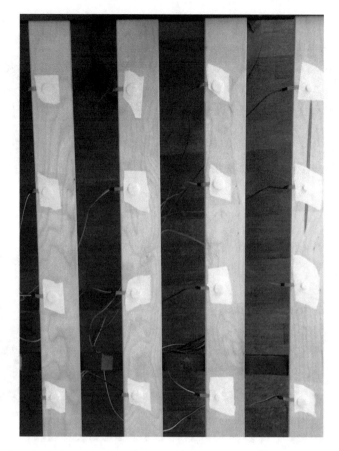

Fig. 11.2 Experimental setup: a grid of force sensing resistor (FSR) sensor nodes placed on the slats

1. For each FSR pressure value p_j, where $j \in N$ (being N the set of the installed FSRs), if $\sum_{j=1}^{N} p_j^{(w)} > \gamma$, where γ is the average pressure, the presence of the patient is ascertained.

2. Only when the patient is detected, for each FSR sensor j, the mean over a W window $P_j^{W} = \frac{1}{W} \sum_{w=1}^{W} p_j$ is evaluated.

3. The difference between two consecutive pressure values $V_j = abs(P_j^{W} - P_j^{W-1})$ is calculated to find significant variations.

4. The variation values are sorted and filtered with a linear weighted filter. The obtained output is a set of sorted and weighted values V_j of significant variations in terms of pressure amplitude.

5. If $\sum_{j=1}^{N} V_j \le \alpha$ (where α is defined as the minimum value for which the pressure variation can be considered as a real movement), the patient is not moving. The α parameter could be defined by leveraging the pressure trace of the day before

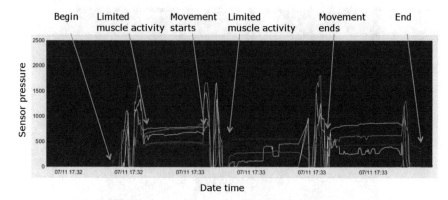

Fig. 11.3 An example of six FSR time series

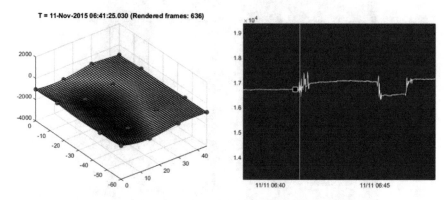

Fig. 11.4 An example of a false positive considering a subset of sixteen FSRs

or by an ad-hoc calibration procedure during the installation of the proposed system.

5. When $\sum_{j=1}^{N} V_j > \alpha$, the movement is detected and the algorithm goes back to step 1.

In particular, γ has been chosen as twice the pressure value of the empty bed, while α was fixed to the 30% of $\sum_{j=1}^{N} V_j$ (i.e. a significant variation).

11.4.2 Sleep Position Detection

Classifying different object means assigning a class to the object considering a set of different possible classes. In the general case, an object is defined by many characteristics (features), providing information about the object class. The information associated to a single characteristic is usually not sufficient to solve the classification problem, so that the correct class can only be inferred by combining all the features. The purpose of our study is to classify users' positions (objects), the features are the FSR signals, and four classes are considered, namely, supine, prone, right lateral, and left lateral.

Many supervised learners are present in literature to solve several classification problems [10]. In our work, the model is made on a set of observations (training set). After this phase, the classifier combines the characteristics and generalizes the learned pattern, by correctly assigning a class to unseen objects. Performance is evaluated applying to the trained model different observations (test set).

In this paper, in order to validate our system, seven different machine learning models have been applied and compared on the task of classifying FSR signals. The goal is to verify that an automatic classification of the user's positions in bed is possible and to carry out a preliminary study in order to choose the best algorithm. The considered models include statistical learning systems (Naive Bayes, Logistic Regression, IbK), ensemble methods (Bagging, HyperPipes), and rule-based learning systems (Decision trees, Decision table). Table 11.1 shows the considered algorithms, along with a raw and short summary of their strengths and weaknesses. Some basic characteristics are investigated for each method: problem type (if the method is able to face classification and/or regression tasks); training speed; prediction speed; automatically feature learning property; if the classifier is parametric or not.

The automatic feature learning property is based on the assumption that not all the features are equal. Some features can be irrelevant and, for example, lead the algorithm to misclassification. On the other hand, some features can be much important than others. A learner can be able to perform automatically the feature selection task, using a scoring method to rank and select the features and, also, to find correlations among them.

Table 11.1 Comparison between the used classification methods

Algorithm	Problem type	Training speed	Prediction speed	Automatically feature learning	Parametric
Decision table	Classification	Slow	Fast	No	No
G. Naive Bayes	Classification	Fast	Fast	No	Yes
Simple logistic	Classification	Fast	Fast	No	Yes
IBk lazy	Class. and Regr.	Fast	Depends on n	No	No
Hyper pipes	Classification	Slow	Fast	No	No
Bagging	Class. and Regr.	Slow	Fast	Yes	No
Random forest	Class. and Regr.	Slow	Moderate	Yes	No

Considering parametric models, we can identify a finite number of parameters. For example, linear models, such as linear regressors, have a finite number of weight coefficients. Vice versa, in non-parametric models, the complexity of the model grows with the number of training data, because the model has not a fixed structure.

In the following, the used classification algorithms are shortly introduced. In our experiments, machine learning methods used are obtained from the WEKA [15] package.

Decision Tables

Decision tables are one of the simplest machine learning techniques [18]. Basically, a decision table consists of a hierarchical table in which each entry in the higher level table gets broken down by the values of a pair of additional features to form another table. Creating a decision table might involve selecting some of the features. The problem is, of course, to decide which features to leave out without affecting the final decision. In our case, we have no a priori information about which FSR must be considered or not. In fact, each sensor, and consequently each feature, can be useful in order to identify a particular user position. Thus, a Decision table approach uses the simplest method of attribute selection: Best First. It searches the space of attributes by greedy hill climbing, augmented with a backtracking facility.

Naive Bayes

Naive Bayes classifiers are a family of simple probabilistic tools based on applying Bayes' theorem. Naive Bayes classifiers employ the class posterior probabilities given a feature vector [17] as the discriminant function. Therefore, approximations are commonly used, such as using the simplifying assumption that features are independent given the class. This assumption of independence is certainly simplistic. However, it is largely adopted in real scenarios and it works very well in many cases, particularly when datasets are filtered with an a priori data selection, in order to avoid redundant records. The Naive Bayes method might not be the best for our scenario because it does not work when an attribute may not occur in the training set in conjunction with every class value.

Logistic Regression

An important class of supervised learning approach is composed by linear regression methods. Into this class an important and well-known method is Logistic regression. Basically, this method uses linear regression in order to provide probabilities [13]. Furthermore, it can be used with binary classification problems. In fact, calculating a linear function as regression function, we use a threshold approach in order to evaluate output in terms of 0 and 1. We can extend this concept involving more than two classes applying a linear function to each class. Output is set to 1 for input that belong to that class. Otherwise we set to 0 for input belonging to others classes. Basically, we provide n regression functions and n different classes. After this separation, when a unseen sample is given, the class with the largest output must be chosen.

Furthermore, we use a sophisticated technique to evaluate probability estimates. Considering classic linear regression method a linear sum is evaluated. In logistic regression a similar sum has to be calculated, but we provide it using an exponential formula:

$$Pr[1|a_1, a_2, \ldots, a_k] = 1/(1 + exp(-w_0 - w_1a_1 - \cdots - w_ka_k)),$$

where $a_1, \ldots, a_k$ are real input features, and $w_0, \ldots, w_k$ are the function parameters. This function is a "logit" transform. $w_0, \ldots, w_k$ parameters are used to minimize the error function on the training set. Through logit function we choose weights to maximize a *log-likelihood function*:

$$\mathscr{L} = \sum_{i=1}^{n}(1 - x^{(i)}) \log(1 - Pr[1|a_1^{(i)}, a_2^{(i)}, \ldots, a_k^{(i)}]) +$$
$$x^{(i)} \log(Pr[1|a_1^{(i)}, a_2^{(i)}, \ldots, a_k^{(i)}]),$$

$x^{(i)}, a_1^{(i)}, \ldots,$ and $a_k^{(i)}$ are, respectively, actual class and features of the i-th pattern of the training set. As mentioned before we try to extend this concept to many classes. At this purpose, we use the probabilities to sum to 1 over the various different classes. This constraint, however, adds computational complexity in order to evaluate joint optimization problem [14].

Lazy Learners
Exploring different supervised approaches, it is enticing to apply a completely different point of view, using Lazy learners, also known as prototype methods. The peculiarity of this class of methods is that they are memory-based and no model is required to be fit [2]. Specifically, we consider the k-nearest neighbors (k-NN) algorithm, a classical non-parametric approach where the function is only locally approximated, whereas all the computations are deferred until classification. The principle behind k-NN is to discover the k (we consider $k = 1$) closest training examples in the feature space with respect to the new sample. The training phase of the k-NN algorithm consists in storing the features and the class label of the training objects. In the classification phase, an unlabeled object is classified by assigning the most frequent label among those of the k training samples nearest to it. During test, new objects are classified based on a voting criteria: the k nearest objects from the training set are considered, and the new object is assigned to the most common class amongst its k nearest neighbors. A variable of this method is represented by the choice of the distance function, used to identify the nearest neighbors. Various distance metrics can be used, the Euclidean distance being the most common. In this work, considering that data were uniformly gathered, we used the most basic settings for the algorithm: Euclidean distance and k set to 1. This means that the class label chosen was the same as the one of the closest training object.

Using k-NN, the target function is approximated locally for each query to the system. These learning systems can simultaneously solve multiple problems, which constitutes, at the same time, their strength and weakness, since for a large input space, they are computationally expensive. These methods usually allow good results when there are not regular separation of the decision boundaries. Our case seems to fit perfectly with this definition.

HyperPipe

A HyperPipe is a fast classifier that is based on simple counts. During the training phase, an n-dimensional (parallel-)pipe is constructed for each class [29]. The pipe will contain all the feature values associated with its class. Test instances are classified according to the category that "most contains the instance". In this way, for each class, a pipe works as a boundary hyper-solid for each numeric feature. At prediction time, the predicted class is the one for which the greatest number of attribute values of the test instance fall within the corresponding bounds.

Bagging

Bagging is a meta-algorithm, that allows to combine and improve the results obtained by other methods. Actually, having a dataset composed by few classes and many samples for each class, classification algorithms may be affected by classical over-fitting problems. The bagging method is known for its capability of avoiding this problem [11]. Basically, the idea is that of creating a set of different training sets, by sampling them from the whole dataset, and combining the different outputs by averaging them or, in our case, voting. As a meta-algorithm, the Bagging method is based on a classification model for the classification phase. In our case, we chose a fast decision tree learner, namely REPTree. This base learner builds a decision and/ or regression tree using information gain or variance and prunes it using reduced-error pruning (with back-fitting).

Considering our data, even taking into account a high number of decision trees, this approach can lead us to a bad overall accuracy. This is due to an intrinsic property of the algorithm that choose, in order to make decision trees, which variable to split in order to minimize the error. In this way, decision trees have a high correlation and a low bias in their own predictions.

Random Forest

A natural step over the bagging approach is represented by the random forest (RF) algorithm. It is also based on decision trees and it is considered as an improvement of the bagging model. Moreover, it allows a decorrelation between trees and, consequently, between their predictions [12]. The idea behind the decision tree methods is quite simple: to make a tree in which each internal node is labeled with an input feature. The arcs from a node representing a particular feature are labeled with each of the possible values of that feature. Each leaf of the tree is labeled with a class or a probability distribution over the classes.

In practice, random forest seems to fit well in our case. In fact, the random forest model is a non-parametric model and, consequently, it does not need any a priori assumption; it is able to face complex input-output relations; it is robust to errors in labels and outliers. This last property is very useful in our case since data are labelled even considering some transition phases from a position to another, in order to make a realistic training set, in comparison with a data acquisition campaign performed throughout the whole night.

11.5   Results

We evaluate the performance obtained by the presented system in two stages. In the first phase, we evaluate the results obtained by the sleep stage detection module, then, we compare, in terms of accuracy, the results obtained by the considered machine learning techniques in the position detection task.

11.5.1   Experimental Setup

In order to evaluate the stage detection performance, we deployed the system in a real test bed, involving a male user of 70 kg of weight and 1.80 m of height. Ground-truth data was gathered using a video camera, able to record in night vision mode, placed in front of the bed. A screenshot of the video recorded is shown in Fig. 11.5. The experiments were carried out for three consecutive nights.

The system was set up with a sampling rate of 10 Hz, that, besides sleep positions detection, can enable a more comprehensive understanding of the human sleep behavior. Indeed, sleep positions represent a fundamental part of the human sleep behavior but, during a sleep session, some sleep disorders (e.g. Restless Legs Syndrome) can also be observed.

From the point of view of the sleep position accuracy evaluation, the gathered dataset along the three nights contains too few samples to apply supervised learning methods. Consequently, a larger benchmark dataset has been constructed. Indeed, few samples can lead us to an ill-posed machine learning problem. For our purpose,

Fig. 11.5 Screenshot of the ground-truth recorded using a video camera in night vision

we prepared an ad hoc test site with a single bed. In particular, we deployed our system into a sleep simulation test bed, involving different users and mattresses with different thicknesses. We have repeated our experiments for 6 days, involving users with different characteristics in terms of weight and height. Each user stayed in the same position for five consecutively minutes, permuting their body position in order to collect data for each class: right, left, prone, supine. As a result, the dataset is composed of 72,000 pressure values, divided into the four different classes, for each user. This approach guarantees us a well-posed benchmark. Furthermore, the three users, different by weight and height, allowed us to gather heterogeneous data in order to test the system adaptability to different users in terms of vital and physiological parameters.

11.5.2   Experimental Results

Regarding the sleep stages detection module, we show the obtained performance in terms of detection of movements. Figure 11.6 shows the output of the module during an entire night. The figure shows that the movements are correctly detected: the pink square dots represent the time instances when the time series (sum of FSR pressures gathered) is strongly variable. The first step of the stage detection algorithm acts as a presence detection filter, resulting particularly useful to prevent a movement detection when the user is not in bed. Indeed, in the figure we can also see how the

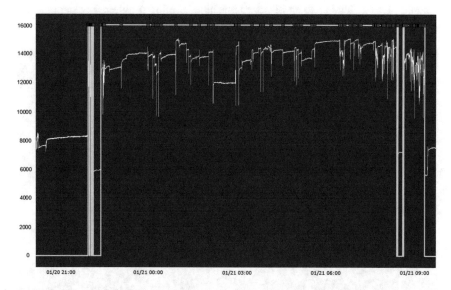

Fig. 11.6 The stage detection algorithm output during an entire night. No movements are detected when the bold line, which represents the presence of a person on the bed, is zero. Otherwise, movements are precisely detected (represented with square dots)

proposed algorithm detects, approximately between 8:00 AM and 8:30 AM, that the user got out of the bed, coming back after few minutes. Moreover, at the beginning of the experiment, the user was asked to get in and out of the bed several time, in order to stress the algorithm. Nevertheless, all the movements were correctly detected, assessing the robustness of the system over the three different nights of tests.

In order to support bedsore risk assessment, the false positive analysis is essential. If the system falsely recognizes the immobility of the user, this impacts on the number of the needed interventions by the caregiver with an overestimation. On the contrary, if the system recognizes a motion of the patient while he/she was motionless, the number of the caregiver interventions will be underestimated. The proposed algorithm shows no false positives, which is key result for its deployment in real test sites.

Regarding the sleep position classification module, we show the obtained performance in terms of detection of correctly classified instances. In order to show the accuracy obtained by each implemented technique, we show the relative confusion matrices. A confusion matrix is a compact representation of the performance of a classifier: each row represents the target class (ground truth positions), while columns represent the predicted classes (predicted positions). A perfect classification method correctly classifies all the instances, so that in the confusion matrix only the diagonal elements are not null. In general, the larger the number in the diagonal cell is, the better the classifier performs. The experimentation has also been designed to assess whether the classifier can be constructed off-line, without adapting its parameters to the user under test.

In order to guarantee an unbiased estimation, training set and test set should ideally be kept separated during the model construction. Successively, the test set can be used to evaluate the obtained model. For this reason, we split the whole dataset into three subsets, each one related to a single user data acquisition for two hours. Then, two experiments were carried out: in the first experiment, a single user dataset is used for training and another single user dataset for testing; in the second experiment, a single user dataset is used for training and the other two (merged together) for the testing.

Tables 11.2, 11.3, 11.4, 11.5, 11.6, 11.7 and 11.8 show the overall average percentage score for each algorithm, considering the two above mentioned configurations for training and test.

As expected, different methods led us to different performances in terms of global accuracy. All the three statistical learning (SL) methods perform well on the tree different scenarios. Tables 11.2, 11.3, and 11.4 show the performance obtained by the algorithms in the SL group with accuracies between 82.2% (worst case) and 95.9% (best case) with classes, in some cases, which are perfectly predicted. The results are promising and suggest that such methods are able to correctly classify the user's postures.

Slightly worse results are obtained when considering methods of the ensemble group. In fact, as shown in Tables 11.5 and 11.6, accuracies are between 68.4 and 83.7%. The worst performance was achieved using the Decision Table method, whose accuracy ranges between 41.6 and 44.3 (see Table 11.7). Actually, such a result is

Table 11.2 Performance evaluation of Naive Bayes method

GT position	User1 – User2 Position predicted				Acc. %	User1 – User3 Position predicted				Acc.%	User1 – User2–3 Position predicted				Acc. %
	S	P	R	L		S	P	R	L		S	P	R	L	
Supine	2964	0	2	583	92.0	12,070	0	18	13	82.2	15,034	0	20	596	84.5
Prone	0	2945	540	8		0	6867	1127	2835		0	9812	1667	2843	
Right	0	0	3554	0		6	1834	9285	592		6	1834	12,839	592	
Left	0	0	0	3551		0	1203	630	10,033		0	1203	630	13,584	

Table 11.3 Performance evaluation of logistic regression method

GT position	User1 – User2					User1 – User3					User1 – User2-3				
	Position predicted				Acc. %	Position predicted				Acc. %	Position predicted				Acc. %
	S	P	R	L		S	P	R	L		S	P	R	L	
Supine	2955	277	317	0	95.8	10,889	0	1212	0	88.4	13,844	277	1529	0	90.2
Prone	0	3493	0	0		1719	9110	0	0		1719	12,603	0	0	
Right	0	0	3554	0		4	8	11,113	592		4	8	14,667	592	
Left	0	0	0	3551		592	602	644	10,028		592	602	644	13,579	

Table 11.4 Performance evaluation of IbK method

| | User1 – User2 | | | | | User1 – User3 | | | | | User1 – User2–3 | | | | |
| | Position predicted | | | | Acc.% | Position predicted | | | | Acc. % | Position predicted | | | | Acc. % |
GT position	S	P	R	L		S	P	R	L		S	P	R	L	
Supine	2966	396	0	187	95.9	12,101	0	0	0	93.3	15,067	396	0	187	93.4
Prone	0	3493	0	0		1109	9624	3	93		1109	13,117	3	93	
Right	0	0	3554	0		2	654	10,469	592		2	650	14,027	592	
Left	0	0	1	3550		8	0	634	11,224		8	0	635	14,774	

Table 11.5 Performance evaluation of Bagging method

GT position	User1 – User2					User1 – User3					User1 – User2–3				
	Position predicted				Acc. %	Position predicted				Acc. %	Position predicted				Acc. %
	S	P	R	L		S	P	R	L		S	P	R	L	
Supine	604	571	71	303	72.5	10,691	783	18	609	68.4	13,295	1354	89	912	69.3
Prone	0	3493	0	0		3	7616	666	2544		3	11,109	666	2544	
Right	604	1186	1764	0		5	1919	9201	592		609	3105	10,965	592	
Left	369	792	0	2390		5619	1329	620	4298		5988	2121	620	6688	

Table 11.6 Performance evaluation of HyperPipes method

GT position	User1 – User2					User1 – User3					User1 – User2–3				
	Position predicted				Acc. %	Position predicted				Acc. %	Position predicted				Acc. %
	S	P	R	L		S	P	R	L		S	P	R	L	
Supine	2487	1062	0	0	83.7	10,187	481	1433	0	77.3	12,674	1543	1433	0	78.8
Prone	0	3493	0	0		1	6755	4060	13		1	10,248	4060	13	
Right	4	0	3550	0		0	4	11,120	593		4	4	14,670	593	
Left	642	265	330	2314		2849	0	1102	7915		3491	265	1432	10,229	

Table 11.7 Performance evaluation of Decision Table method

GT position	User1 – User2					User1 – User3					User1 – User2–3				
	Position predicted				Acc. %	Position predicted				Acc. %	Position predicted				Acc. %
	S	P	R	L		S	P	R	L		S	P	R	L	
Supine	3172	371	6	0	44.3	12,100	0	1	0	41.6	15,272	371	7	0	42.2
Prone	2635	858	0	0		7430	354	2826	219		10,065	1212	2826	219	
Right	1599	0	1955	0		5738	348	5631	0		7337	348	7586	0	
Left	1759	726	785	281		9189	942	470	1265		10,948	1668	1255	1546	

Table 11.8 Performance evaluation of RF method using 100 trees and no-replacement

GT position	User1 – User2					User1 – User3					User1 – User2–3				
	Position predicted				Acc. %	Position predicted				Acc. %	Position predicted				Acc. %
	S	P	R	L		S	P	R	L		S	P	R	L	
Supine	2963	579	0	0	95.7	11,960	0	141	0	89.6	14,924	579	147	0	91.0
Prone	0	3493	0	0		610	10,218	1	0		610	13,711	1	0	
Right	0	0	3554	0		2	8	11,115	592		2	8	14,669	592	
Left	0	0	23	3528		2844	9	635	8378		2844	9	658	11,906	

Table 11.9 Performance evaluation of random forest according to different number of trees

GT position	User1 – User2-3 – 200 Trees					User1 – User2-3 – 500 Trees					User1 – User2-3 800 Trees				
	Position predicted				Acc. %	Position predicted				Acc. %	Position predicted				Acc. %
	S	P	R	L		S	P	R	L		S	P	R	L	
Supine	14,696	583	371	0	95.4	14,875	583	192	0	95.7	14,742	583	325	0	95.4
Prone	0	14,322	0	0		0	14,322	0	0		0	14,322	0	0	
Right	0	11	14,668	592		0	11	14,668	592		0	11	14,668	592	
Left	0	14	1222	14,181		0	1	1226	14,190		39	0	1230	14,148	

Table 11.10 Classification performance of random forest with 500 trees, after downsampling with no-replacement

Sample size (%)	User1–User2	User1–User3	User1–User2–3 (%)
20	95.8	94.9	95.1
10	95.8	91.2	92.3
5	91.9	89.5	90.1

expected, since the main advantage of approaches based on decision tables lies in their simplicity and low computational cost, whereas the classification performance is usually lower with respect to other machine learning methods.

The Random Forest (RF) method, instead, can be considered the best method in our application as shown in Table 11.8. In terms of global performance, it shows an accuracy between 89.6 and 95.7%.

In order to justify such a result, we need to recall some notions about random forests. The Random forest approach belongs to the class of ensemble methods, based on a combination of tree predictors. Each tree is composed using a sub-sampling of the training set. Combining outputs from each tree the algorithm is able to improve the generalization performance and avoid the over-fitting problem. Eventually, when the number of trees goes to infinity, the Strong Law of Large Numbers always guarantees that the RF accuracy converge to that of the optimal predictor.

Table 11.9 shows the global accuracy values, obtained by Random Forest, running the algorithm with different number of trees, and seeds fixed to 1.

The Random Forest algorithm reaches an accuracy of 95.4% for User1–User2–User3 case, better than all the other methods previously shown. Table 11.10 shows how a different percentage of the original dataset, with no-replacement and fixed number of trees equal to 500, impacts in terms of accuracies. Random Forest, considering the number of features involved in our scenario, needs approximately 100 s for the learning phase, considering all the position together. Instead, a real-time prediction can be performed. A down-sampling strategy can be useful in the case in which the model construction is performed on-board at microcontroller-level or on some other resource-constrained device. Indeed, the overall complexity of RF, in terms of computational speed, depends on several factors, such as the number of trees, features, and instances. RF, trying to find an optimal predictor scanning several levels of possibilities, can require a good deal of computing power and memory available.

11.6 Conclusions

This work presents an unobtrusive sleep monitoring system, suitable for long term monitoring, that exploits a sensing method based on force sensor resistors. The system is able to detect movements and sleep patterns. The high versatility of the proposed system allows its use in several application scenarios, such as to assess the risk of pressure ulcer, to monitor bed exits or to observe the influence of medication on sleep behavior.

In this paper, we have compared several supervised learning algorithms, in order to obtain the most suitable solution in this context. Comparative experimental results from seven different approaches demonstrate that we can infer about user positions, sleep duration, and routines using an unobtrusive setting by using the IbK method. Indeed, the IbK method achieves about 95% of accuracy. A similar performance has been obtained by the Random Forest technique. However, since the Random Forest is a non-parametric model, characterized by an automatic feature learning technique, it should be preferred to detect the user's positions. Despite of the good accuracy, this model could not immediately apply in more realistic scenarios. Further analysis, considering different users in terms of gender, height, weight and features in general, will be performed for a more robust system validation.

Acknowledgements This work is supported by the INTESA "Servizi ICT integrati per il benessere di soggetti fragili" project, under the APQ MIUR MISE Regione Toscana (DGRT 758 del 16/09/2013) FAR–FAR 2014 Program.

References

1. Acebo C, Sadeh A, Seifer R, Tzischinsky O, Wolfson A, Hafer A, Carskadon M (1999) Estimating sleep patterns with activity monitoring in children and adolescents: how many nights are necessary for reliable measures? Sleep 22(1):95–103
2. Aha DW, Kibler D, Albert MK (1991) Instance-based learning algorithms. Mach Learn 6(1):37–66
3. Alfeo AL, Barsocchi P, Cimino MG, La Rosa D, Palumbo F, Vaglini G (2017) Sleep behavior assessment via smartwatch and stigmergic receptive fields. In: Personal and ubiquitous computing, pp 1–17
4. Association American Sleep Disorders et al (1995) Practice parameters for the use of actigraphy in the clinical assessment of sleep disorders. Sleep 18(4):285–287
5. Baranyi P, Csapo A (2012) Definition and synergies of cognitive infocommunications. Acta Polytech Hung 9(1):67–83
6. Baranyi P, Csapo A, Sallai G (2015) Cognitive infocommunications (CogInfoCom). Springer
7. Barbon G, Margolis M, Palumbo F, Raimondi F, Weldin N (2016) Taking Arduino to the internet of things: the ASIP programming model. Comput Commun
8. Barsocchi P (2013) Position recognition to support bedsores prevention. IEEE J Biomed Health Inf 17(1):53–59
9. Barsocchi P, Bianchini M, Crivello A, Rosa DL, Palumbo F, Scarselli F (2016) An unobtrusive sleep monitoring system for the human sleep behaviour understanding. In: 2016 7th IEEE international conference on cognitive infocommunications (CogInfoCom), pp 000,091–000,096
10. Bishop CM (2006) Pattern recognition. Mach Learn 128:1–58
11. Breiman L (1996) Bagging predictors. Mach Learn 24(2):123–140
12. Breiman L (2001) Random forests. Mach Learn 45(1):5–32
13. Friedman J, Hastie T, Tibshirani R et al (2000) Additive logistic regression: a statistical view of boosting (with discussion and a rejoinder by the authors). Ann Stat 28(2):337–407
14. Hair JF Jr, Anderson RE, Tatham RL, Black WC (1995) Multivariate data analysis with readings, 4th edn. Prentice-Hall Inc, Upper Saddle River, NJ, USA

15. Hall M, Frank E, Holmes G, Pfahringer B, Reutemann P, Witten IH (2009) The weka data mining software: an update. SIGKDD Explor Newsl 11(1):10–18. https://doi.org/10.1145/1656274.1656278
16. Iwasaki M, Iwata S, Iemura A, Yamashita N, Tomino Y, Anme T, Yamagata Z, Iwata O, Matsuishi T (2010) Utility of subjective sleep assessment tools for healthy preschool children: a comparative study between sleep logs, questionnaires, and actigraphy. J Epidemiol 20(2):143–149
17. John GH, Langley P (1995) Estimating continuous distributions in bayesian classifiers. In: Proceedings of the eleventh conference on Uncertainty in artificial intelligence. Morgan Kaufmann Publishers Inc., pp 338–345
18. Kohavi R (1995) The power of decision tables. In: Proceedings of the European conference on machine learning, Springer, pp 174–189
19. Neubauer DN (1999) Sleep problems in the elderly. Am Family Physician 59(9):2551–2558
20. Palumbo F, Barsocchi P, Furfari F, Ferro E (2013) AAL middleware infrastructure for green bed activity monitoring. J Sens
21. Perlis ML, Smith LJ, Lyness JM, Matteson SR, Pigeon WR, Jungquist CR, Tu X (2006) Insomnia as a risk factor for onset of depression in the elderly. Behav Sleep Med 4(2):104–113
22. Roepke SK, Ancoli-Israel S (2010) Sleep disorders in the elderly. Indian J Med Res 131:302–310
23. Rus S, Grosse-Puppendahl T, Kuijper A (2014) Recognition of bed postures using mutual capacitance sensing. In: European Conference on ambient intelligence, Springer, pp 51–66
24. Sadeh A, Hauri PJ, Kripke DF, Lavie P (1995) The role of actigraphy in the evaluation of sleep disorders. Sleep 18(4):288–302
25. Seelye A, Mattek N, Howieson D, Riley T, Wild K, Kaye J (2015) The impact of sleep on neuropsychological performance in cognitively intact older adults using a novel in-home sensor-based sleep assessment approach. Clin Neuropsychol 29(1):53–66
26. Townsend DI, Holtzman M, Goubran R, Frize M, Knoefel F (2011) Measurement of torso movement with delay mapping using an unobtrusive pressure-sensor array. IEEE Trans Instrum Meas 60(5):1751–1760
27. Waltisberg D, Arnrich B, Tröster G (2014) Sleep quality monitoring with the smart bed. In: Pervasive health, Springer, pp 211–227
28. Wiggs L, Montgomery P, Stores G (2005) Actigraphic and parent reports of sleep patterns and sleep disorders in children with subtypes of attention-deficit hyperactivity disorder. Sleep—New York then Westchester 28(11):1437
29. Witten IH, Frank E, Hall MA, Pal CJ (2016) Data mining: practical machine learning tools and techniques. Morgan Kaufmann

Chapter 12
Electroencephalogram-Based Brain-Computer Interface for Internet of Robotic Things

Jozsef Katona, Tibor Ujbanyi, Gergely Sziladi and Attila Kovari

Abstract Several papers focus on the IoT ranging from consumer oriented to industrial products. The IoT concept has become usual since the beginning of the 21st century and was introduced formally in 2005 [45, 53]. IoT gives the possibility for lots of uniquely addressable "things" to communicate and exchange information with each other over the existing network systems and protocols [1, 10, 15]. The IoT enables to make information detected by these objects transmittable, and the objects themselves controllable, by using the current network infrastructure [13, 18]. This provides the opportunity to integrate the physical world and IT systems in an even greater scale, which leads to the enhancement of efficiency, accuracy, and economics by minimal human intervention.

12.1 Introduction

Several papers focus on the IoT ranging from consumer oriented to industrial products. The IoT concept has become usual since the beginning of the 21st century and was introduced formally in 2005 [45, 53]. IoT gives the possibility for lots of uniquely addressable things to communicate and exchange information with each other over the existing network systems and protocols [1, 10, 15]. The IoT enables to

J. Katona (✉) · T. Ujbanyi · G. Sziladi · A. Kovari
University of Dunaujvaros, Tancsics M. u. 1/A, Dunaujvaros 2401, Hungary
e-mail: katonaj@uniduna.hu

T. Ujbanyi
e-mail: ujbanyit@uniduna.hu

G. Sziladi
e-mail: sziladig@uniduna.hu

A. Kovari
e-mail: kovari@uniduna.hu

© Springer International Publishing AG, part of Springer Nature 2019
R. Klempous et al. (eds.), *Cognitive Infocommunications, Theory and Applications*,
Topics in Intelligent Engineering and Informatics 13,
https://doi.org/10.1007/978-3-319-95996-2_12

make information detected by these objects transmittable, and the objects themselves controllable, by using the current network infrastructure [13, 18]. This provides the opportunity to integrate the physical world and IT systems in an even greater scale, which leads to the enhancement of efficiency, accuracy, and economics by minimal human intervention. In this paper a BCI system based human-robot test environment is implemented using TCP/IP communication, where the latency of human actuation has been analyzed.

12.1.1 Internet of Robotic Things and BCI

The IoT technology provides several possibilities for expanding opportunities of robots, for example the usage of intelligent actuated devices that can impact the physical world as a robot. ABI Research has created and introduced the Internet of Robotics Things (IoRT) expression in its Report of 2014. The Internet of Robotic Things is a new concept [52] based on IoT for supporting robotic systems including industrial, home robots or other complex automatic systems with humanlike skills, where partly autonomous systems can communicate with each other. These devices use ad-hoc, local, distributed or cloud service based intelligence to optimize acts and motions in the physical world taking into account several factors for example cooperative, flexible, safety, production, and logistics aspects using information exchange, and data sharing.

When the information scanned by the robot is not adequate for the optimal operation the robot can collect additional information from the environment and/or can use additional cloud services to determine the appropriate action. To achieve this functionality several technologies have to be applied based on the knowledge of robotics, mechatronics, cyber-physics, artificial intelligence, bio-engineering, information exchange network or cooperation (Fig. 12.1).

The IoRT concept dealing with the extension of IoT and robotic devices to provide advanced, adaptive, more intelligent, collaborative, and heterogeneous robotic capabilities using the technologies, among others, Networked Robots, Cloud Robotics, Robot as a service. In order to achieve the mentioned aims a number of important aspects have to be considered, such as standardization, interoperability, common architecture/infrastructure design including time-varying network latency, and security.

In the IoRT system, the robot is integrated into the smart environment. IoT technology, the cooperative robots, and the communication of the equipment significantly contribute to the automation and optimization of the systems, but in several cases, for instance security-critical cases, human supervisory of these systems needs to be ensured. Human-Machine Interfaces (HMI) enabling communication between IT-systems and humans are emphasized, in which field there are many researches in progress, such as processes based on observation of gestures, eye moves or brain activity. There are several methods available for brain activity, but in terms of mobility and price, the application of EEG based devices worn on the head promises

Fig. 12.1 Main fields of
Internet of Robotics Things

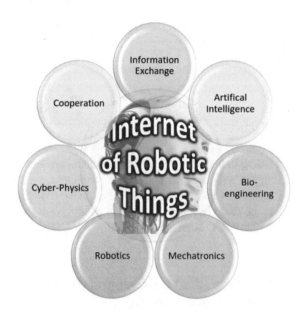

the most opportunities. EEG device measuring human brain activity was invented
by Hans Berger in 1929, searching for the answer by using this device, What kind
of connections between differential psychological and physiological conditions can
be shown. This technique has been improved in the past decades [50], and several
changes occurring during the activity of different brain lesions have been observed.
By today, this technology has enabled the design of mobile headsets capable of reg-
istering EEG signals. The brain-computer interface (BCI) is one of the most rapidly
developing multidisciplinary research fields nowadays.

BCI system, which can transmit different information (commands) based on brain
activity by processing signals arriving from the human brain, creates a channel
between the brain and an external device (for example a mobile robot). The goal
of the first BCI researches was to enhance life quality of patients suffering from different neurological disorders. But by nowadays [44, 46], research fields have already
embraced other application fields, such as the application of control features. BCI
systems based on the operation of EEG type devices, have a relatively simple design,
they are portable, safe, and their operation is also quite simple [3, 4, 26, 30, 32, 44,
46, 47].

The article examines the design opportunity of supervisory achieved by moving
of mobile robotapplying IoT technology via brain-computer interface, which can
utilize some cognitive human skills and features by a BCI system (Fig. 12.2).

To design a brain-computer interface system, a headset applying a cost-efficient
EEG biosensor, and a self-built mobile robot have been used. In the testing environment built-up by the help of these devices, users brain activity is examined by
an EEG headset, which transmits data to a computer for process and evaluation.
BCI application running on the computer controls the speed of a WiFi mobile robot

Fig. 12.2 Knowledge base
of Internet of Robotics
Things

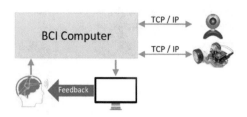

according to the processed information, and enables remote observation of the robot
through a web camera. The article examines interventional latency in the created
testing environment in that case, when the user can directly see the mobile robot, and
when the robot is observed through a web camera, while the speed of the robot can
be influenced by changing attention.

12.2 Theoretical Background of EEG Based Brain Computer Interfaces

Human brain activity induces minor voltage changes, which can be measured on the
head skin, and this information may be registered by EEG devices. The registered
signals can be interpretable by the BCI, which can be transformed to command lines.

12.2.1 Role of PFC in Controlling Concentration Strength

Frontal lobe of human brain is responsible for several duties. Among others, it controls conscious moves, and our thinking and social behavior are also managed by it.
It also plays a major role in problem solving, planning, speaking and controlling of
skeletal muscles, and the neuron network is responsible for these complex functions.
The first part of the lobe is called Prefrontal Cortex—PFC), which is the relevant
control centre of our brain, furthermore, nerves enter to the lobe from all of our
senses, and this region is an important regulator of attention abilities as well [2, 16,
24, 27, 49].

During anatomical examination of the brain, it has been proved, that frontal lobe
consists of the primer motoric lobe, and the PFC in front of it, in which ventral and
dorsal regions responsible for different functionalities can be observed [31, 42, 51].
The biological complex role of PFC is presented by multiple exchange connections
between fore-brain and brain-stem [28, 31, 42], furthermore, it plays a key role in
such cognitive processes, like ongoing maintenance of attention, learning, remembering, and autonomous operations, so PFC regions are significant in controlling the
strength of concentration [7, 11, 12, 19, 29, 37, 39, 40, 43]. Figure 12.3 shows the

Fig. 12.3 The role in PFC
area in each function [39]

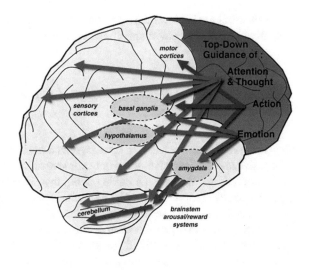

location of PFC in the frontal lobe, and its brain regions responsible for different functions.

From the Fig. 12.3, the following three functions can be read: attention, emotional reaction, behavior and decision making, furthermore, the location of brain regions responsible for these functions are also visible. Among these areas, regarding brain-computer interface, for achieving the goals of profession, attention is the most important. For examination of attention, among the later presented measuring areas, the so-called Frontopolar1 (Fp1) position is the most suitable, which is an important perspective regarding the system to be established.

It has been observed, that such patients, whose PFC had got injured, were hardly, or even not able to acquire new knowledge, adapt to different tasks, solve problems, or execute tasks requiring high-level concentration [6, 11, 43]. PFC, beside the controlling of the value of attention, plays a major role in the optimal operation of short term memory, while for temporary store of sensor information, and for connecting reactions on external stimuli, flawless operation of PFC is inevitable [14, 34].

For monitoring the PFC activity of human brain, several technologies are applied, primarily in medical technologies are applied, primarily in medical and laboratory environment, such as Computed [Axial] Tomography (CT), Magnetic Resonance Imaging (MRI), Functional Magnetic Resonance Imaging (fMRI), Single Photon Emission Tomography (SPECT), Proton Emission Tomography (PET), and EEG [8, 17, 25, 38, 41].

The above listed different imaging devices provide quite accurate image about the structural and functional operation of the brain, but these devices are not mobile. By the development of EEG devices, mobile EEG based devices worn on the head have become available, which depending on the number of sensors, give image about functional operation of each areas of the brain [54]. The wear designed EEG based headsets is not too inconvenient even after longer period of time, so it can be applied not only in laboratory environment, but also in everyday use.

12.2.2 EEG Registering and Signal Processing Devices of BCI Systems

Registration of brain signals can be categorized according to several methods, such as invasive and non-invasive. If the detector sensor is placed inside the human skull, in the brain tissue, then we are talking about invasive process, whose benefit is that high-frequency components can be measured much clearer and more accurate, but due to health risks and several ethical aspects, they are primarily used in animal experiments. In case of non-invasive method, electrodes are placed outside the skull, according to the 10–20 international standard (Fig. 12.4). This application method is much more likely used on humans, because it does not endanger them due to the implantation, but it has the disadvantage, that the measured signals are noisier [9, 33, 48].

In the past years, several EEG headsets have been developed not only for medical use, which are operated from own batteries in order to ensure mobile use. Now some widespread EEG headsets are being introduced, which are also suitable for achieving unique-developed BCI.

Fig. 12.4 The International 10–20 standard, Fp1 measurement position relevant for the research is well-shown on the left of the frontal lobe

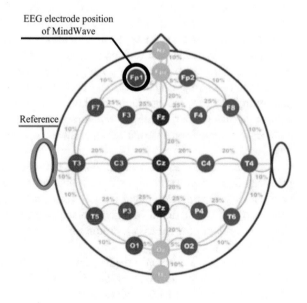

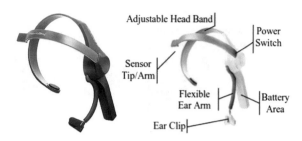

Fig. 12.5   MindWave EEG headset

12.2.2.1   Neurosky MindWave Mobile

The MindWave Mobile EEG headset (Fig. 12.5) is the development of Neurosky. Several universities cooperated in the development of NeuroSky ThinkGear EEG technology, such as Stanford, Carnegie Mellon, University of Washington, University of Wollongong and the Trinity College. In the device, dry EEG sensor made of stainless alloy is applied in Fp1 position, and a reference sensor, by a detector connecting to the earlobe. The device transmits pre-processed data through Bluetooth wireless connection. EEG signals are sampled with 512 Hz frequency, and digitalizes with a 12 bit A/D converter [55, 56]. The device, thanks to ThinkGear technology is suitable for calculating an attention value [55, 56], this value is used in the examinations.

12.3   IoRT Ready Mobil Robot

The mobile robot is a self-made unit assembled from main parts capable of WiFi wireless communication, which enables easy attachment with virtually all computer based control units, and even achievement of IoRT functionality.

12.3.1   Design of the Mobile Robots Hardware

At the design of the mobile robot, simple design, upgrade-ability, and IoT based on wireless communications were the main aspects. To achieve the experiments, simple control feature, the speed control had to be ensured. In case of the mobile robot, the module based on ESP8266 chip manufactured by Espressif Systems was chosen, in which the micro controller and a WIFi communication unit were achieved in an integrated circuit.

The device is available in severally designed modules, for development purposes, the ESP-12E containing integrated aerial making important outputs of the circuit available was chosen. To the module, a 2nd generation LoLin developer chip was

Fig. 12.6 LoLin ESP8266 developer chip and I/O adapter module

Fig. 12.7 Assembled
mobile robot

chosen, which contains the 3.3 V voltage regulator required for mains supply via
USB port, and the USB-UART serial converter based on CH340G circuit, required
for programming via USB port was chosen. The connection of other external modules
to the LoLin developer chip is supported by a separate I/O module (Fig. 12.6).

For the purpose of the mobile robot, primarily the simple design, however universally applicable design providing upgrade-ability, enabling chassis moving mechanism was preferred. Regarding the moving mechanism, the two driven front wheels
and one free rear wheel design was preferred. The mobile robot kit contains a remov-
able surfaced, transparent plexi chassis, two driven wheels, an engine unit containing
two 1:48 gear direct current motors, and other supplements required for assembly.

For driving the direct current motors, a module based on L9110S motor controller
circuit was chosen. In the front of the robot, three distance detectors were attached
to detect the environment and avoid collisions. The mobile robot containing the
assembled modules placed on the chassis, and the motor and mains inputs are shown
on Fig. 12.7.

12.3.2   Design of Mobile Robots Software

For ESP8266 micro-controller capable of WiFi communication, several development environments are available, such as the Arduino development environment.
For programming the developer chip, driver for CH340G integrated circuit enabling
USB-UART connection has to be installed, which creates a virtual serial port on
the computer. The programming of WiFi communication connection is supported by
the ESP8266WIFI function library. The BCI system sends the data varying between
0 and 100, defining the mobile robots speed, to the mobile robots controller unit,
joining as a client to the server via TCP/IP connection. Flow-chart of the program
enabling speed control of the mobile robot is shown on Fig. 12.8.

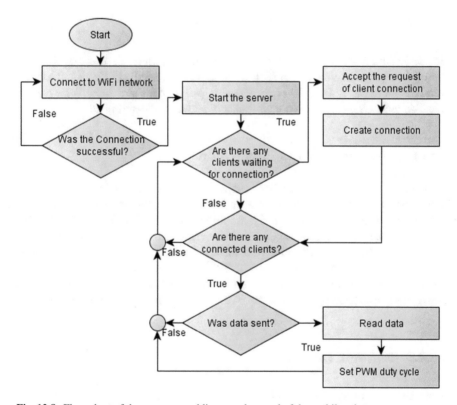

Fig. 12.8 Flow-chart of the program enabling speed control of the mobile robot

12.4 Implementation of the Software of BCI-Based System

For managing data measured by the MindWave EEG headset, such a software is required, which is able to process the detected brain bioelectric signals in real time, at the site of the measurement, and also enables the graphical visualization of these information, and their storage for posterior processes.

The software, beside the process of EEG signals, is suitable for the visualization of a remote IP camera, and by its support, apart from the location, enables the observation of the move of a mobile robot, and also allows the definition of latencies related to the control of the robot.

12.4.1 Planning of BCI-Based System

In the so-called overview phase of the development, different aspects of the BCI-based testing system, and its connections have been revealed. Beside that, the structural design of the system, the elements participating in the structure, their communication technologies, and also their transmit media were determined.

12.4.1.1 Components of BCI-Based System

Following the measurement, digitalization, pre-process, normalization, filtering of bioelectrical signals coming from the brain activity, primarily the time periodic and/or frequency range features of EEG signals are performed. According to the features of the EEG signal, classification of signals is performed, which information may be displayed, or by their utilization, even further functions can be achieved, such as speed control of a remote mobile robot using TCP/IP connection, and the display of the online screen of an IP camera, which means a feedback for the user regarding the move of the robot. Components of the BCI-based system are shown on Fig. 12.9.

12.4.2 Software Design

During the software design, software structures, data structures and algorithm descriptions to be implemented have been defined. During the planning, the logical plan of the system has been created, as well as the static, dynamic and case models, which provide the main part of the systems implementation.

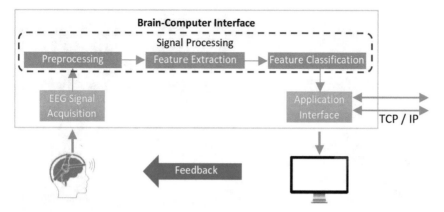

Fig. 12.9 Components of a BCI-Based system

12.4.2.1   Algorithm Flowchart

Figure 12.10 describes the main logical plan of the implemented system by a flow-chart. On the flow-chart, main steps of the programs operation are well-shown; data process, data communication required for speed control of the mobile robot, and data storage have been implemented in the line 1.

12.5   Results

The software of the BCI-based system was realized on the basis of the planning models introduced in the previous sub-chapter. The developed BCI interface is appropriate test environment for the examination of BCI interface latency. In this section the test environment, the examination and the test results have been introduced.

12.5.1   BCI Control

The implemented application, such as the BCI interface, realizes the adaptation of EEG headset, the process of information sent by the headset, their presentation, and the control of devices capable of remote communication via TCP/IP protocol, as well as the monitoring of the process via a web camera. The application is capable of simultaneous presentation of several processed information, displaying both the actual values of brain wave strengths defined by the spectrum-analysis of brain bio-electrical signals, and their changes by time. The system set-up this way achieves the evaluation of some cognitive factors, like attention level, by using the strength of brain waves defined by brain activity.

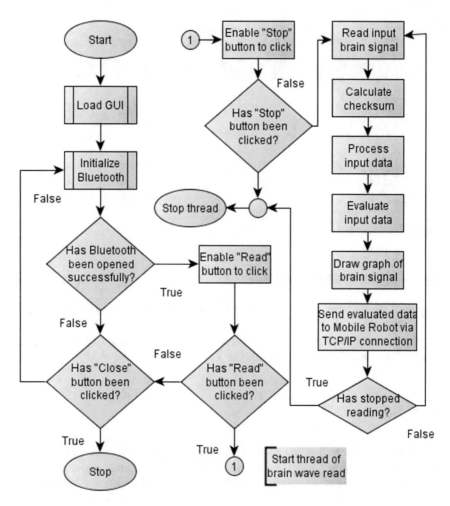

Fig. 12.10 Overview of logical plan of BCI system software

Following the throughout study of dynamic planning models, the application may be divided into five main function parts: data process, display of processed data, forwarding required data for the mobile robot, storage of processed data, and the display of the IP cameras view for the user. The data processing unit performs the reading, conversion and process of data sent by the EEG headset via wireless communication, while the data display unit performs the display of defined signals, by a pre-defined bar graph and timing diagram. Forwarding of evaluated data to the mobile robot, and the display of the IP cameras view have been performed by applying TCP/IP protocol. The evaluated and processed data have been stored in a pre-defined structure. Figure 12.11 shows the applications user interface during operation.

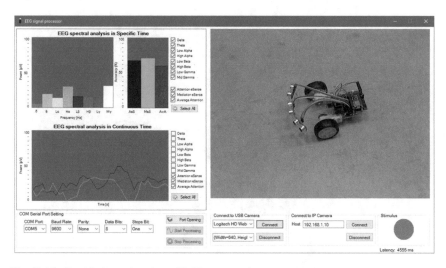

Fig. 12.11 Graphical user interface of BCI system software

12.5.2 Test Environment

For testing the control of the robot, a testing environment has been realized. The testing environment is suitable for performing supervisory or control functions of the robot, based on attention level defined by processing EEG headset data. The structure of the system is shown on Fig. 12.12

Figure 12.13 shows the structure of the designed BCI based system, whereas the BCI application runs on the computer. The computer and the MindWave EEG headset connect to each other via Bluetooth technology, while the network device supporting remote control, applies wire transfer. Remote devices, the mobile robot controlled during the test, and the IP camera monitoring it, communicate with the network device via wireless TCP/IP connection.

12.5.3 Examine the Response Latency of BCI Based Control

By applying the testing environment, with the co-operation of testing subjects, it has been examined, with what latency the testing subjects were able to give a command depending on attention level. The task of testing subjects was to increase their attention level above 60% when the light in the bottom-right corner of the BCI application turned red, and in case of exceeding that level, the mobile robot received via TCP/IP connection a speed level divergent from zero making the robot move. The mobile robot provided a feedback of launching the robot to the BCI application. During the test, the period between the light turning red to the exceeding of 60% attention level,

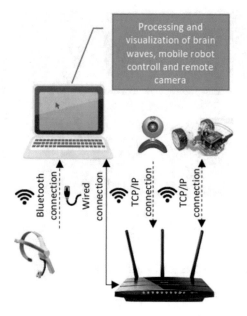

Fig. 12.12 Structure of BCI-based system

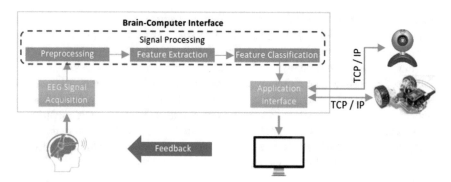

Fig. 12.13 Components of BCI-based system

as well as the mobile robots feedback have been logged. This time period provides the latency of communication between the BCI system and the mobile robot. According to these, two tests have been performed, in the first case (Fig. 12.14) the BCI system and the robot were in the same sub-network, and the testing subjects could directly see the robot. In the second case (Fig. 12.15), the robot was in another sub-network, and the testing subjects could see the robot only via a web camera, so remote control was performed.

In both cases, the robot had to be launched three times in a row, increasing their attention level following the light turning red. In the test, 10 secondary school stu-

Fig. 12.14 BCI based direct robot control

Fig. 12.15 BCI based remote robot control

dents, at ages varying between 15 and 17 participated, particularly 8 boys and 2 girls. Results of the two tests are shown in Tables 12.1 and 12.2.

It is shown in the results of the first test, that the average time from the appearance of the signal to the launch of the robot, is almost 8 s, most of the time was spent for reaching higher attention level, while robot communication required only some milliseconds, depending on network latency. In case of the second test, performance of the remote control was similar to the first test, in this case, average latency was also almost 8 s, and of course conscious influence of attention level required some seconds. As a conclusion, however, we can declare, that conscious influence of attention level requires some seconds. During the performance of the tests, it was experienced, that testing subjects had to practice the use of devices for 10–20 min, before they became able to consciously influence their attention levels in that extent, which could be detected by the device.

Table 12.1 Time-varying control latency Test1 (T1)

Test subjects	SR[a]latency [ms[c] ± SD[d]]	N[b]latency [ms[c] ± SD[d]]	Total latency [ms[c] ± SD[d]]
1	5812.42 ± 1723.32	0.97 ± 0.04	5813.39 ± 1723.36
2	10722.33 ± 2652.15	1.02 ± 0.04	10723.35 ± 2652.19
3	9756.19 ± 2152.88	0.99 ± 0.04	9757.19 ± 2152.92
4	6917.11 ± 2017.56	0.90 ± 0.04	6918.01 ± 2017.60
5	7012.86 ± 1352.98	0.95 ± 0.04	7013.80 ± 1353.02
6	11121.96 ± 2822.42	0.91 ± 0.04	11122.87 ± 2822.46
7	4872.77 ± 1248.73	0.95 ± 0.04	4873.72 ± 1248.77
8	7786.84 ± 1992.89	0.98 ± 0.04	7787.82 ± 1992.93
9	12123.56 ± 3465.39	0.96 ± 0.04	12124.51 ± 3465.43
10	6992.72 ± 2053.56	0.91 ± 0.04	6993.63 ± 2053.60
Average	8311.88 ± 2148.19	0.95 ± 0.04	8312.83 ± 2450.53

[a]*SR* Speed reference
[b]*N* Network
[c]*ms* millisecond
[d]*SD* Standard Deviation

Table 12.2 Time-varying control latency Test2 (T2)

Test subjects	SR[a]latency [ms[c] ± SD[d]]	N[b]latency [ms[c] ± SD[d]]	Total latency [ms[c] ± SD[d]]
1	7913.26 ± 2013.19	2.83 ± 0.15	7916.10 ± 2013.34
2	8124.94 ± 2482.64	3.01 ± 0.17	8127.95 ± 2482.81
3	7372.58 ± 1923.81	2.72 ± 0.16	7375.30 ± 1923.97
4	6123.54 ± 1876.66	3.25 ± 0.17	6126.79 ± 1876.82
5	9456.22 ± 2913.17	2.79 ± 0.16	9459.01 ± 2913.33
6	10011.68 ± 3178.43	2.82 ± 0.15	10014.51 ± 3178.58
7	5456.39 ± 1642.68	2.98 ± 0.16	5459.37 ± 1642.84
8	6234.39 ± 1988.16	3.08 ± 0.16	6237.46 ± 1988.32
9	11045.70 ± 3411.42	3.07 ± 0.16	11048.77 ± 3411.58
10	5923.33 ± 1781.14	2.92 ± 0.16	5926.26 ± 1781.29
Average	7766.20 ± 2321.13	2.95 ± 0.16	7769.12 ± 1907.10

[a]*SR* Speed reference
[b]*N* Network
[c]*ms* millisecond
[d]*SD* Standard Deviation

Figure 12.16 shows the latency of reaching higher attention level and consequently higher speed reference while the test subjects directly see the robot.

Figure 12.17 shows this latency when the test subjects control the robot remotely and see the robot via a web camera only.

In case of T2 the latency was slightly less than Test1 T1 probably because the second test took place after more exercises. On the basis of the test results, it can

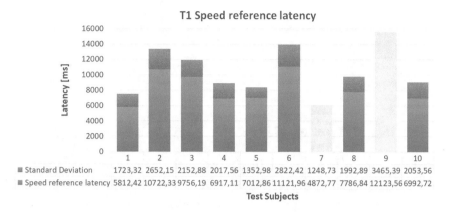

Fig. 12.16 Speed reference latency T1

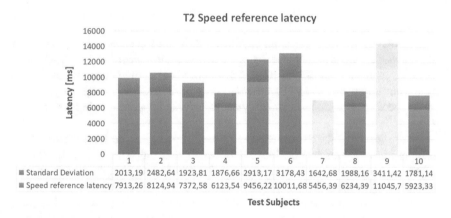

Fig. 12.17 Speed reference latency T2

be stated that, in the case of the test subjects examined in the current research, the control of the attention level required for moving the robot can be improved.

According to the figures, the best and the worst results performed by the test subjects can be compared. In the case of the first test, when the users could directly see the mobile robot, the testing subject 7 after receiving the target stimulus was able to achieve the concentration level required for the starting of the mobile robot within 4872.77 ± 1248.73 ms. This value in the case of the testing subject 9 was 12123.56 ± 3465.39 ms, which means 7250.79 ms difference.

In the case of the second test, when the robot could be seen only via a web camera, the testing subject 7 after receiving the target stimulus was able to achieve the concentration level required for the starting of the mobile robot within 5456.22 ± 1642.68 ms, and this value in the case of the testing subject 9 was 11045.70 ± 3411.42 ms, which means 5589.3 ms difference.

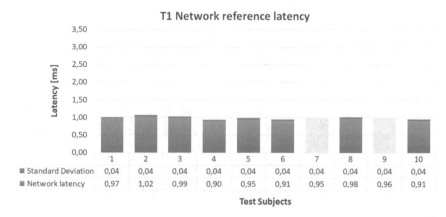

Fig. 12.18 Network latency T1

Between the two test cases no significant difference is shown, so the visibility of the mobile roboteither physically or in virtual environment has no considerable influence on the period of time needed to achieve the required attention level. Less significant improvement can be observed only in the case of the test subject with weaker performance, and it is probably due to the practice gained during the previous tests.

On the ground of the examination mentioned above, it can be established that the examined test subjects are not able to influence their attention level in the same way; difference can be seen so the starting time of the mobile robot in the case of the test subjects is different.

Figure 12.18, On the basis of the whole examination, the first test case shows the latency results took place in the network. In this case the used devices were connected

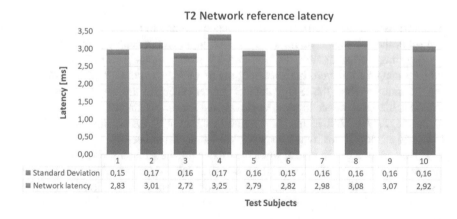

Fig. 12.19 Network latency T2

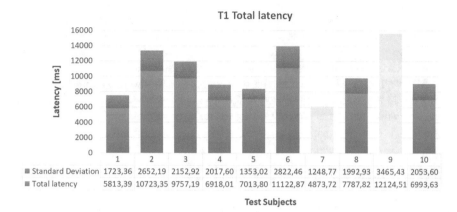

Fig. 12.20 Total latency T1

to an identical subnet; they were placed physically close to each other in the same laboratory room.

Figure 12.19, In the case of the second test, the latency results are shown based on each test subject, where the used devices were connected to different subnets, placed physically more remote from each other than in the case of the first test. At the first test the value of the latency was average 0.95 ± 0.04 ms while this value in the second test case increased to 2.95 ± 16 ms, which is primarily due to the data delivery of the network devices and the latency caused by the network flow.

Figure 12.20, On the basis of the total examination, in the case of the first test, the total latency of the devices connected to the same subnet can be seen, which is the required time duration to achieve the appropriate attention level, and also the latency results took place in the network are shown on the basis of each test subject.

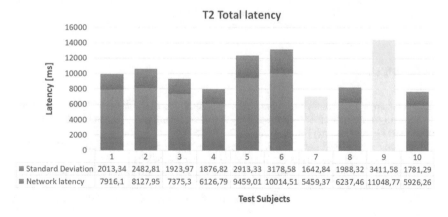

Fig. 12.21 Total latency T2

Figure 12.21, On the basis of the total examination, in the case of the second test, the total latency of divergent subnets can be seen, which is the required time duration to achieve the appropriate attention level, and also the values of the latency in the network took place on the basis of each test subject are shown.

All in all, it can be declared that differences can be shown among the people involved in the examination related to the speed of the influence on attention, and that the realization of remote controlling has no significant impact on it.

12.6   Conclusion

In the article, the BCI system designed for human-computer based control of IoT based robot (IoRT) unit has been introduced, which supports modern robots with the opportunities of technology on the basis of IoT. Thanks to the performed BCI system and an IoRT device, such testing environment has been set up, which is suitable for the realization of controlling local and remote robots. In the testing environment, performance of human intervention and its latency due to the BCI system have been examined. According to the experiences of the performed tests, can be stated, that for proper operation of BCI system, testing subjects had to practice the use of the device first, in order to reach appropriate results. On the other hand, testing subjects were able to perform human intervention only with some seconds latency, although this latency did not depend on whether the testing subjects controlled the robot directly or by remote control. The BCI system gives proper basis to test the technology, and on the ground of the gained results, determination of the courses for further improvements. The IoRT and BCI can be used well in education also to apply in innovative, disruptive, cooperative learning environment [20–23] and using modern ICT technology [5, 35, 36].

Acknowledgements The project is sponsored by EFOP-3.6.1-16-2016-00003 founds, Consolidate long-term R and D and I processes at the University of Dunaujvaros.

References

1. Adelmann R, Langheinrich M, Floerkemeier C (2006) A toolkit for bar code recognition and resolving on camera phones—jump-starting the internet of things. In: Hochberger C, Liskowsky R (eds) GI Jahrestagung (2). LNI, GI, vol 94, p 366–373
2. Arnsten Amy FT, Berridge Craig W, McCracken James T (2009) The neurobiological basis of attention-deficit/hyperactivity disorder. Prim Psychiatry 16:47–54
3. Baranyi P, Csapo A (2012) Definition and synergies of cognitive infocommunications. Acta Polytech Hung 9(1):67–83
4. Baranyi P, Csapo A, Gyula S (2015) Cognitive infocommunications (CogInfoCom). Springer, Heidelberg, p 378

5. Benedek A, Molnar G (2014) Supporting the m-learning based knowledge transfer in university education and corporate sector. In: Arnedillo Sánchez I, Isaías P (eds) Proceedings of the 10th international conference on mobile learning 2014, Madrid, Spain, Feb 2014, pp 339–343

6. Brown VJ, Bowman EM (2002) Rodent models of prefrontal cortical function. Trends Neurosci 25:340–343

7. Cardinal RN, Parkinson JA, Hall J, Everitt BJ (2002) Emotion and motivation: the role of the amygdala, ventral striatum, and prefrontal cortex. Neurosci Biobehav Rev 26:321–352

8. Cauda F, Cavanna AE, Dágata F, Sacco K, Duca S, Geminiani GC (2011) Functional connectivity and coactivation of the nucleus accumbens: a combined functional connectivity and structure-based meta-analysis. J Cognit Neurosci 23:2864–2877

9. Chen F, Jia Y, Xi N (2013) Non-invasive EEG based mental state identification using nonlinear combination. In: 2013 IEEE international conference on robotics and biomimetics (ROBIO). https://doi.org/10.1109/robio.2013.6739789

10. Christian F et al (2008) The internet of things. In: Floerkemeier C, Langheinrich M, Fleisch E, Sarma SE (eds) IOT 2008: first international conference, Zurich, Switzerland, March 2008, vol 4952. Springer, Heidelberg, p 378

11. Dalley JW, Cardinal RN, Robbins TW (2004) Prefrontal executive and cognitive functions in rodents: neural and neurochemical substrates. Neurosci Biobehav Rev 28:771–784

12. Feenstra M, Botterblom M, Uum JV (2002) Behavioral arousal and increased dopamine efflux after blockade of NMDA-receptors in the prefrontal cortex are dependent on activation of glutamatergic neurotransmission. Neuropharmacology 42:752–763

13. Fortino G (2016) Agents meet the IoT: toward ecosystems of networked smart objects. IEEE Syst Man Cybern Mag 2:43–47

14. Freedman M, Oscar-Berman M (1986) Bilateral frontal lobe disease and selective delayed response deficits in humans. Behav Neurosci 100:337–342

15. Friedemann M, Christian F (2010) From the internet of computers to the internet of things. In: Sachs K, Petrov I, Guerrero P (eds) From active data management to event-based systems and more, Papers in Honor of Alejandro Buchmann on the Occasion of His 60th Birthday, vol 6462. Springer, Heidelberg, pp 242–259

16. Friganovic K, Medved M, Cifrek M (2016) Brain-computer interface based on steady-state visual evoked potentials. In: 2016 39th international convention on information and communication technology, electronics and microelectronics (MIPRO). https://doi.org/10.1109/mipro.2016.7522174

17. Gallo DA, Mcdonough IM, Scimeca J (2010) Dissociating source memory decisions in the prefrontal cortex: fMRI of diagnostic and disqualifying monitoring. J Cognit Neurosci 22:955–969

18. Hakiri A, Berthou P, Gokhale A, Abdellatif S (2015) Publish/subscribe-enabled software defined networking for efficient and scalable IoT communications. IEEE Commun Mag 53:48–54

19. Heidbreder CA, Groenewegen HJ (2003) The medial prefrontal cortex in the rat: evidence for a dorso-ventral distinction based upon functional and anatomical characteristics. Neurosci Biobehav Rev 27:555–579

20. Horvath I (2016) Disruptive technologies in higher education. In: 2016 7th IEEE international conference on cognitive infocommunications (CogInfoCom), Wroclaw, Poland, Oct 2016. pp 347–352. https://doi.org/10.1109/CogInfoCom.2016.7804574

21. Horvath I, Kvasznicza Z (2016) Innovative engineering training—today's answer to the challenges of the future. In: 2016 International education conference, Venice, Italy, June 2016, pp 647-1–647-7

22. Horvath I (2016) Innovative engineering education in the cooperative VR environment. In: 2016 7th IEEE international conference on cognitive infocommunications (CogInfoCom), Wroclaw, Poland, Oct 2016, pp 359–364. https://doi.org/10.1109/CogInfoCom.2016.7804576

23. Horvath I (2016) Digital life gap between students and lecturers. In: 2016 7th IEEE international conference on cognitive infocommunications (CogInfoCom), Wroclaw, Poland, Oct 2016. pp 353–358. https://doi.org/10.1109/CogInfoCom.2016.7804575

24. Jadidi AF, Davoodi R, Moradi MH, Yoonessi A (2014) The impact of numeration on visual attention during a psychophysical task; an ERP study. In: 2014 21th Iranian conference on biomedical engineering (ICBME). https://doi.org/10.1109/icbme.2014.7043947
25. Jansma JM, Ramsey NF, Slagter HA, Kahn RS (2001) Functional anatomical correlates of controlled and automatic processing. J Cognit Neurosci 13:730–743
26. Kaysa WA, Widyotriatmo A (2013) Design of brain-computer interface platform for semi real-time commanding electrical wheelchair simulator movement. In: 2013 3rd international conference on instrumentation control and automation (ICA). https://doi.org/10.1109/ica.2013.6734043
27. Kennerley SW, Dahmubed AF, Lara AH, Wallis JD (2009) Neurons in the frontal lobe encode the value of multiple decision variables. J Cognit Neurosci (21)6:1162–1178
28. Kita H, Oomura Y (1981) Reciprocal connections between the lateral hypothalamus and the frontal cortex in the rat: electrophysiological and anatomical observations. Brain Res 213:116
29. Kolb B (1991) Chapter 25 animal models for human PFC-related disorders. In: Progress in brain research the prefrontal its structure, function and cortex pathology, p 501–519
30. Laar BVD, Gurkok H, Bos DP-O, Poel M, Nijholt A (2013) Experiencing BCI control in a popular computer game. IEEE Trans Comput Intell AI Games 5(2):176–184
31. Lacroix L, Spinelli S, Heidbreder CA, Feldon J (2000) Differential role of the medial and lateral prefrontal cortices in fear and anxiety. Behav Neurosci 114:11191130
32. Lebedev MA, Nicolelis MA (2006) Brainmachine interfaces: past, present and future. Trends Neurosci 29(9):536–546
33. Lew E, Chavarriaga R, Zhang H, Seeck M, Millan JRD (2012) Self-paced movement intention detection from human brain signals: invasive and non-invasive EEG. In: 2012 annual international conference of the IEEE engineering in medicine and biology society. https://doi.org/10.1109/embc.2012.6346665
34. Milner B (1982) Some cognitive effects of frontal-lobe lesions in man. Philos Trans R Soc B Biol Sci 298:211–226
35. Molnar G Modern ICT based teaching and learning support systems and solutions in higher education practice. In: Turcani M, Drlik M, Kapusta J, Svec P (eds) 10th international scientific conference on distance learning in applied informatics, Sturovo, Slovakia, May 2014, pp 421–430
36. Molnar G Szuts Z (2014) Advanced mobile communication and media devices and applications in the base of higher education. In: Aniko S (ed) SISY 2014: IEEE 12th international symposium on intelligent systems and informatics. Subotica, Serbia, Sept 2014, pp 169–174
37. Morgane P, Galler J, Mokler D (2005) A review of systems and networks of the limbic forebrain/limbic midbrain. Prog Neurobiol 75:143–160
38. Murase N, Rothwell JC, Kaji R, Urushihara R, Murayama N, Igasaki T, Sakata-Igasaki M, Shibasaki H (2007) Acute effect of subthreshold low-frequency repetitive transcranial magnetic stimulation over the premotor cortex in writer's cramp. In: 2007 IEEE/ICME international conference on complex medical engineering. https://doi.org/10.1109/iccme.2007.4382100.
39. Nagy B, Szabó I, Papp S, Takács G, Szalay C, Karádi Z (2012) Glucose-monitoring neurons in the mediodorsal prefrontal cortex. Brain Res 1444:38–44
40. Nagy B, Takács G, Szabó I, Lénárd L, Karádi Z (2012) Taste reactivity alterations after streptozotocin microinjection into the mediodorsal prefrontal cortex. Behav Brain Res 234:228–232
41. Nordahl CW, Ranganath C, Yonelinas AP, Decarli C, Fletcher E, Jagust WJ (2006) White matter changes compromise prefrontal cortex function in healthy elderly individuals. J Cognit Neurosci 18:418–429
42. Peers PV (2005) Attentional functions of parietal and frontal cortex. Cereb Cortex 15:1469–1484
43. Ramnani N, Owen AM (2004) Anterior prefrontal cortex: insights into function from anatomy and neuroimaging. Nat Rev Neurosci 5:184–194
44. Ullah K, Ali M, Rizwan M, Imran M (2011) Low-cost single-channel EEG based communication system for people with lock-in syndrome. In: 2011 IEEE 14th international multitopic conference. https://doi.org/10.1109/inmic.2011.6151455

45. Vermesan O, Friess P (eds) (2013) Internet of things: converging technologies for smart environments and integrated ecosystems. Aalborg, Denmark
46. Wolpaw JR et al (2002) Braincomputer interfaces for communication and control. IEEE Commun Mag 13(6):7356–7382
47. Wolpaw JR, Wolpaw EW (2012) Brain-computer interfaces: something new under the sun. In: Wolpaw EW, Wolpaw EW (eds) Brain-computer interfaces: principles and practice. Oxford University Press, New York, pp 3–12
48. Yeon C, Kim D, Kim K, Chung E (2014) Sensory-evoked potential using a non-invasive flexible multi-channel dry EEG electrode with vibration motor stimulation. In: IEEE SENSORS 2014 proceedings. https://doi.org/10.1109/icsens.2014.6985049
49. Yoshida H, Tanaka Y, Kikkawa S (2012) EEG analysis of frontal lobe area in arousal maintenance state against sleepiness. In: 2012 annual international conference of the IEEE engineering in medicine and biology society. https://doi.org/10.1109/embc.2012.6346578
50. Yu Ishikawa Y, Takata M, Joe K (2012) Constitution and phase analysis of alpha waves. In: The 5th international conference on biomedical engineering, 2012. https://doi.org/10.1109/bmeicon.2012.6465482
51. Zald DH (2006) Neuropsychological assessment of the orbitofrontal cortex. The Orbitofrontal Cortex 449–480
52. ABI Research (2014) The Internet of Robotic Things. In: Technology analysis report. Available via DIALOG. https://www.abiresearch.com/market-research/product/1019712-the-internet-of-robotic-things. Accessed 28 Jan 2017
53. International Telecommunication Union (ITU) (2005) The Internet of Things. In: ITU internet reports. Available via DIALOG. https://www.itu.int/net/wsis/tunis/newsroom/stats/The-Internet-of-Things-2005.pdf. Accessed 28 Jan 2017
54. NeuroSky Inc (2009) Brain wave signal (EEG) of NeuroSky. Available via DIALOG. http://www.frontiernerds.com/files/neurosky-vs-medical-eeg.pdf. Accessed 28 Jan 2017
55. NeuroSky Inc (2015) MindWave Mobile: User Guide. Available via DIALOG. http://download.neurosky.com/support_page_files/MindWaveMobile/docs/mindwave/_mobile_user_guide.pdf. Accessed 28 Jan 2017
56. NeuroSky Inc (2015) "store.neurosky.com". Available via DIALOG. http://store.neurosky.com/pages/mindwave. Accessed 28 Jan 2017

Chapter 13
CogInfoCom-Driven Surgical Skill Training and Assessment

Developing a Novel Anatomical Phantom and Performance Assessment Method for Laparoscopic Prostatectomy Training

László Jaksa, Illés Nigicser, Balázs Szabó, Dénes Ákos Nagy, Péter Galambos and Tamás Haidegger

Abstract The systematic assessment and development of human learning capabilities is one of the biggest challenges in applied sciences. It can be observed within the medical domain how evidence-based paradigms are gradually gaining space. In this chapter, the development process of a laparoscopic box trainer is introduced. A simulator including a phantom for prostatectomy is described, which feeds into medical staff training and skill assessment. An overview of laparoscopic surgical simulators is provided. Based on the state of the art and our previous experience, a clear need was formulated to develop a partially physical, partially computer-integrated simulator. To gain a better understanding of the cognitive load and physical stress,

L. Jaksa (✉) · B. Szabó · D. Á. Nagy · P. Galambos · T. Haidegger
Antal Bejczy Center for Intelligent Robotics, Óbuda University, 82 Kiscelli Street,
Budapest 1032, Hungary
e-mail: laszlo.jaksa@irob.uni-obuda.hu

B. Szabó
e-mail: balazs.szabo@irob.uni-obuda.hu

D. Á. Nagy
e-mail: denes.nagy@irob.uni-obuda.hu

P. Galambos
e-mail: peter.galambos@irob.uni-obuda.hu

T. Haidegger
e-mail: tamas.haidegger@irob.uni-obuda.hu

I. Nigicser
University of Surrey, Guildford GU2 7XH, UK
e-mail: nigicseri@gmail.com

T. Haidegger
Austrian Center for Medical Innovation and Technology, 2 Viktor Kaplan Street,
2700 Wiener Neustadt, Austria

© Springer International Publishing AG, part of Springer Nature 2019
R. Klempous et al. (eds.), *Cognitive Infocommunications, Theory and Applications*,
Topics in Intelligent Engineering and Informatics 13,
https://doi.org/10.1007/978-3-319-95996-2_13

277

force measurement was used in the test environment. The force and time data were used to evaluate the performance of the participant. A new assessment method was described, which can be used to point out the weak aspects of surgical technique, and the participants can do this on their own. Computer-integrated assistive technologies for surgical education are believed to rapidly become the gold standard on a global scale.

13.1   Introduction

The improvement of assistive technologies in surgical device development has taken a rapid turn where high level support systems like the da Vinci system are gaining importance. The foundation of these were established by the emergence of cognitive sciences and human factor engineering. In the development process of surgical devices, bringing invasiveness to a minimum has become an important focus, as it reduces the damage done to tissues, improves surgery and recovery time, increases precision and makes a wider scope of surgeries feasible. As a first level, there has been the introduction of laparoscopic surgery, in which, small incisions are enough for tool introduction, and an endoscopic camera is used as a digital sensor. The next, second level goes over the first one, featuring the invention of the Computer Assisted Laparoscopic Robot [4], most significantly the da Vinci System, where an artificial entity is introduced that bridges the cognitive control of the doctor and the procedural execution during surgeries, resulting in better accuracy, safety and efficiency. This then serves as a ground for a whole new area of development of artificial and natural cognitive systems communication in the field of medical technology. The development of assistive technologies for surgery has reached a point where even the cognitive abilities of surgeons can be supported and enhanced by artificial cognitive systems. Decision support systems are routinely employed in image-guided surgeries, and low level autonomous functions are also on the development horizon. Currently one of the most developed assistive technologies of such is embodied by the da Vinci Surgical System. The da Vinci System is a teleoperation system where the surgeon navigates the surgery from a distance through a refined system as a medium resulting in a very high level support system (Fig. 13.1).

Nevertheless, the real need in Hungary and the region is currently more towards low level support and teaching devices, that by itself may make a big difference in the way junior medical staff acquires and maintains surgical skills. As a reason this chapter elaborates on the first mentioned level, the enhancement of the manual technique for laparoscopic surgery, as a crucial and central step in the surgical device development process. More in particular, it focuses on the way an increasing level of training equipment fidelity enhances the cognitive learning process of doctors on laparoscopic surgeries. For assessing this phenomenon, an own laparoscopic trainer box was further developed and implemented, featuring a training phantom for prostatectomy with high fidelity tissue models and anatomy.

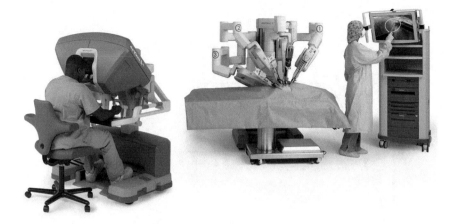

Fig. 13.1 The da Vinci surgical system consisting of the surgeon console, a patient-side cart with endo-wrist instruments and the vision system. Photo courtesy by Intuitive Surgical, Inc.

13.2 Theory

Laparoscopic surgery requires a relatively high level of cognitive capability. Fine psychomotor skills and dexterity are indispensable for the surgeon. Hospitals and medical centers have started applying a wide array of teaching devices, since these abilities can be and must be developed. The assessment of learning process efficiency can be conducted through observing the invested energy and time in a particular learning environment against skill improvement. This can be visualized by learning curves, which graphically displays the required time to complete a task against practice volume. These curves are mainly used in the field of cognitive psychology. The test results on performing a wide range of both mental or cognitive and practical tasks that are based on dexterity assessment, show that these learning curves usually follow a pattern called the power law of practice. It is characterized by a steep start and a plateau after reaching a certain amount of practice volume (Fig. 13.2) [23].

The slope of the curve i.e. the derivative before the plateau indicates learning process quality, thus the steeper the slope the faster the learning. Such learning curves were taken over from industrial applications and they are used in economics to a great extent in relating unit costs to accumulated volume [28]. This term began gaining importance in medical education and training and more significantly in minimally invasive surgeries as that is a technically and cognitively demanding process of learning [14]. In a recent review article a thorough literature search about high-fidelity medical simulators was conducted (109 journal articles) in order to characterize the features and aspects of simulation that lead to effective learning. These included the availability of feedback, room for repetitive practice, curriculum integration into medical school education, a range of difficulty levels, individualized learning and defined outcomes as the most important educational features among others [16].

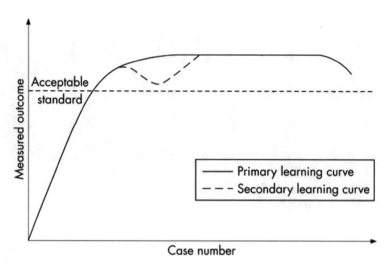

Fig. 13.2 A learning curve that shows the power law of practice where the initial steep slope gradually reaches a plateau. Also, it represents an ideal surgical learning curve [14]

The aspect of feedback including haptic, visual or auditory, indicates that the more senses are involved in a learning process the steeper the learning curve will get. Research shows that the presence of the haptic force feedback as a sense of touch at laparoscopic surgical training resulted in improved training transfer [18]. Accurate visual and haptic fidelity of tissues on laparoscopic simulators may further improve learning outcomes. However, many research articles failed to find evidence of a linear correlation between training effectiveness and high model fidelity [12, 18]. In spite of this, a paper states that an ideal laparoscopic training model should be physiologically and anatomically similar to a real human patient, which means fidelity still has great importance in surgical education [35].

13.3 Background

The best way to gain proficiency in laparoscopic operations is to observe, assist and then perform surgeries on real patients. However this involves high risk factor for the patient in the skill acquisition of junior medical staff and thus moving from observer to assistant and finally to performing surgeon is a slow and regulated process. In order to accelerate this skill acquisition, outside-operating theater learning facility is required.

Nowadays laparoscopic surgery is widely spread in the clinical practice which is why appropriate training and assessment of skills in laparoscopic procedures became an important issue. These coordination skills significantly differ from those of traditional open surgeries. Due to the flat screen the depth perception is reduced and the tactile information is lost [21]. To the development of these appropriate skills there is a clearly defined learning curve associated, where the teaching methods used in conventional surgery cannot guide the students with full satisfaction and without risk [1]. Surgical simulation may be the solution to this problem, the usability of which is confirmed by several studies completed in laparoscopic training [22]. A review article about low cost, suitable-for-home-use laparoscopic simulators highlighted many very low cost, not on the market, prototype solutions which can be built easily. These devices obviously did not undergo any validation thus their practicality is opposed by their dubious educational value [21]. A study examined the maintainability of laparoscopic surgical skills of inexperienced subjects. After a short training on a simulator followed by a test the gained skills were still significantly present and maintained upon re-examination even after one year [10]. The simulation grants the possibility for the student to practice laparoscopic surgical technique outside the operating theater with less stress and without the risk of damaging the patient [13]. A number of approaches and solutions to such a simulation were developed ranging from simple and cheap, but low fidelity box trainers to high fidelity virtual reality simulators. Box trainers are usually simply constructed, portable, allowing for basic laparoscopic skills development: for example maneuvering and manipulating with the tools, cutting and suturing. The computerized surgical simulators use computer graphics and sometimes even haptic feedback to create a virtual anatomic environment enabling the practice of specific surgical procedures in addition to general surgical skills [8]. According to a study the laparoscopic virtual reality simulators improve the surgical skills more than box trainers. It is not clear however how this additional development will appear in everyday surgical practice [36]. The virtual reality simulators are expensive which makes it impossible for many educational institutions to purchase them making it difficult to access for students [5, 27]. Nowadays box trainers are much more commonly used than virtual reality simulators [1]. The educational integration of such simulation equipment vastly varies for different disciplines, for example there is much less emphasis on practice on simulators in urology where there is more emphasis on practice with senior supervision, as opposed to in other endoscopic surgical disciplines [34].

Before autopsies became allowed in medical schools, students and medical staff used wax models for centuries in order to learn traditional open surgery [26]. For the purpose of learning and education human cadaver dissection has been widely used although their high cost and limited availability led to the application of slightly more accessible animal cadavers as a first step in teaching. Laparoscopic surgeries have been performed on animal models that represented an in vivo training environment that could provide skill acquisition opportunity of hand-eye coordination, depth perception and knot-tying [35]. Mostly pigs were chosen for these animal models since they resemble the human anatomy very well thus they are used widely and successfully for laparoscopic cholecystectomy training [19]. However, the fact that access to animal cadavers have still been limited resulted in a need for creating artificial simulation tools and environments. Medical schools and hospitals have started using simple and low-cost box trainers for laparoscopic surgery training that provide exercises for depth-queuing, hand-eye coordination and triangulation. In addition, some of these trainers can be built at home and may be used for individual practice sessions and thus the already mentioned restrictions concerning medical working hours spent on training can be avoided [9, 20, 26]. These are simple, practical and cheap solutions although they do not model real anatomy.

For more accurate anatomical simulations, especially from a visual standpoint, virtual reality (VR) and augmented reality (AR) simulators have been developed. VR simulation creates a 3D virtual environment where various operations can be repeatedly simulated without any need of external assistance, although this tool does not provide haptic or tactile feedback of the procedure thus limiting the slope of the learning curve. AR differs from VR mainly in providing real time force feedback thus improving the quality of training [26]. However, AR and VR software both have high costs and thus their price-efficiency makes purchasing these devices infeasible for certain institutions [4]. Because of the limitations of all the aforementioned educational environments and tools, simple laparoscopic box trainers upgraded with realistic anatomical models can provide affordable, practical but still high-quality means for training, featuring realistic tactile and haptic feedback and fill this market gap. Such box trainers are already present on the market although with limited scope. Such a phenomenon is visible in other surgical disciplines that use laparoscopic instruments. A recent study [17] presents a training dummy for pediatric Nissen fundoplication.

Commercially available box trainers were examined in a thorough overview. During the research for mapping the laparoscopic trainers the Google search engine was used with the following keywords: laparoscopic simulator, laparoscopic trainer, laparoscopic box trainer, virtual reality surgical simulator, laparoscopic surgical simulator. Additionally, overview articles and case studies were searched in Google Scholar with a similar method.

The chosen skill assessment method will greatly affect the qualifying surgery skill level for students so the choice of the appropriate method of measurement is key in ensuring the quality of surgical education [32]. For subjects with minimal laparoscopic experience, neither did the box trainer practice nor did the LapVR virtual reality simulator practice show any significant correlation with the scores achieved in the GOALS (Global Assessment of Laparoscopic Operative Skills) evaluation system. This is a warning sign that these methods are not necessarily suitable for rapid skill assessment or selection [29]. There is a program aiming to unify laparoscopic skill assessment which is called the FLS program (Fundamentals of Laparoscopic Surgery) which is established by the SAGES (Society of American Gastrointestinal and Endoscopic Surgeons). Their practical skill enhancement and assessment module consists of five tasks: peg transfer, cutting out a pattern, loop suturing, suturing with intracorporeal knot and suturing with extracorporeal knot [26]. The FRS (Fundamentals of Robotic Surgery) is the surgical robotic FLS embodiment.

The features of the laparoscopic box trainers found during the research are detailed in Tables 13.1 and 13.2. In case of both tables, in the "Curriculum" column there is no difference in content between the "FLS" exercises and the "basics". This is because for such products that do not have validated FLS compatibility the term "FLS" could not be used but an alternative term had to be referenced. The virtual reality simulators

Table 13.1 Laparoscopic box trainers I

Manufacturer	Product	Price range	FLS compliance	Portability	Forceps	Curriculum	Resource (webpage)
3-Dmed	LapTab trainer	400–1000 $	Yes	Excellent	Included	FLS	3d-med.com
	T3 plus	2500–3000 $	Yes	Good	No data	FLS, camera handling	
	T5	2500–3500 $	Yes	Good	No data	Basics, camera handling, ultrasound	
	T9	3000–3500 $	No data	Good	No data	FLS, camera handling	
	T12	4000–5000 $	Yes	Acceptable	No data	FLS, camera handling	
	ForceSense	8500–9000 $	No data	Undefined	Undefined	Force measurement	
CamTronics	LapStar	No data	Yes	Good	Not included	Basics	laparoscopic-trainer.com
Delletec	Laparoscopy simulator	No data	No data	Acceptable	Not included	Basics, camera handling	delletec.com
eoSurgical	eoSim	700–4000 $	No data	Excellent	Included	Basics	eosurgical.com
Ethicon	TASKit	400–600 $	Yes	Excellent	Not included	FLS	ethicon.com
Grena	Laparoscopic trainer system	200–300 $	No data	Good	No data	Basics	grena.co.uk

Table 13.2 Laparoscopic box trainers II

Manufacturer	Product	Price range	FLS compliance	Portability	Forceps	Curriculum	Resource (webpage)
Hospiinz	Endo trainer	No data	No data	Good	Included	Basics	hospiinz.com
Inovus surgical solutions	Pyxus	500–2200 $	No data	Acceptable	Included	Basics	inovus.org
iSurgicals	iSim2	3000–4800 $	No data	Excellent	Included	Basics	isurgicals.com
	iSim SMART	1500–2000 $	No data	Excellent	Included	Basics	
Lagis endoSurgical	Laparoscopy simulator	No data	No data	Excellent	Not included	Basics	lagis.com.tw
Limbs and things	FLS system	1000–6200 $	Yes	Good	No data	FLS	limbsand-things.com
	Helago HD laparoscopic trainer	8500–9300 $	No data	Acceptable	No data	Basics	
Nahl medical	HY-01 laparoscopic trainer box	No data	No data	Good	Included	Basics	nahlmed.com
Pharmabotics	BodyTorso laparoscopic trainer BTS300D	600–1000 $	No data	Acceptable	Not included	Basics, camera handling	pharma-botics.com
Samed	Laparoscopie-Trainer	No data	No data	Acceptable	No data	Basics	samed-dresden.com
Simulab	LapTrainer	2000–2200 $	Yes	Excellent	Not included	FLS, camera handling	simulab.com

are playing an increasingly important role in surgical education. The features and details of the virtual reality simulators available on the market are summarized in Table 13.3. The application of force feedback through built-in haptic forceps is one of the most significant features of virtual reality simulators since enabling the sense of tactile properties of only virtually existing tissues greatly enhances the fidelity and the training value.

In recent years the rapid development of telerobotic systems made it possible for such robots to not only be used in industrial settings but in clinical applications too. This way the surgeon is able to perform an operation far from the patient by controlling a robotic interface. The medical teleoperation systems are very successful in increasing precision. Each year, more than 1.5 million patients benefit from surgery performed by telerobotic systems worldwide [31]. The design and development of such medical teleoperation systems apart from establishing full motion control is also facing not-everyday-challenges because of the strict regulations stemming from the operational environment and from the unique mechanical properties of soft tissues [30]. Apart from the engineering challenges the education of the appropriate use of such robotic systems is a complex issue. The robotic surgical simulators (Table 13.4) greatly facilitate the acquisition of movement coordination skills needed for robotic surgery. Additionally, the simulators aiming to enhance cooperation skills of assisting medical staff is a new branch of surgical education (Table 13.4).

Apart from the robotic surgical simulators, none of the mentioned simulators contain 3D image display. According to a study [11] the 3D image display results in a significant improvement over 2D visualization in the skills for beginner and experienced laparoscopic surgeons. The skills of a beginner surgeon viewing a 3D image would compete with an experienced surgeon's skills viewing a 2D image only. Another article stated that students using laparoscopic simulators with 3D visualization are able to perform specific tasks significantly quicker with the same amount of practice compared to students practicing on simulators with 2D visualisation with a flat screen [24]. The laparoscopic surgical simulators are not only good for education but also for warm-up before surgery as this type of warm-up improves the time of performing particular tasks [7]. According to a study, learned and practiced skills on laparoscopic simulators also impact and improve arthroscopic skills [2]. Another rising discipline within laparoscopic training is gamification [6]. A recent study [25] indicates that playing with specific video games is an effective way for surgeons to warm up before actual surgeries. Devices that are designed for home use may add another dimension to skill training, as forming training curricula for home circumstances is a specific challenge [33]. Pure mental training also proved to be an effective and cheap supplement to regular skill training regarding laparoscopic cholecystectomy [15]. These results can indicate the future directions of the development of laparoscopic skill training simulators.

Table 13.3 Laparoscopic virtual reality simulators

Manufacturer	Product	Price range	Portability	Forceps	Curriculum, FLS compliance	Resource (webpage)
CAE healthcare	LapVR	85000–90000 $	Difficult	Built-in, haptic	Virtual surgery	caehealthcare.com
	ProMIS3	No data	Difficult	Real	Virtual surgery, Augmented reality, FLS	
Medical-X	Lap-X VR	55000–60000 $	Good	Real	Virtual surgery	medicalx.com
	Lap-X Hybrid	55000– 60000 $	Good	Real	Augmented reality	
	Lap-X II	72000–105000 $	Difficult	Real	Virtual surgery	
Nintendo	Underground the game	200–300 $	Good	Built-in	Gamified FLS tasks	underground–thegame.com
Open simulation	LapKit	40–60 $	Excellent	Real	Augmented reality	opensimulation.org
Simbionix	LapMentor	No data	Difficult	Built-in, haptic	Virtual surgery	simbionix.com
Simendo	Simendo Pro 3	19000–20000 $	Good	Built-in	Virtual surgery	simendo.eu
Simsurgery	SEP	No data	Difficult	Built-in	Virtual surgery	simsurgery.com
	D-box	No data	Acceptable	Built-in	Virtual surgery	
Surgical science	LapSim	30000–45000 $	Difficult	Built-in, haptic	Virtual surgery	surgical–science.com
Touch surgery	Touch Surgery	No data	Excellent	Undefined	Virtual surgery	touchsurgery.com

Table 13.4 Robotic surgery and operation room simulators

Manufacturer	Product	Price range	Resource (webpage)
BBZ medical technologies	Actaeon	30000–32000 $	bbzmedical-technologies.com
Intuitive surgical	da Vinci Skills Simulator	No data	intuitivesurgical.com
Mimic simulation	dV-Trainer	60000–100000 $	mimicsimulation.com
	FlexVR	No data	
ORZONE	ORCAMP MIS setup	No data	orzone.com
Simbionix	Robotix mentor	No data	simbionix.com
	Team mentor	No data	
Sim surgery	SEP robot	No data	simsurgery.com
Simulated surgicals	Robotic surgery simulator	No data	simulatedsurgicals.com
Surgical science	TeamSim	No data	surgical-science.com

13.4 Motivation and Design Requirements

A novel box trainer design for radical prostatectomy simulation was chosen as a starting point, since this procedure is one of the most common routine surgeries performed both with manual laparoscopic technique and with robotic assistance by the da Vinci System. Because of this an educational need for laparoscopic prostatectomy is clearly on demand. This project was a continuation of a first box trainer prototype, created by Barcza [4]. For creating the second, upgraded prototype the development aspects focused on the pelvic phantom which required the complete remodeling of the pelvic floor muscles (PFM) and the rectum and improving fidelity of the bladder and the Connective Tissue models.

In the first prototype only craftsman solutions were used for modeling the rectum and the PFM. The rectum was made by making cuts on a hollow insulation foam tube of appropriate diameter to form a realistic rectum shape. As a next step, transparent silicone was poured on this tube, after which the foam tube was removed and the interior of the hollow shape was painted red to give a realistic appearance. The PFM was fabricated from smaller foam sheets by taping them to the inner pelvic cavity to form a seemingly continuous surface which was then also painted red.

3D-modeling the rectum and the PFM was a very time-consuming and tedious task, which reduced efficiency in replication. Even though a mould for silicone moulding was already available for the bladder in the first prototype, the mould had to be constantly rotated until the given amount of two-component silicone solidified since the bladder had to be hollow. This, made the manufacturing process slow, and also resulted in a very uneven surface for the bladder. Finally, the Connective Tissue was just indicated with painted gauze pieces which were very far from real anatomy both in texture and appearance (Fig. 13.3).

The motivation behind the creation of the second prototype was finding a totally new and much more efficient technology for reproduction that also provides more realistic phantoms for both texture and appearance, which can be easily optimized for mass-production. In order to take the robotic surgical training environment into consideration, another design requirement was set to achieve a high quality dexterity training environment that simulates a real-life surgery where the gripping, tearing and cutting of tissues feel realistic, with some limited bleeding authenticity included.

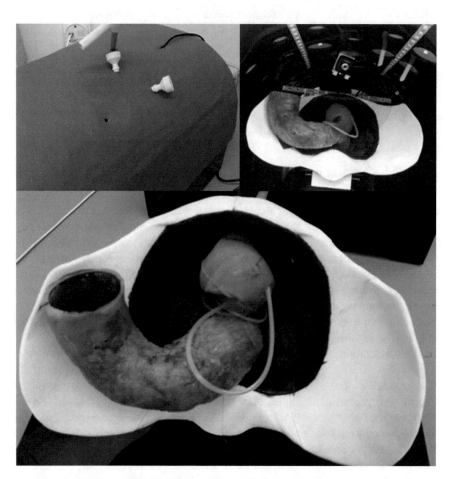

Fig. 13.3 The first prototype of the laparoscopic training box with the pelvic phantom

13.5 Anatomical Phantom Design and Implementation

As a first step in formulating new rectum design concept, a 3D model of the rectum was created using Solid Edge and Blender 3D modeling software. The goal was to make an outer mould for silicone fill-up and then a thin inner mould for a paraffin core. This paraffin body would then be inserted and fixed in the outer mould. This way a gap would be left between the paraffin and the outer mould with a wall thickness equivalent to the real rectum. The gap would then be filled up with silicone after which the paraffin could be melted and removed from the model resulting in a realistic, hollow rectum phantom.

The 3D model was constructed using the Solid Edge CAD software, based on known size parameters and anatomical appearance. The muscle-like, smooth, wavy surface of the rectum was then sculpted on the 3D model in Blender. The negative of the rectum was taken after returning to Solid Edge again, resulting in the mould files, ready for 3D printing from PLA material. For the paraffin core, a thinner mould was created in a similar fashion. After assembling the paraffin core with the outer mould, the gap was filled up Rubosil SR-20, a commercial two-component silicon. For the desired stiffness and color, silicone oil and paint was also added (Fig. 13.4).

The pelvic floor muscle (PFM) was designed similarly. For the new concept of the PFM, our main goals were that it should be manufactured and replaced with ease while it still fits as tight as possible. Therefore, the base of the 3D model was the existing Pelvis model. Based on reference pictures, the pelvic model was complemented in Blender, then the unwanted parts of the model were deleted so that only the ones that determine the PFM remained. Considering the possibility of dis-, and reassembly of the mould, it had to be sectioned into pieces. The planes

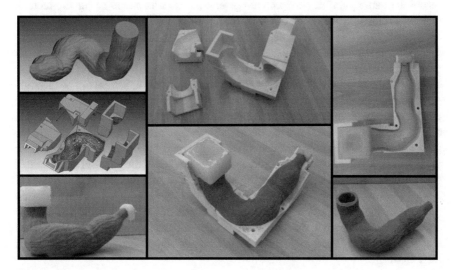

Fig. 13.4 The design steps and the final result of the rectum model

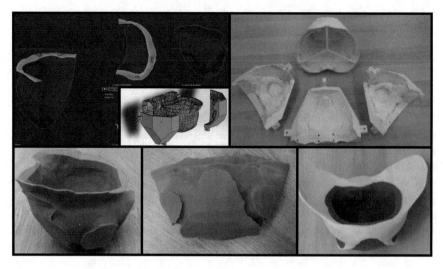

Fig. 13.5 The design steps and the final result of the pelvic floor muscle

and surfaces used to section the mould was also made in Blender. Next, the PFM model file was imported to SolidWorks as a solid body, then it was subtracted from a solid cuboid body and cut into three separate bodies, using the previously generated surfaces. After optimizing for 3D printing, the mould was printed using PLA. After manufacturing and assembly, the silicone PFM was created using the same method as before (Fig. 13.5).

The design of the new bladder was actually an improvement of the first prototype, where the same paraffin core technological concept was added as for the rectum, making production faster. This was a crucial step, as the bladder was often damaged during practice sessions, especially with inexperienced users. The Connective Tissue became an additional feature compared to the first prototype. A composite was created from painted gauze and gel candle, resulting in both texturally and visually plausible properties that simulate a realistic dissection experience. The bleeding simulation by a simple silicone tube was the other extra feature. It was filled with painted water, held under pressure by a syringe, and applied to the plexus santorini only because of its size. However, during a operation surgical knotting is applied to prevent bleeding. Bleeding simulation would be more important during the dissection of blood vessels around the pediculus. In that area, bleeding should be prevented by electro-surgical instruments but the simulation of this was not on the scope of this project. Another important improvement was the installation of load cells on the base of the pelvic phantom for sensing reaction forces in x, y and z directions during practice. The final result became a precise and anatomically realistic phantom that is simple to replicate and assemble, hence may be suitable for mass production (Fig. 13.6).

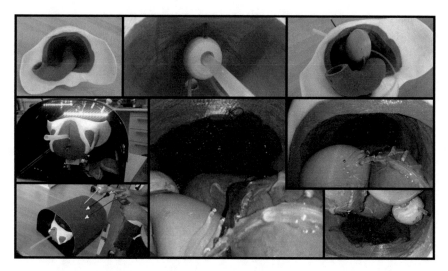

Fig. 13.6 The new, second prototype of the pelvic phantom with extra features of higher fidelity of connective tissue and bleeding effect when cutting the Plexus Santorini

13.6   Experimental Setup

In the box trainer there are three TAL220 load cells thus at the attachment of the pelvic phantom the forces in three perpendicular directions could be measured (Fig. 13.7). The load cells were not suitable to measure the forces exerted by each laparoscopic tool during the procedure but only suitable to measure the sum of the reaction forces caused by the laparoscopic tools together. The signals from the load cells were sampled at a frequency of 10 Hz and converted using three analog-digital converters which was then fed to the Arduino Nano microcontroller. This was screened and tracked by Arduino IDE serial monitor as the data arrived from the microcontroller to the PC. The calibration of the load cells was achieved by placing five different masses on top of the phantom reading off the signal of the load cell in the z direction parallel to the gravitational direction. Knowing the masses and the signal strengths the sensor constant was determined showing the unit change of the signal corresponding to a given force.

The improvement and quality of the described second prototype was assessed by experimental surgery simulation using the device. The tests were conducted identically to the methods used in [4]. The participants were asked to perform a simplified prostatectomy on the laparoscopic box trainer that consisted of five tasks: the cutting of the plexus santorini, the right pediculus, the left pediculus, the urethra at the prostate base and the urethra at the prostate apex (Fig. 13.8). The measured variables were the reaction forces at the suspension point of the phantom during the experiment, the time of completion for each task and the time of the entire procedure. A survey [4] was filled by the users along with both prototypes. It assessed the previous

Fig. 13.7 The force measurement system under the phantom

Fig. 13.8 The measurement setup

surgical experience, and included questions regarding the fidelity and usability of the box trainer and phantom for laparoscopic surgery simulation and training. After each user, the phantom was prepared for the next test with the meltable gel candle and the various sizes of silicone tubes cut to unit lengths. This way only a few minutes were required to save all the force and time data, and to reset the phantom for the next user after each trial.

13.7 Results

The tests were conducted on a total of 13 participants: 7 laymen, 3 surgery residents and 3 expert surgeons. In the first prototype tests a total of 23 participants, 9 laymen, 7 medical students, 5 residents and 2 specialists were tested thus the results of both tests are comparable [4]. The average time spent on each task and for the entire test against experience groups are summarized in Table 13.5 also including the overall times from the first prototype tests for comparison. The average value of the resident/specialist group excludes one specialist's time results as it was unreliable due to a known experimental setup error.

It is clearly visible here that the overall time for the whole operation decreases as the medical experience of the participant increases resulting in a significant 29% difference between the time of the laymen and of the residents and specialists. It is also shown that the average time decreased for the second prototype at each individual task except for the first two. Cutting of the plexus santorini and the right pediculus required somewhat more time in case of the second prototype. The compared overall time of procedure on groups of laymen and resident and specialist for both tests are illustrated on Fig. 13.9.

This shows that the average time of procedure generally decreased for the second prototype compared to the first, but this decrease was rather small for participants with more experience. The subjective evaluation of the usability and fidelity of the phantom and box trainer for laparoscopic surgical training and simulation are summarized in Table 13.6 including the results from the first prototype tests for comparison.

This shows that the weakest fidelity remained for the preparation in both prototypes, but the overall suitability of the trainer box became reasonably high scoring 89% on a 1–5 scale. It was also observable that all of the individual scores showed a slightly decreasing tendency for the second prototype since nine of the average scores went below 4 out of 5. A summary for the obtained force values with respect to time scores are presented in Table 13.7.

Once the measured force data was imported into Microsoft Office Excel for every subject the respective three directional force and the resultant force pattern (calculating with the constant found in the calibration phase) with respect to time was available. These maximum force values gave a range of 4.26–12.46 N. Knowing the force and time results the next task was to objectively analyze the performance of the subjects. For this such a measuring variable was needed to be introduced that would be proportional to the time data and the measured highest force data. The lower the force (smooth movement) and the shorter the time needed during the tasks, the better the performance of the subject will be considered. Based on this the Time-Force Product (TFP) was introduced which is the multiplication of the completion time [s] and the measured highest force [N] which gives units of Ns. The physical significance of the TFP value is not interpreted but it is used as a performance indicator only. Since it is beneficial to be quick and apply as little force as possible during the procedure, the lower the TFP the higher the operation performance. Here the average of TFPs for different groups showed that the laymen had a 60% higher TFP compared to the residents/specialists in average (Table 13.7). In the dataset a two sampled T-test and Welch test was applied too. By 95% significance level both tests

Table 13.5 Average time (s) and standard deviation of the tasks in groups (Avg. +/− Dev.)

	Cutting of plexus santorini	Cutting of right pediculus	Cutting of left pediculus	Cutting of urethra at the apex of prostate	Cutting of urethra at the base of prostate	Overall time of procedure
Laymen	95.9	80.3	111.4	127.8	90.0	506.9
	(+/− 43.8)	(+/− 41.1)	(+/− 57.6)	(+/− 80.1)	(+/− 47.7)	(+/− 197.9)
Residents/specialists	48.8	96.8	90.2	90.8	70.2	395.2
	(+/− 50.2)	(+/− 44.5)	(+/− 143.9)	(+/− 87.9)	(+/− 45.7)	(+/− 143.9)
Summarization of the second prototype	75.7	87.4	103.2	111.9	81.5	450.1
						(+/− 175.1)
Summarization of the first prototype	71.1	80.8	150.2	131.6	84.2	517.9
						(+/− 281.1)

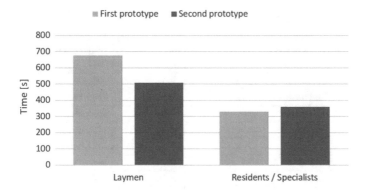

Fig. 13.9 Compared overall time results of the first and second prototype tests

Table 13.6 Subjective evaluations of the laparoscopic box trainer averaged over the entire population of experimental participants. Evaluations taken on a scale of 1–5 where 5 represents the best score

Question	Average score of the second prototype	Average score of the first prototype
Suitableness of the experiment for laparoscopic training	4.43	4.85
Similarity of		
• Task for the camera navigation	4.36	4.50
• Task for the preparation	4.04	4.21
• Task for the cutting of Plexus Santorini	4.43	4.50
• Task for the cutting of pediculus	4.29	4.28
• Task for the cutting of urethra	4.07	4.57
Fidelity of the phantom	4.21	4.43
Suitability of the box trainer for laparoscopic environment simulation	4.43	4.64

null hypothesis was discarded and concluded that the 60% difference of the average TFP value of the laymen and the experienced medical staff was significant. Based on the measurement results and the statistical analysis the TFP proved to be applicable for reflecting the level of training of medical staff and thus may even be used for exam evaluation and marking. However the fine-tuning of the weights used should be done following advice from experienced surgeons.

Table 13.7 The results of the force measurements

Participant	Medical experience	Time (s)	Maximal force (N)	Time-force product [TFP] (Ns)	TFP (Ns) average
GT	Layman	334	8.71	2907.92	3614.66
TÁ	Layman	452	6.52	2945.44	
JL	Layman	381	7.90	3010.20	
TM	Layman	814	4.26	3466.05	
ER	Layman	783	4.67	3657.61	
NI	Layman	355	12.46	4422.73	
NT	Layman	580	8.44	4892.71	
JK	Resident	268	4.29	1148.41	2253.91
NDA	Resident	394	4.37	1723.59	
RÁ	Resident	481	6.47	3109.79	
SZJ	Specialist	314	5.02	1576.56	
MS	Specialist	339	10.95	3711.18	

13.8   Discussion

According to the results of the trials and measurements the novel anatomical phantom proved to be successful by reaching very high scores on the subjective evaluation provided by the users. None of the aspects fell below a score of 4 out of 5. This shows that not only the laymen but also the more experienced medical staff agreed on that the device is suitable for laparoscopic surgical simulation. The phantom received a score of 4.21 out of 5 for fidelity which is proof for the successful accomplishment of the main focus of the design (Table 13.6). The 29% reduction of the overall procedure time for the residents and specialists compared to the laymen indicates that the prior laparoscopic experience and medical knowledge affected performance significantly (Table 13.5). This clearly validates the suitability of the prototype since it shows that the laparoscopic skills and experience were transferable to the box trainer environment. These findings firmly support the assumption and statement that practice sessions done on this box trainer will result in transferable skills to real laparoscopic surgeries. The newly introduced metric named Time Force Product proved to be a suitable performance indicator since successful surgery requires both swift (little time) and safe (little force) procedures. In average the TFP results of the laymen were 60% higher compared to the residents/specialists which is a clear and significant difference. It forms firm evidence that the new box trainer is suitable for skill assessment with the force measurement and evaluation features (Table 13.7).

13.8.1   Comparison of the Results of the First and Second Prototype

The comparison of subjective evaluations of the first and second prototype showed a slight general decrease, though this was not significant (Table 13.5). These comparison results are not reliably quantifiable since the participants were not the same for each prototype and thus the participants could only execute absolute judgment and could not give a valid comparison which would have been more reliable. The conclusion of these subjective results were rather about the fidelity of the second design bringing similar results concerning quality as the first one, while also being much better suited for mass production. The comparison of time results possibly indicated the effect of improved anatomical phantom fidelity. In average, all the individual tasks were completed slightly quicker on the second prototype except for the first two tasks: cutting of the plexus santorini and the right pediculus (Table 13.5). The reason behind this could be that the increased fidelity of the Connective Tissue affected these first two tasks mostly because some tissue preparation was necessary. The improvement in fidelity also affected the average overall time of procedure for the laymen and residents/specialists (Fig. 13.9). Results show that this improved fidelity influences laymen significantly more than experienced medical staff. This indicates that at the beginning of the learning curve, i.e. in the initial stages of learning, fidelity is a more important factor while for experienced medical staff it does not make a significant difference after a certain point [12, 18]. It is to be added however, that this study did not consist of repeated tests of the same individuals thus an analysis of learning curve differences between the models could not be conducted.

13.8.2   Discussion of Improvement Outlooks

Since the development of the box trainer and the prostatectomy phantom is still in the research phase there are a number of required improvements and future goals. The second prototype made significant advancement regarding reproducibility and anatomical fidelity giving the research new directions of improvement. During testing clinicians reported the need to improve the quality of the endoscopic camera suggesting that it would significantly improve the quality of the simulation experience. They have also noted that the design of the bladder needs to be improved. First, in order to establish secure connection to the urethra the implementation of urinary catheters was suggested. Second, the modeling of a deflated or collapsed bladder was phrased which may be achieved by using softer silicone composition. Additionally the fidelity of the bladder could be further approximated by modeling a wavy, smooth muscle surface. The prostate gland preparation, separation from other tissues and dissection of blood vessels and nerves may be better modeled with potentially adding extra tasks of clipping, suturing and using electrical surgical tools suspending bleeding. The modeling of membrane like connective tissue may also be added for improved preparation simulation. Beyond these simpler modifications a more accurate improvement in anatomical shape fidelity would be the use of 3D models from CT scans with the appropriate software but using these technologies may be beyond the design requirements of this product. Finally after having created a third prototype the research testing and validation process needs to be more in depth and extensive. This could be achieved by significantly increasing the number of participants of professional clinicians in the experiment, including the previous prototypes in the testing for comparison feedback and conducting repeated experiments on participants for the examination of learning curves. Lastly additional metrics may be included for surgical performance testing such as visual 3D tracking and observations of tissue deformation apart from the already implemented force sensing.

13.8.3 Evaluation of Force Data

It is important to mention that the TFP does not show which component of the subject's skills are missing in case of bad performance. In this case it is advisable to observe the subject's time and force data independently (Fig. 13.10). From the measurements there are three main categories the subjects can grouped into independent of the level of surgical training: exerting big forces but working quickly, exerting low forces but also being slow, or exerting low forces and being quick. The latter is the ultimate goal of the practice since a surgeon needs smooth movements and quick performance within sensible limits.

Here a few individual cases will be analyses. Analyzing the data of a subject grouped in the first category (exerting big forces but working quickly) it can be seen

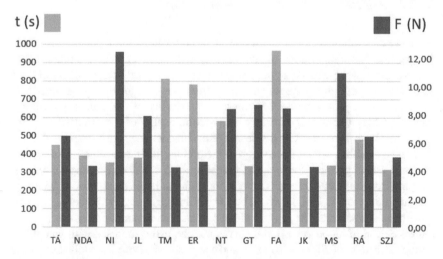

Fig. 13.10 The completion time and maximal resulting force of each participant

Fig. 13.11 Case analysis
I—fast completion, great
forces

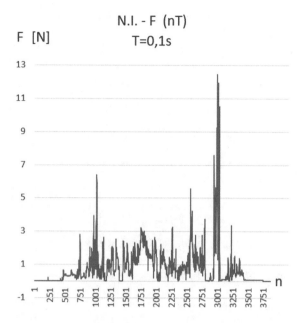

that the highest force mostly determining the high TFP value is only observable in a few occasions (Fig. 13.11). This indicates possible errors. Based on the time data associated to the force values and based on the video recording it can be found in which part of the simulated surgery did the error happen that so significantly determined the TFP results. Such uncontrolled motion is dangerous because in real surgical settings such an abrupt movement can likely cause injury to the patient. Based on this information the student can identify their areas of weakness by themselves and work on improving them.

Analysing the results of another subject from the second category (exerting low forces but also being slow) it is seen that the patient would not be exposed to danger of injury but the slow work can lead to fatigue of the surgeon which can increase the probability of the occurrence of an unwanted mistakes, and also the time the patient needs to spend under anesthesia (Fig. 13.12).

Based on these it can be stated that in ideal cases the measured highest force would minimally deviate from the average force values. This would mean complete absence of any abrupt movement. The real significance of this and the possibility of achieving this would be a subject of future measurements and a question of argument. In ideal cases the subject would perform the tasks as quick as possible. Further question may arise whether the abrupt movements would decrease the overall procedure time or not. Further analysis of this question is not detailed in this study.

Examining the results of another subject from the third category (exerting low forces and being quick) it is clear that the subject achieved the lowest maximum force result and performed the procedure the quickest (Fig. 13.13). It can be seen that

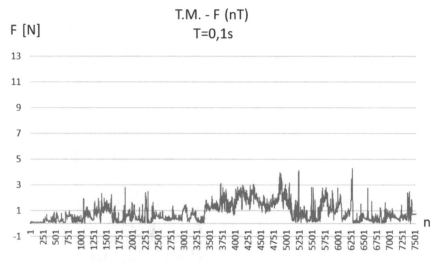

Fig. 13.12 Case analysis II—slow completion, low forces

Fig. 13.13 Case analysis III—fast completion, low forces (ideal case)

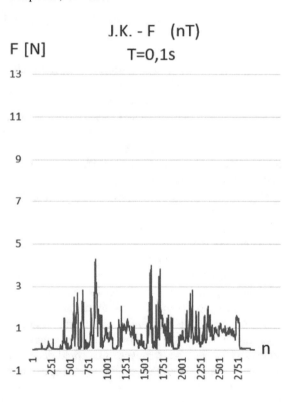

even in this case the TFP value is determined by a few spiking and instantaneous force values.

13.8.4 Developmental Context

The box trainer may meet a number of clinical, educational and surgical technological needs that feed into the research of inter-cognitive communication between human and artificial cognitive systems in the field of surgical robotics [3]. First of all it meets educational needs since medical students could gain relevant practice and dexterity using the box trainer. Second, it may be used for skill assessment, for which primarily the product would need to be validated. This validation is aimed to be achieved by analyzing and quantifying the performance of specialists of laparoscopic surgeries. This analysis and performance quantification on the box trainer can be obtained from visual data of the endoscopic camera and from force data of the load cells (e.g. TFP values (Table 13.7)). Once the performance of a number of specialists are quantified the validation process can start and be completed. Skill assessment can be then achieved with the box trainer using objective standards that may become a crucial educational asset for standardized examination. Finally once a quantified successful surgical procedure is already achieved it can feed into another technological advancement namely the robotic surgery automation for the da Vinci System. Therefore the finalized box trainer can potentially lead to the future of medical robotics where the da Vinci System with increasingly automated surgical capabilities embodies the artificial cognitive system in surgical technology.

13.9 Conclusion

In this chapter a new laparoscopic box trainer was introduced which has been designed to accommodate several types of procedures, and therefore can be used in medical education to practice different types of procedures. One of these procedures is radical prostatectomy, for which we presented an anatomical phantom with high fidelity silicone tissue models. With the model the goal was to create a low cost, easily reproducible phantom which can be mass produced. The development is still in the research phase but early results from tests with clinicians proved that the phantom can be used for medical training and could become an important platform for surgical education.

Apart from developing a new anatomically relevant pelvic phantom a new measurement method was introduced with which objective performance may be measured. In case of insufficient performance the in-depth analysis of the force and time results can be used to identify the causing errors. This error diagnosis can be independently conducted by the subject which increases the efficiency of practice without the need of supervision. The next step of the development may be to place the force sensors at the tip of the laparoscopic tools so that the force data of each tool may be analyzed independently. This would allow a more detailed characterization of the user's insufficiently developed skills. The TFP could also be introduced in haptic virtual-reality systems, where it would be easy to measure the forces within each instrument separately. A virtual environment would also eliminate the need for single-use organ phantoms. However, it would be much more expensive to develop a sufficiently realistic virtual reality software.

The separate analysis of individual tasks may also be edifying. An algorithm for the introduced method can easily be developed so that in the future the assessment methods may be accompanied with a developed software which will supposedly enhance the assessment of the performance of residents and specialists practicing laparoscopic surgical procedures. The extension of this method to other surgeries may increase the effectiveness of the system which would however require the development of further anatomical phantoms.

The phantom is planned to be updated to move from anatomical correctness to the exact modeling of the surgical field. For this the surgical procedure needs to be better examined and understood. It is planned that the surgical environment will be better involved in the phantom, for which one example is the use of urinary catheters, and the modeling of softer collapsible bladder. Further research is planned to examine the procedure and phases of radical prostatectomy, find metrics to measure surgical performance and evaluate the progress of surgical skill development for surgery residents. The first step towards this goal was to implement force sensing into the phantom, but later research intends to expand the range of measured parameters with 3D tracking for example and observations on tissue deformation. Lastly, the presented box trainer and phantom will later be validated on the da Vinci surgical robot, examining the performance of robotic surgery and developing a curriculum for robotic surgery education.

Acknowledgements The research was supported by the Hungarian OTKA PD 116121 grant. This work has been supported by ACMIT (Austrian Center for Medical Innovation and Technology), which is funded within the scope of the COMET (Competence Centers for Excellent Technologies) program of the Austrian Government. T. Haidegger is supported through the New National Excellence Program of the Ministry of Human Capacities. Partial support of this work came from the Hungarian State and the European Union under the EFOP-3.6.1-16-2016-00010 project.

References

1. Aggarwal R, Moorthy K, Darzi A (2004) Laparoscopic skills training and assessment. Br J Surg 91(12):1549–1558. https://doi.org/10.1002/bjs.4816
2. Akhtar K, Sugand K, Wijendra A, Sarvesvaran M, Sperrin M, Standfield N, Cobb J, Gupte C (2016) The transferability of generic minimally invasive surgical skills: is there crossover of core skills between laparoscopy and arthroscopy? J Surg Educ 73(2):329–338. https://doi.org/10.1016/j.jsurg.2015.10.010
3. Baranyi P (2012) Definition and synergies of cognitive Infocommunications. Acta Polytech Hung 9(1):67–83

4. Barcza SM (2016) Surgical skill assessment with robotic technology. PhD thesis, Budapest University of Technology and Economics, Faculty of Electrical Engineering and Informatics
5. Bernier GV, Sanchez JE (2016) Surgical simulation: the value of individualization. Soc Am Gastrointest Endosc Surg
6. Cowan B, Sabri H, Kapralos B, Cristancho S, Moussa F, Dubrowski A (2011) SCETF: serious game surgical cognitive education and training framework. In: 2011 IEEE international games innovation conference (IGIC), pp 130–133. https://doi.org/10.1109/IGIC.2011.6115117
7. Cruz JA, Reis ST, Frati R, Duarte RJ, Nguyen H, Srougi M, Passerotti CC (2016) Does warm-up training in a virtual reality simulator improve surgical performance? a prospective randomized analysis. J Surg Educ 73(6):974–978. https://doi.org/10.1016/j.jsurg.2016.04.020
8. Dunkin B, Adrales GL, Apelgren K, Mellinger JD (2007) Surgical simulation: a current review. Surg Endosc 21(3):357–366. https://doi.org/10.1007/s00464-006-9072-0
9. Duren B, Boxel G (2014) Use your phone to build a simple laparoscopic trainer. J Minim Access Surg 10(4):219–220
10. Fitzgerald JEF, Caesar BC (2012) The European working time directive: a practical review for surgical trainees. Int J Surg 10(8):399–403. https://doi.org/10.1016/j.ijsu.2012.08.007
11. Folaranmi SE, Partridge RW, Brennan PM, Hennessey IA (2016) Does a 3d image improve laparoscopic motor skills? J Laparoendosc Adv Surg Tech 26(8):671–673. https://doi.org/10.1089/lap.2016.0208
12. Friedman Z, Siddiqui N, Katznelson R, Devito I, Bould MD, Naik V (2009) Clinical impact of epidural anesthesia simulation on short- and long-term learning curve: high-versus low-fidelity model training. Reg Anesth Pain Med 34(3):229–232
13. Gaba DM (2004) The future vision of simulation in health care. Qual Saf Health Care 13(suppl 1):i2–i10. https://doi.org/10.1136/qshc.2004.009878
14. Hopper AN, Jamison MH, Lewis WG (2007) Learning curves in surgical practice. Postgrad Med J 83(986):777–779
15. Immenroth M, Bürger T, Brenner J, Nagelschmidt M, Eberspächer H, Troidl H (2007) Mental training in surgical education: a randomized controlled trial. Ann Surg 245(3):385–391. https://doi.org/10.1097/01.sla.0000251575.95171.b3
16. Issenberg SB, McGaghie WC, Petrusa ER, Lee Gordon D, Scalese RJ (2005) Features and uses of high-fidelity medical simulations that lead to effective learning: a BEME systematic review. Med Teach 27(1):10–28
17. Jimbo T, Ieiri S, Obata S, Uemura M, Souzaki R, Matsuoka N, Katayama T, Masumoto K, Hashizume M, Taguchi T (2017) A new innovative laparoscopic fundoplication training simulator with a surgical skill validation system. Surg Endosc 31(4):1688–1696. https://doi.org/10.1007/s00464-016-5159-4
18. Kim HK (2004) Virtual-reality-based laparoscopic surgical training: the role of simulation fidelity in haptic feedback. Comput Aided Surg. DWRMAS 9(5):227–234. https://doi.org/10.3109/10929080500066997
19. Kirwan WO, Kaar TK, Waldron R (1991) Starting laparoscopic cholecystectomy the pig as a training model 10(4):219–220
20. Lee M, Savage J, Dias M, Bergersen Ph, Winter M (2015) Box, cable and smartphone: a simple laparoscopic trainer. Clin Teach 12(6):384–388. https://doi.org/10.1111/tct.12380
21. Li MM, George J (2016) A systematic review of low-cost laparoscopic simulators. Surg Endosc 1–11. https://doi.org/10.1007/s00464-016-4953-3
22. Nagendran M, Gurusamy KS, Aggarwal R, Loizidou M, Davidson BR (2013) Virtual reality training for surgical trainees in laparoscopic surgery. Cochrane Database Syst Rev (8). https://doi.org/10.1002/14651858.CD006575.pub3
23. Ritter FE (2002) The learning curve. In: International encyclopedia of the social and behavioral sciences, pp 8602–8605
24. Romero-Loera S, Cárdenas-Lailson LE, de la Concha-Bermejillo F, Crisanto-Campos BA, Valenzuela-Salazar C, Moreno-Portillo M (2016) Skills comparison using a 2d vs. 3d laparoscopic simulator. Cirugáa y Cirujanos (English edn) 84(1):37–44. https://doi.org/10.1016/j.circen.2015.12.012

25. Rosser JC, Liu X, Jacobs C, Choi KM, Jalink MB, Ten Cate Hoedemaker HO (2017) Impact of super monkey ball and underground video games on basic and advanced laparoscopic skill training. Surg Endosc 31(4):1544–1549. https://doi.org/10.1007/s00464-016-5059-7
26. Sándor J, Lengyel B, Haidegger T, Saftics G, Papp G, Nagy A, Wéber G (2010) Minimally invasive surgical technologies: challenges in education and training. Asian J Endosc Surg 3(3):101–108. https://doi.org/10.1111/j.1758-5910.2010.00050.x
27. Schijven M, Jakimowicz J (2004) Virtual reality surgical laparoscopic simulators—how to choose. Res Gate 17(12):1943–50. https://doi.org/10.1007/s00464-003-9052-6
28. Spence AM (1981) The learning curve and competition. Bell J Econ 12(1):49–70
29. Steigerwald SN, Park J, Hardy KM, Gillman L, Vergis AS (2015) The fundamentals of laparoscopic surgery and LapVR evaluation metrics may not correlate with operative performance in a novice cohort. Med Educ Online 20:30024
30. Takács A, Kovács L, Rudas I, Precup RE, Haidegger T (2015) Models for force control in telesurgical robot systems. Acta Polytech Hung 12(8):95–114
31. Takács A, Nagy D, Rudas I, Haidegger T (2016) Origins of surgical robotics. Acta Polytech Hung 13(1):13–30
32. Thinggaard E, Bjerrum F, Strandbygaard J, Gögenur I, Konge L (2016) Ensuring competency of novice laparoscopic surgeons-exploring standard setting methods and their consequences. J Surg Educ 73(6):986–991. https://doi.org/10.1016/j.jsurg.2016.05.008
33. Thinggaard E, Konge L, Bjerrum F, Strandbygaard J, Gögenur I, Spanager L (2017) Take-home training in a simulation-based laparoscopy course. Surg Endosc 31(4):1738–1745. https://doi.org/10.1007/s00464-016-5166-5
34. Undre S, Darzi A (2007) Laparoscopy simulators. J Endourol 21(3):274–279. https://doi.org/10.1089/end.2007.9980
35. van Velthoven RF (2006) Methods for laparoscopic training using animal models. PubMed 7(2):114–9
36. Zendejas B, Brydges R, Hamstra SJ, Cook DA (2013) State of the evidence on simulation-based training for laparoscopic surgery. Syst Rev Res Gate 257(4):586–93. https://doi.org/10.1097/SLA.0b013e318288c40b

Chapter 14
Cognitive Cloud-Based Telemedicine System

Ábel Garai, István Péntek and Attila Adamkó

Abstract Telemedicine instruments and e-Health mobile wearable devices are designed to enhance patients' quality of life. The adequate man-and-machine cognitive ecosystem is the missing link for that. This research program is dedicated to deliver the suitable solution. This exploratory study focuses specifically on the universal interoperability among the healthcare domains and its actors. This research's goal is the establishment of adaptive informatics framework for telemedicine. This is achieved through the deployed open telemedicine interoperability hub-system. The presented inter-cognitive sensor-sharing system solution augments the healthcare ecosystem through extended interconnection among the telemedicine, IoT e-Health and hospital information system domains. The research team's motivation is to deliver advanced living condition for patients with chronic diseases through remotely accessible telemedicine and e-Health solutions relying upon the open telemedicine interoperability hub-system. Industry-wide accepted technologies are applied to achieve this goal. The general purpose of this experiment is building an augmented, adaptive, cognitive and also universal healthcare information technology ecosphere.

14.1 Introduction

Remote monitoring systems trace back more than 100 years ago [10], the first appearance of a tele-stethoscope, a forerunner of personalized medical care. The significance of remote monitoring systems increased by ranks of magnitude upon

Á. Garai (✉) · I. Péntek · A. Adamkó
Faculty of Informatics, University of Debrecen, Kassai Str. 26, Debrecen, Hungary
e-mail: garai.abel@inf.unideb.hu

I. Péntek
e-mail: pentek.istvan@inf.unideb.hu

A. Adamkó
e-mail: adamko.attila@inf.unideb.hu

the achievements in life and medical sciences, in sensory and executing mechanisms approaching treatments in chronic diseases [37, 40]. Treatment of chronic diseases implies regular physician visits and related decisions. Both the number of visits and the related decisions are eased in time and follow-up by personalized telecare provisions. Fitting those said before, patients under medical care are no more passive targets rather, should have an active share in communication, instead [1]. Nowadays, the developing way of telemedicine is paved by the speed up of translating technology inventions into medicine and, by the urgent need for healthcare data collection, store and processing, (digital content management) in safe condition. Advances reporting capabilities is a basic need for the eHealth industry. Telemedicine instruments can emit huge files, meanwhile healthcare professionals should reach the information at the appropriate moment and location. Therefore, adequate conversion, transformation, interpretation and presentation are required for health-related bulk-data captured by telemedicine appliances. The study's goal is to create a flexible telemedicine interoperability hub system to extend the options of classical hospital systems. To achieve this goal, industry-wide accepted technologies have been applied [4]. Our motivation is to deliver advanced living condition for patients with chronic diseases through remotely accessible telemedicine and e-Health solutions relying upon the open telemedicine interoperability hub-system. The general purpose of this experiment is building a universally applicable information technology solution interconnecting the telemedicine instruments, the e-Health smart devices and hospital IT systems.

Future telemedicine systems are to be accessed from multiple end-points: tablets, smart-phones, integrated systems, cars or hospital IT systems [34]. The best solution for these systems is the Cloud architecture. Within the Cloud we separate Infrastructure-as-a-Service, Platform-as-a-Service and Software-as-a-Service solutions. The most important features of these solutions are the availability, security and scalability. These are also the needs of the modern healthcare IT systems. There are numerous different data exchange standards and formats within the health-IT industry. Modern systems need to support different ways of communication. There are only very few stand-alone systems. Most of the healthcare systems operate in a diversified architectural landscape. The elements of this whole landscape connect with interfaces to each other. The arising Internet-of-Things (IoT) technologies can be inserted into the healthcare IT domain, if it supports the standard communication methods. In practice, these items have their own specific data representation and encryption. Therefore, these cannot be inserted into the healthcare ecosystem automatically. In practice, the eHealth IoT devices intercept bio-sensory data and send it via Bluetooth to an end-device. However, it generally works only with its own factory-designed app or Bluetooth-enabled instrument [32]. The spread of IoT devices call for attention: the communication methods and data coding should be standardized in order to build an interoperable landscape and to exploit this technology. The IoT eHealth's market penetration could positively influence the overall healthcare supply chain. However, there are roadblocks to be solved before it can bring its benefit. The research focuses on the Cognitive Infocommunication-doctrines [6, 7] for building a human-machine interaction-based healthcare ecosystem. There are many international healthcare projects running parallel. There are also many new electronic devices on the other hand: w-lan scales, electronic blood pressure monitors or fitness equipments. These tools have built-in sensors, and these emit the measurement data. These data are forwarded with a wireless-connection to a smartphone. The end-user can access the data on a smartphone app.

The cloud services are important for the telemedicine and clinical information services. These healthcare equipments usually have Bluetooth, Near Field Communication (NFC) or wireless chips. Therefore, it is possible to capture the data from these devices without physical connection. However, these emit generally unstructured data. Therefore, we need methods to standardize this data-flow. On the other hand, we need to store the received data. Technically we need to establish a generally applicable healthcare transaction processing mechanism. There are four keywords to be considered for successful healthcare data processing: atomicity, consistency, isolation or durability. There are two further keywords for effective transaction processing: scalability and distribution. The Cloud architecture can handle scalability efficiently. The recent in-memory data mining systems (VoltDB) are also important for real-time healthcare prognosis. These systems, once they are established correctly, can provide healthcare forecasts on an individual basis and also on populational level.

14.2 Adaptive eHealth Content Management

We can classify telemedicine and e-Health adaptive content management at three different levels:

(a) technical,
(b) semantic,
(c) process.

The technical level refers to the technical capability of the interoperability of the connected systems. The semantic level refers to the ability to interpret the exchanged information beyond pure technical data-exchange. The process level targets the overall organizations: the different processes need to be harmonized in order to gain the maximum efficiency. The healthcare industry is generally process-oriented. Semantic interoperability also implies that complex telemedicine systems as a whole are harmonized with each other [35].

14.2.1 Technical Level

Technical interoperability refers to the basic ability to send and receive data. XML-schemas assist modern transaction processing. These files and their skeletons provide

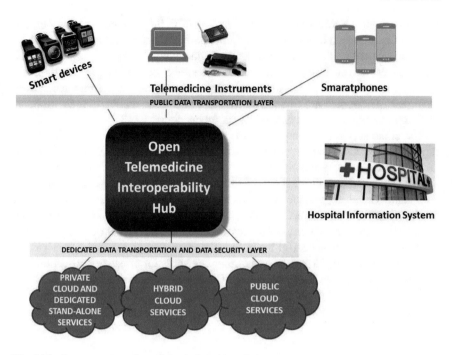

Fig. 14.1 Common open telemedicine hub and interface recommendation

the necessary data structure for the interoperability. Before correct XMLs are built, the two or more organizations need to agree on the data to be exchanged and their structure. The receiving and sending systems also should be adjusted in order to build a stable data-exchange framework. Another was for ensuring standardized content management is using the Common Object Request Broker Architecture (CORBA) [29]. This is a standard owned by the Object Management Group (OMG) [9]. The different telemedicine systems are generally written in different programming languages on different software platforms. These different systems generally use their own data-representation. The CORBA standard provides interoperability among such different platforms and system landscapes. There is another way to establish interoperability among different, distributed software components for Internet-enabled computers. This is the Distributed Component Object Model (DCOM) [38] technology enables communication among software components distributed across networked computers. Telemedicine and healthcare systems are connected into complex IT-landscapes. For this reason, DCOM offers the suitable solution for such distributed systems and networked subcomponents in a distributed framework. Figure 14.1 shows the Common Open Telemedicine Hub-Software (OTI-Hub).

14.2.2 Semantic Level

Semantic content management means, that the shared information is interpretable on both or on all interconnected systems. Within the healthcare supply chain the SNOMED-CT [12] is the most accepted standard worldwide. The International Health Terminology Standards Development Organisation (IHTSDO) [33] owns and upgrades it. A general standard is necessary for a content management framework. Furthermore, a centrally and generally acknowledged one is required for dynamic adaptivity. There is another internationally acknowledged standard within the healthcare industry: the HL7 [11]. The Health Level Seven International ANSI-accredited organization is responsible for this. The HL7 controls the seventh OSI-layer. This is the application layer. This is critical for the interruption-free messaging among different medical systems. This is also called the Reference Information Model (ANSI/HL7 RIM R3-2010 and ISO 21731). This is a part of the HL7 standard. Within the HL7 we distinguish between the version 2 and the version 3. The RIM is applicable not only for health care communication but also for the documentation. The World Health Organization (WHO) issued the International Statistical Classification of Diseases and Related Problems (ICD-10) [42]. This is a medical classification list. These structures and decodes the diseases, their symptoms, the associated complaints and other factors like external causes or social circumstances. The ICD-10 is generally internationally applicable and already introduces in most of the developed countries: UK (1995), Canada (2000), China (2002), France (2005) or USA (2015). The international acknowledgement of this nomenclature is critical for the healthcare industry generally, and for the telemedicine in particular. This is the missing link between the different and heterogenous systems. This influences not only the interoperability between the different systems, but also the effective access to the Electronic Health Records (EHR) [26]. There are EHRs, where different standardization methodologies took place. Such an example is the OpenEHR, which uses combinations of different values, data types and structures.

14.2.3 Process Level

For process interoperability there are different institutions and national bodies. These organizations are responsible for elaborating the basic rules for the successful interoperability among healthcare systems from different service providers. One of the biggest and most influential bodies is the Healthcare Information Technology Standards in the USA, which was founded in 2010. Another significant body is the National Library of Medicine. In Canada, the most important organization is the Health Infoway founded in 2010. The UK introduced the Connecting for Health Strategy in 2005. The International Health Terminology Standards Development Organization (IHTSDO) was established in the EU (headquarters in Denmark). This organization is committed to enhance the digital content management of the interconnected health systems. There is a possibility for customized content management. This plays an important role for hospice and palliative care. The process level inter-operability is also important on social, political, organizational and technological level. Artificial intelligence and computer decision support system (CDSS) can use the medical inputs coming from different devices and from the medical staff. The level of quality assurance is critical, because it influences the applicability of the healthcare information base. Root-problems can be incompleteness or inconsistency of the data.

14.3 Dynamic Cloud Architecture

The healthcare information systems are required to continuously exchange their information with internal and external systems, instruments. The way how they do this influences the quality of the output. Encryption and authentication also play significant role for the securing patient information. The different interconnected systems share sensitive information. They generally use open channels for that, like the Internet. Therefore, adequate authentication procedures are required for safeguarding the personal information. There are several pros and cons for cloud architecture. In same cases it is safer, than stand-alone servers. However, healthcare decision-makers still hesitate to transform the hospital information systems into cloud-based distributed solutions.

Telemedicine instruments and eHealth devices emit personal and confidential information. Therefore, the suitable authentication methodology and encryption mechanism is critical for the overall security framework. Business continuity and disaster recovery planning also important features. Systems, healthcare personal and patients share sensitive and private health-care information through these systems. As a result, information security measures are critical parts. The safety and security should be planned, monitored, controlled and reported on a timely manner. In order to bridge the different healthcare technologies, we have created a Common Open Telemedicine Hub-Software (COTH) [35]. Blood-sugar monitoring gadgets, blood-pressure monitors and smart watches are connected to cell-phone apps via wireless data-exchange methodology. These instruments are connected to the host systems through the internet. Technically this means GSM, WLAN, Bluetooth, LAN or USB connection. Traditional and dedicated healthcare information systems are generally linked through a dedicated WAN to the Internet or to other organizations, like ministries or social security organs. There are at least two systems for data exchange: a recipient and the sender [25]. We embedded our COTH in Cloud-infrastructure assisted by ICT-services. We apply different ICT services: wireless and landline network connectivity. This means also connection to the Internet. We rely on services of international ICT providers. These provide the infrastructure for transporting the data packages from one system to another. The most influential Cloud Service Providers provide not only data storage capabilities, but also analytical and forecasting capabilities. The aim is to use the capabilities of the Cloud architecture including global availability. Healthcare data consists of text, images, videos and data-streams. All these need to be stored in a structured manner. The cloud architecture provides the necessary scalability for this.

The overall healthcare system landscape is heterogenous and complex. It includes different devices and technologies: telemedicine instruments, smart devices and hospital information systems [3, 19]. These mentioned systems use different internal logic, different data-structure and stand for different purpose [13]. Our COTH is specified to be able to establish communication with a wide variety of different data structures coming from and transferred to alternative devices at staying at different locations. The COTH is open for other, new healthcare instruments, smart devices and different type of sources. The proposed and demonstrated dynamically adaptive content management framework stands for the correct representation of the standardized healthcare data.

Market-leader eHealth and ICT corporations [14] optimize their business models utilizing the technical possibilities of the Cloud infrastructure. Besides that, compliance remains a significant cornerstone: accessibility routines and legal prerequisites are even more important, than physical premises. Safeguarding the sensitive healthcare information covers physical and non-physical safety and security measures. There are different factors taken into consideration in the course of the integration of traditional healthcare systems and modern telemedicine system landscapes. Nowadays only a small part of the available bio-sensory information is used in the everyday healthcare supply chain. The length of the patient-visits is generally very limited, and the overall logic of the healthcare systems does not really support the implementation of the technically available options, like integration of e-Health devices [31]. On the other hand, hospital information systems rely upon reliable and validated data. This should be the same for telemedicine instruments and e-Health gadgets. However, the overall quality of the data from these devices does not reach the level of the traditional healthcare systems [41]. Healthcare smart-devices should be certified first, in order to be applied and inserted into the healthcare supply chain. This is a critical part of the general applicability of these devices, and also substantial for the COTH-research. The general aim of the study is to provide, that extended or real-time data is displayed on the doctor's monitor.

A couple of managers from international ICT providers were questions, whether the currently available cloud technology and infrastructure would be able to handle the real-time transfer and processing of continuous bio-sensory information. Interestingly we received the answer, that these managers see it as a real possibility. This is a significant step, and it also means, that our COTH-research is justified. As in-memory real-time analysis tools evolve, general population-level healthcare information analysis and forecasting is possible. Now Cloud architectures can provide the necessary bandwidth, memory, storage- and calculation-capacity. The global wireless, landline and satellite infrastructures are critical to utilize the e-Health capabilities on an international basis. The interconnection of these devices will require extended storage capacities for the exponentially increasing data volumes. Therefore, there is a demand for dynamically adaptive e-Health content management. The COTH gives the basis and flexibility to reach this target.

14.4   Open Telemedicine Interoperability Hub Research Program

Recent studies drafted general software platform for telemedicine instruments clinical systems integration [27, 43]. However, these cited researches remain on theoretical basis. On the other hand, the available studies focus either on telemedicine instrument integration or on mobile wearable body-sensory appliances linkage. We aimed to establish the empirical basis for universal Internet of Things (IoT) Clinical Systems interconnection.

Our research is a synthesis of classical healthcare information system architecture, telemedicine instrument landscape, and mobile wearable bio-sensory technology. We have developed an Open Telemedicine Interoperability Hub (OTI-Hub, Fig. 14.2), put into operation and functionally evaluated. Below follows the summary on the international healthcare interoperability standards, nomenclatures, further, an enhancement proposal through common open telemedicine interface standard recommendation is demonstrated.

Sensor-based smart devices capture, store and transmit healthcare data in semi-continuous data series. However, traditional hospital systems transmit and store static patient-information. Technical solutions bridging classical clinical information systems and eHealth smart devices, remain still a missing link. The hospital information

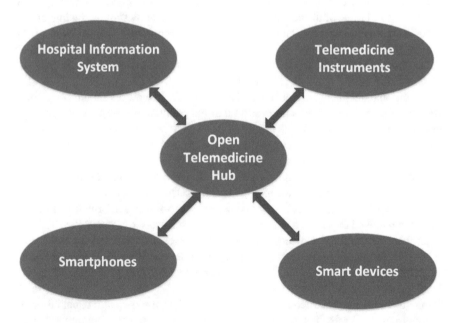

Fig. 14.2 Open telemedicine interoperability hub (OTI-Hub) interconnecting the segregated medical domains

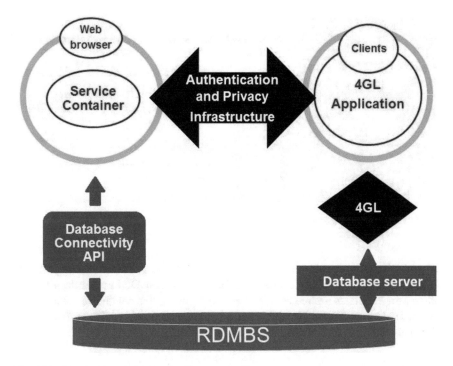

Fig. 14.3 Classical hospital information system outline

systems, telemedicine instruments and e-Health devices store, process and transmit healthcare data in a different way.

It is unresolved, how clinical systems with e-Health smart devices will flaw-lessly communicate. This is an important interoperability issue. It means technical and methodical interoperability among these wearable smart e-Health devices, the telemedicine landscape and the traditional hospital information system domain. This study utilizes the test- and acceptance-system of a selected hospital information system of an internationally renowned service provider (Fig. 14.3). MedSol information system operates in sixty hospitals over Europe and serves forty thousand users.

Analyses are given on enhanced leading international clinical interoperability standard application, on Systematized Nomenclature of Medicine (SNOMED [39]) and HL7. Notwithstanding that the latest HL7 v3 standard covers almost all inter-nationally applicable medical information fields instead of the previous HL7 v2.x version [22], the HL7 v2.x is the de facto international clinical information sys-tems interoperability standard. Therefore, the HL7 v2.3 was applied in the research program for the clinical spirometer-healthcare information system interconnection.

A locale of the research so far has been the Semmelweis University 2nd Pedi-atric Clinic Department of Pulmonology. During the research the PDD-301/shm spirometer is connected to the clinical information system (MedSol and eMedSol, Fig. 14.4). The clinical information system is accessed with tablets through local

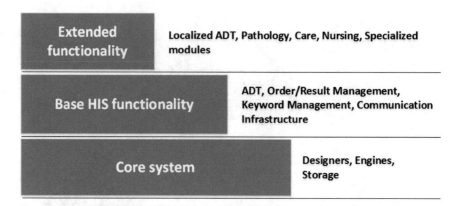

Fig. 14.4 Concept of modular-hierarchical hospital information system

clinical WLAN by the medical staff. The research demonstrates how the mobile spirometer and the also mobile clinical information system GUI cooperate with each other. This simulates mobile telemedicine deployment, like healthcare solutions for remote underpopulated regions (Norway, Sweden and Canada). The different HL7 v2.x and HL7 v3 standards are also demonstrated.

14.5   Examples for Clinical Targets in Service

14.5.1   Diabetes

Diabetes remains global endemic disease number one. Insulin pumps developed for diabetes with sensors implanted under the skin require stable data-circulation. This solution supplemented with cloud-based interface and data-processing programs transfer the actual blood sugar level to the designated diabetes center through telecare and telemedicine software systems. The actual patient blood sugar information is displayed directly at the diabetes center, and the blood sugar timelines, periods are monitored customized to the emerging patient complaints.

14.5.2   Asthma

The telemedicine instrument landscape (Fig. 14.2) is represented by the PDD-301/shm medical spirometer installed. This mobile device is connected through Health Level Seven [HL7 [18, 20]]-based interface to the hospital information system.

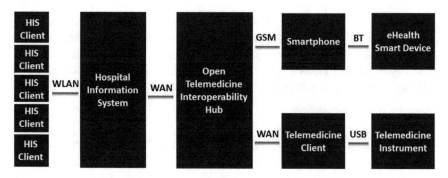

Fig. 14.5 Telemedicine and eHealth systems interoperability landscape with OTI-Hub

The mobile telemedicine device is blown by the patient at home and the results are automatically uploaded to the asthma clinical center through the Internet. The specialist analyzes the results during the subsequent personal patient visit or conducts a remote determination of the necessary medical intervention. The medical professional evaluates the spirometry test results based on criteria system standard.

14.5.3   Electrocardiogram ECG

Principles above apply to ECG tests: a heart rate monitoring smart wearable bracelet represents eHealth smart device technology connected to the hospital information system. Similar principles apply for smart blood pressure monitoring device of patients with cardiovascular disease.

The Research Program has been conducted with the aim to establish service architecture through the OTI-Hub enabling bi-directional syntactic interoperability among HIS, TI and IoT eHealth devices. Therefore we raised a cloud-based system landscape (Fig. 14.5) based on the OTI-Hub and the HL7 interoperability standard. The novelty of this architecture is that it integrates the classical healthcare information system architecture, the standard telemedicine environment and the IoT eHealth technology [15, 36].

Cloud architectures are classified as private, public and hybrid clouds (community cloud is excluded in this regard due to associated reliability and availability concerns). The private cloud offers feasible technical solution for sensitive personal patient data; the public cloud delivers the necessary scalability. Therefore, the final technical architecture belongs to hybrid architecture category. The applied cloud-based Healthcare Information System (HIS), eMedsol, operates in sixty hospitals and serves forty thousand users in Hungary, Romania, Czech-Republic and Bosnia-Herzegovina. The eMedSol is equipped with UNIX-based (Linux, AIX HP-UX, SCO) WebSphere Application Server (V5, V6) interconnected with Oracle (10gR2) and Progress (V10 OpenEdge) relational database management systems (RDBMS).

Each installation (for clinical institution under three hundred beds) is equipped with a primary and a secondary virtual server, two quad-CPUs and 16 GB Memory. The OTI-Hub is embedded in Google public cloud (Google Cloud Platform), and sensitive patient personal data are stored in the M?nchen-based Open Telekom Cloud datacenter, therefore patient data remains within the EU.

14.6 Research Methodology and Software Technology

Clinical systems interoperability reaches beyond plain data-exchange: it constitutes interoperability at technical, semantic and at process level. In the empirical model of the research the OSI model (ISO/IEC 7498-1:1994 [21]) is mapped against the aforementioned interoperability levels, therefore these three interoperability modalities are interpreted also at the corresponding information technology abstraction layer. Technical and semantic interoperability is targeted within the presented research. Among the technical interoperability modalities instead of the TCP/IP the file-based interface connection has been elected, since this option offered significantly more flexibility during the research [2].

The following instruments have been selected and allocated to the research program: Spirometer PDD-301/shm as clinical telemedicine instrument, Microsoft Band I and Microsoft Band II smart wristband as eHealth sensory devices, Nokia Lumia 930 smartphone (Windows 10 Mobile operating system), Dell Latitude E6520 (Windows 10 32 bit operating system, i5-2520M chipset, 4 GB RAM and 256 GB HDD) primary laptop, Dell Latitude E6220 (Windows 7 64 bit operating system, i5-2520M chipset, 4 GB RAM, 128 GB SDD) secondary laptop, three Lenovo MIIX 300-10IBY tablets and an ACER SWITCH SW3-013-12CD tablet. Each tablet is equipped with 10,1 display (WXGA and HD IPS), 2 GB memory, 64 GB internal storage and Windows 10 operating system. All laptops and tablets fit to the 802.11g WLAN and Bluetooth 4.0 standards. The spirometer is USB-enabled. The selected smart wristbands are manufactured with built in- Bluetooth 4.0 communication chipsets. Each instrument of the lab equipment package has been individually tested prior to the experiment.

A specific private cloud was established for the research. This ran on stand-alone x86-64 architecture equipped with Intel i5 processor, 256 GB SSD and 4 GB RAM. The operating system for the private Cloud is Red Hat Enterprise Linux 7.0 3.10.0-229, the virtualization is provided by VMware Workstation v6.5.0 and the relational database management system is supplied by MySQL v5.6. The cloud-based version of the hospital information test system runs in a commercial private cloud (Telekom Cloud). The Open Telemedicine Interoperability Hub data transmission module is embedded in a commercial public cloud (Microsoft Azure).

The HIS runs on J2EE WebSphere Application server V6, relying upon Oracle RDBMS 10gR2 and Progress V10 OpenEdge RDBMS. The HIS is hosted on Unix operating system. Floating licenses were made available for reaching the online, cloud-based edition of the selected HIS through the research tablets. The

Open Telemedicine Interoperability Hub development environment consisted of the Universal Windows Application Development Tools (1.4.1), Windows 10 Software Development Kit 10.0.25431.01 Update 3 and Microsoft .NET Framework Version 4.6.01038. The OTI-Hub internal database was developed by SQL Server Data Tools 14.0.60519.0. The OTI-Hub App was developed with Visual Studio Tools for Universal Windows Apps 14.0.25527.01. The OTI-Hub middleware was settled in Microsoft Azure Mobile Services Tools 1.4. Red Hat Enterprise Linux 7.0 3.10.0-229 provided the operating system for the private cloud established specifically for the research.

The spirometry desktop program has been installed on a standalone Dell Latitude E6520 laptop equipped with Windows 10 operating system. The spirometer has been calibrated by the manufacturer for the research. Forced vital capacity spirometry test has been undertaken with a healthy individual. Having the test results stored in the spirometry desktop software, the HL7 v2.3.1 interface file has been exported. This interface file has been processed by the cloud-based OTI-Hub. The OTI-Hub appended the spirometry information with the previously transformed cardio body-sensor information captured by the L18 Smart Bluetooth Wristband. The generated HL7 interface file is imported after parameterization into the factory acceptance test instance of the MedSol hospital information system (Fig. 14.6). Both the imported spirometry and cardio test results are retrieved and displayed by the patient report query of the hospital information system. The information technology results are validated by the Department of Information Technology, University of Debrecen and by

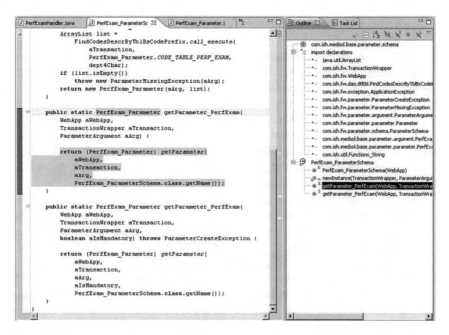

Fig. 14.6 Hospital information system acceptance instance for OTI-Hub interconnection

T-Systems Healthcare Competence Center Central and Eastern Europe. The clinical results are validated by the Semmelweis University 2nd Department of Paediatrics.

The implemented system is a distributed, cloud based and scalable. In case of load increase, the system can be scaled up by automatic allocation of new resources into the OTI-Hub cluster.

14.7 Clinical System Interoperability Improvement Proposal and Standard Recommendation

The software development environment for the hub has been selected and concretized (the selected technologies are available on Microsoft stack, and the OTI-Hub has been implemented with Microsoft stack and further open source technologies [4]). The OTI-Hub relies upon the international HL7 standard and provides bi-directional interoperability among the different healthcare platforms and domains [8].

The following technologies are applied within the OTI-Hub (Fig. 14.7) and in its modules:

(a) Receiver module handles the received measurement data through http web request. This module uses a Web API library and associated markup language (represented by Microsoft Web API, JSON and XML in the research). This module uses open authentication protocol (OAuth in this research) to authorize and authenticate the users and devices. The REST principles need to be adapted as far as possible [16, 24];

(b) Transformation module operates with data collected by the receiver module and the main task of this module is data transformation to the chosen interoperability standard format (HL7 and C#-based Windows service in the research);

(c) Data storage module is responsible for building data warehouse from the collected and transformed data. (This module was implemented in Apache Hadoop software library in the research);

(d) Interpreter module works with the data warehouse module and its main task is to interpret the collected and aggregated data. (This module works with Apache Hadoop software library in the research as well);

(e) The most important module is the Hubs integration module. This module offers export data feature from the telemedicine hub into external systems, e.g.: hospital information systems. This module uses REST API endpoints to transport data into external systems; and it applies the technologies summarized at the receiver module (Microsoft Web API, OAuth, JSON and XML in the research [28]);

(f) Real-time web communication module library operates to maintain an open socket between the Hub and the devices (SignalR in the research). The module is critical when a device is recording data frequently and the received data have to be available promptly on the Hub. When lower priority measurement data is not necessarily needed to be available immediately on the Hub, this module can be excluded from the processing chain. The socket gives the ability of

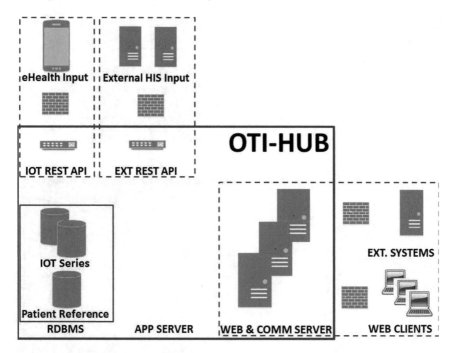

Fig. 14.7   OTI-Hub service architecture

using a channel between the Hub and the device without reconnection and re-authentication. The only significant latency is the network latency in this case. The Hub is prepared for full duplex channel, so the data can be transmitted in both directions;

(g) The Hub is embedded [5] in Hybrid Cloud Architecture (Fig. 14.8). The Hub component, responsible for patient-related master data, is embedded in private cloud architecture (German Telekom Private Cloud in the research). The Hub-components, accountable for the interoperability logic and routing, are implanted in public cloud (Google Cloud Platform in the research).

14.8   Hospital Information System Optimization with Asynchronous Cloud-Based Telemedicine Services

In order to make the communication ways as short as possible between the system elements, direct messaging channels are recommended among the system components instead of the legacy tree structure system components communication model. The adaptive telemedicine services are designated to realize the direct messaging channels between system components integrating the bio-sensory data-flow [30].

Fig. 14.8 Optimized
hospital information system
embedded in hybrid-cloud

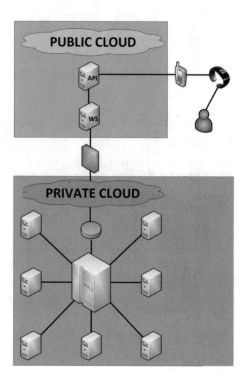

Our proposed Open Telemedicine Interoperability Hub is the manifestation of the
adaptive telemedicine services during the research program.

The given heterogenous hospital information system landscape was modelled by
directed graph, and the graph was mapped into a corresponding adjacency matrix.
The research team thematically outlined the optimized system landscape candidates,
and mapped each of these into separate adjacency matrices. The availability matrices
have been calculated for both the already deployed system landscape and also for
the optimized system landscape candidates relying upon the following formula:

$$S_k = sign \sum_{n-1}^{k} C^n \qquad (14.1)$$

where C is the (square) adjacency matrix and k is the adjacency matrixs dimension.

All the computed availability matrices have been compared against each other to
select the optimal one. The optimum criteria was set to maximize the communication
channels quantity among the system components within the optimized system landscape. For the optimizations sake, the resulting availability matrices were compared,
and the one with the highest overall available link number was selected:

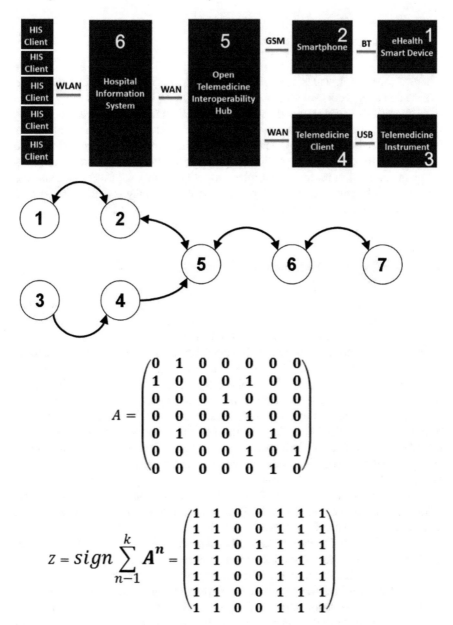

Fig. 14.9 Simplified system architecture optimization methodology

$$x_m = \sum_{i=1}^{n} \sum_{j=1}^{n} a_{i,j} \qquad (14.2)$$

where x is the system landscape solutions identifier, i is the row, j is the column and a is the actual elements value of the square n × n availability matrix.

Consequently, x represents the number of the maximum interconnection among the system components within the proposed system landscape. As more components communicate with each other directly, the overall information exchange performance of the entire system landscape delivers significantly better performance. From optimizations point of view, the higher interconnection grand total (xm) represents the higher level of optimization. This optimization methodology is demonstrated on a simplified system design on Fig. 14.9.

14.9   Testing and Evaluation

The OTI-Hub was interconnected to the mirrored HIS industry test system. The test plan included individual, cluster and integration tests. The individual tests concerned the single research software environment element: the receiver, transformational, storage, interpreter and integrational module of the OTI-Hub. The cluster tests focused both on the eHealth smart device and on the telemedicine instrument thread of the OTI-Hub. The clinical spirometer emits elementary data. However, the smart wearable eHealth device produces continuous time-series. Therefore, a clustertest was carried out. During this clustertest primary data both from the spirometer and from the wearable eHealth device was successfully processed. The integration test provided the overall quality assurance for the OTI-Hub. The telemedicine instrument and the eHealth smart device measured real bio-sensory signals of anonymized individuals, and sent it to the OTI-Hub. The OTI-Hub interpreted, saved, transformed and sent these data to the mirrored HIS industry test system. The allocated tablets were used to load the Cloud-based HIS graphical user interface. The tablets were connected via dedicated WLAN to the HIS industry test system. The results were validated through the GUI on the tablets by clinical professionals. However, the OTI-Hub module, which is responsible for the eHealth wearable device signals interception, proved to be instable due to the regular mandatory operational system upgrade.

A separate load test was performed regarding the automatic cloud architecture scaling. For this validation, exponentially increasing number of parallel input was delivered to the dedicated cloud system. This test was successful as the virtual cloud infrastructure scaled up automatically to process the significantly increased workload. The load test was started with five compute-optimized virtual machines. These virtual machines were predefined with the following parameters: 16 virtual CPU cores, 32 GB allocated RAM and 256 GB allocated disk space. These tests simulated up to 100 000 concurrent wearable eHealth device data flows and up to 10 000 simultaneous simplified medical information system data flows. During these load tests the virtual wearable eHealth devices sent the test measurement values to the OTI-Hub, which processed, transformed and transmitted the captured measurement values into the simulated simplified medical information systems. The load test was successful, as the system successfully transmitted the previously specified number of transactions. A daily total 8 500 000 000 simulated heart rate transaction volume was processed without error during the load test.

14.10   Threats to Validity

The research applied selected medical devices and instruments. As different manufacturers provide different data formats, the OTI-Hub should be enhanced from time to time, instead of offering a generally applicable single solution. The other weak points are the international healthcare standards. As several overlapping healthcare data exchange standards coexist (HL7 v2, HL7 v3, GDT, etc.), the OTI-Hub's output module should also be enhanced as these standards evolve.

14.11   Research Impact

The spirometry (Fig. 14.10) and cardio sensory HL7 test result data transferred over OTI-Hub has been successfully imported, interpreted and presented in the target HIS. As expected, spirometry and cardio test results were correctly reflected through the HIS' patient result query. The cardio information in the HIS query shows the values transformed by OTI-Hub. The results show that HL7-based health data interchange among different Information Technology architectures is completely feasible. The OTI-Hub successfully provides seamless interoperability among Android-, UNIX- and Windows-platforms. Both cloud and standalone architecture components were effectively interconnected during the research.

The OTI-Hub patient-data-related components run on Private Cloud Infrastructure-as-a-Service environment. The other OTI-Hub components run on public cloud Platform-as-a-Service environment. Consequently, the OTI-Hub is embedded in hybrid-cloud.

The HL7-based information exchange among the healthcare information acceptance system, the dedicated standalone spirometry system (Fig. 14.11) and the OTI-Hub was stable and trackable (Fig. 14.12). The combination and reconciliation of the different HL7 versions are mostly handled successfully within the OTI-Hub logic. The conversion of the body-sensory smart device output data stream into meaningful HL7 interface information was a significant challenge. While the spirometry output interface file was emitted and processed by the OTI-Hub correctly, healthcare smart-device manufacturers unfortunately do not provide well-specified output format in most of the cases. These smart device output formats are typically readable for humans, but not exact enough for automatic processing [17].

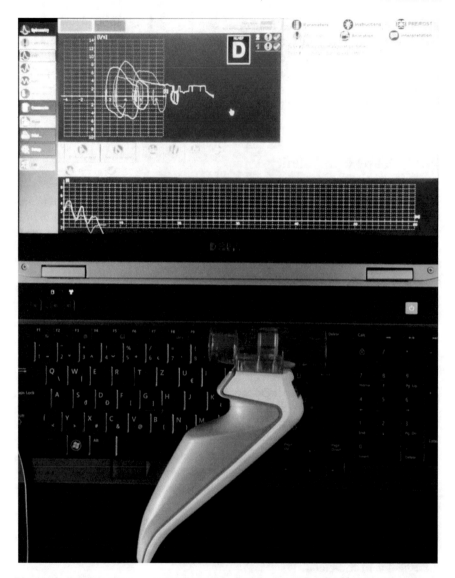

Fig. 14.10 Data interchange over OTI-Hub with clinical spirometer PDD-301/shm

Also, the healthcare smart device output data streams are not ready to be processed by HIS unless they are transformed into meaningful static values. However, in this unique case our OTI-Hub successfully imported and interpolated the randomly chosen healthcare smart devices (smart bracelet) data stream into interpretable HL7 values. Of course, this result calls for further examinations. Since healthcare smart devices of different origin in manufacture use diverse output data-format and data

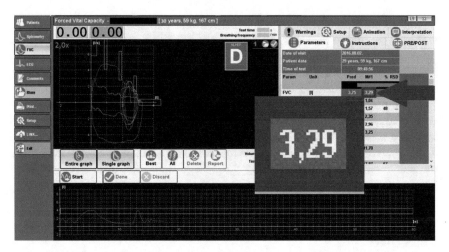

Fig. 14.11 Spirometry clinical test result data-export to OTI-Hub in HL7 v2.3.1 format

```
MSH|^~\&|PXP||||20160802095510|||ORU^R01|20160802095510|P|2.3.1|||NE|AL|HUN|
PID||65488965|18|||Patient^Anonymised^^^^Mr.||19870513|F||||||||||||||||||||
OBR|1||20160802094856^PXP|94011|||20160802094856|||||||||||||||2016080209485
OBX|1|ST|0^FVC^99MKW||3,29|1|3,75||||F
OBX|2|ST|2^FEV*0.5^99MKW||1,04|1||||||F
OBX|3|ST|3^FEV*1.0^99MKW||1,57|1|3,27||||F
OBX|4|ST|5^FEV*0.5/FVC^99MKW||31,70|%||||F
```

Fig. 14.12 Clinical spirometer HL7 v2.3.1 output interface file section through OTI-Hub

emission frequency [23], there is room for overall standardization reaching a single format in the field. Research results so far have met the requirements, however, significant lessons to learn are generated as described next.

14.12 Estimation in Brief

The OTI-Hub optimized and established serves as a methodological basis and software solution in interconnecting the world of IoT and classic healthcare technology, based on standards. We successfully demonstrated and tested the interoperability built on our hybrid cloud-based architectural solution with the OTI-Hub. Its important to understand that healthcare smart device technology generates unprecedentedly huge amount of body-sensory data made available in the foreseeable future. This bulk information serves inter are as important input for epidemic control and set new targets for pharmaceutical development. Big Data analytics methodologies foster pattern and trend analysis based on the captured body-sensory healthcare information base opening a new way for crowd-sourced information handling.

Our proposed hybrid cloud architecture assures the essential scalability for the OTI-Hub in order to bear with the necessarily robust transaction processing capacity. The illustrated architectural topology and systems integration provides technological solution for the integration of bi-directional international body-sensory, telemedicine and classical healthcare data exchange. The illustrated results offer some optimism, however, current national healthcare data-related legal prerequisites need international harmonization to reach the required break-through. The illustrated OTI-Hub solution provides international eHealth data-exchange. Furthermore, the demonstrated cloud-based eHealth interoperability solution conveys the framework for additional application areas, like cloud-based implementation of the da Vinci Surgical System.

References

1. Ackerman MJ, Filart R, Burgess LP, Lee I, Poropatich RK (2010) Developing next-generation telehealth tools and technologies: patients, systems, and data perspectives. Telemed J E Health 16:9395. https://doi.org/10.1089/tmj.2009.0153
2. Adamko A, Arato M, Fazekas G, Juhasz (2007) Performance evaluation of large-scale data processing systems. In: Proceedings of the 7th international conference on applied informatics, Eger, Hungary, 28–31 January 2007, vol 1, pp 295–301
3. Adamko A, Garai A, Pentek I (2016) Common open telemedicine hub and infrastructure with interface recommendation. In: 11th IEEE international symposium on applied computational intelligence and informatics, Timisoara, Romania, 12–14 May 2016, pp 385–390
4. Adamko A, Kollár L (2008) MDA-based development of data-driven Web applications. In: Proceedings of the fourth international conference on web information systems and technologies, Funchal, Madeira, Portugal, 4–7 May 2008, vol 1, pp 252–255
5. Adenuga OA, Kekwaletswe RM, Coleman A (2015) eHealth integration and interoperability issues: towards a solution through enterprise architecture. Health Inf Sci Syst 3:1–8
6. Baranyi P, Csapo A (2012) Definition and synergies of cognitive infocommunications. Acta Polytechnica Hungarica 9(1):67–83
7. Baranyi P, Csapo A, Sallai Gy (2015) Cognitive infocommunications (CogInfoCom). Sprigen International Publishing
8. Bouamrane MM, Tao C, Sarkar IN (2015) Managing interoperability and complexity in health systems. Methods Inf Med 54:1–4
9. COLOMB, Robert et al (2006) The object management group ontology definition metamodel. In: Ontologies for software engineering and software technology. Springer, Berlin, Heidelberg, pp 217–247
10. Dinesen B, Nonnecke B, Lindeman D, Toft E, Kidholm K, Jethwani K, Young HM, Spindler H, Oestergaard CU, Southard JA, Gutierrez M, Anderson N, Albert NM, Han JJ, Nesbitt T (2016) Personalized telehealth in the future: a global research agenda. J Med Internet Res 18:e53. https://doi.org/10.2196/jmir.5257
11. Dolin RH, Alschuler L, Boyer S, Beebe C, Behlen FM, Biron PV (2006) HL7 clinical document architecture, release 2. J Am Med Inf Assoc 13(1):30–39
12. Donnelly K (2006) SNOMED-CT: the advanced terminology and coding system for eHealth. Stud Health Technol Inform 121:279
13. Eren H, Webster JG (2015) The E-medicine, E-health, M-health, telemedicine, and telehealth handbook. Two volume set. CRC Press
14. Fong B, Fong ACM, Li CK (2011) Telemedicine technologies: information technologies in medicine and telehealth. Wiley

15. Garai A (2010) Methodology for assessment validation of platform migration of robust critical IT-systems. In: 8th international conference on applied informatics, Eger, Hungary, 27–30 January 2010, pp 445–448
16. Garai A, Pentek I (2015) Adaptive services with cloud architecture for telemedicine. In: 6th IEEE conference on cognitive infocommunications, Gyor, Hungary, 19–21 October, 2015, pp 369–374
17. Giachetta R, Fekete I (2015) A case study of advancing remote sensing image analysis. Acta Cybernetica, Szeged 22:57–79
18. Health Level Seven Standard Version 2.3.1. An application protocol for electronic data exchange in healthcare environments, Health Level Seven International. http://www.hl7.org/. Accessed 3 March 2017
19. Huang Y, Kammerdiner A (2013) Reduction of service time variation in patient visit groups using decision tree method for an effective scheduling. Int J Healthcare Technol Manag 14(1/2):3–21
20. International Health Terminology Standards Development Organisation, Systematized nomenclature of medicine clinical terms, SNOMED-CT. http://www.ihtsdo.org/snomed-ct. Accessed 3 March 2017
21. ISO/IEC 7498-1:1994 Information Technology (2016) Open systems interconnection—basic reference model: the basic model, OSI-model, International Organization for Standardization, ISO. http://www.iso.org/iso/catalogue_detail.htm?csnumber=20269. Accessed 3 March 2016
22. ISO/HL7 10781:2015, HL7 electronic health records-system functional model, Release 2, HER FM. http://www.iso.org/iso/iso_catalogue/catalogue_tc/catalogue_detail.htm?csnumber=57757. Accessed 3 March 2017
23. Kartsakli E, Antonopoulos A, Alonso L, Verikoukis C (2014) A cloud-assisted random linear coding medium access control protocol for healthcare applications sensors. Spec Iss Sens Data Fus Healthcare 9628–9668
24. Kota L, Jarmai K (2014) Efficient algorithms for optimization of objects and systems. Pollack Periodica 9:121–132
25. Mahmud K, Lenz J (1995) The personal telemedicine system. A new tool for the delivery of health care. J Telemed Telecare 1:173–177
26. Mantas J (2001) Electronic health record. Stud Health Technol Inform 65:250–257
27. Martinez L, Gomez C (2008) Telemedicine in the 21st century. Nova Science Pub Inc
28. Neelakantan P, Reddy ARM (2014) Decentralized load balancing in distributed systems. Pollack Periodica 9:15–28
29. Otte R, Patrick P, Roy M (1996) Understanding CORBA: common object request broker architecture. Prentice Hall PTR
30. Pandi K, Charaf H (2015) Mobile resource management load balancing strategy. Acta Cybernetica, Szeged 22:171–181
31. Rossi RJ (2010) Applied biostatistics for the health sciences. Wiley
32. Sallai Gy (2013) Chapters of future Internet research. In: IEEE 4th international conference on cognitive infocommunications (CogInfoCom), pp 161–166
33. SNOMED C. User Guide January 2007 release. http://www.ihtsdo.org/fileadmin/user_upload/Docs_01Technical_Docs/snomed_ct_user_guidepdf
34. Szabo R, Farkas K, Ispany M, Benczur A, Kollar L, Adamko A (2013) Framework for smart city applications based on participatory sensing. In: IEEE 4th international conference on cognitive infocommunications (CogInfoCom), pp 295–300
35. Toman H, Kovacs L, Jonas A, Hajdu L, Hajdu A (2012) Generalized weighted majority voting with an application to algorithms having spatial output. In: 7th international conference on hybrid artificial intelligence systems, HAIS, Lecture notes in computer science, vol 7209, pp 56–67. https://doi.org/10.1007/978-3-642-28931-6-6
36. Varshney U (2007) Pervasive healthcare and wireless health monitoring. Mob Netw Appl 12:2–3
37. Vigneshvar S, Sudhakumari CC, Senthilkumaran B, Prakash H (2016) Recent advances in biosensor technology for potential applications an overview front. Bioeng Biotechnol 4:11. https://doi.org/10.3389/fbioe.2016.00011

38. Wang YM (2006) U.S. Patent No. 7,082,553. U.S. Patent and Trademark Office, Washington, DC
39. Warren S, Craft RL, Parks RC, Gallagher LK, Garcia RJ, Funkhouser DR (1999) A proposed information architecture for telehealth system interoperability. In: Conference: Toward an electronic patient record, Orlando, USA, 2–6 May, pp 1–10
40. Wootton R (2012) Twenty years of telemedicine in chronic disease management an evidence synthesis. J Telemed Telecare 18:211220. https://doi.org/10.1258/jtt.2012.120219
41. Wootton R (1998) Telemedicine in the national health service. J R Soc Med 91:289–292
42. World Health Organization (1993) The ICD-10 classification of mental and behavioural disorders: diagnostic criteria for research, vol 2. World Health Organization
43. Zarour K (2016) Proposed technical architectural framework supporting heterogeneous applications in a hospital. Int J Electron Healthc 9:19–41

Chapter 15
Pilot Application of Eye-Tracking to Analyze a Computer Exam Test

Tibor Ujbanyi, Gergely Sziladi, Jozsef Katona and Attila Kovari

Abstract From human aspect, several cognitive factors can be determined by different measuring methods. The human visual attention and some hidden cognitive processes may be revealed and examined by Eye-tracking. Using the parameters of eye-tracking, human activity, the visual attention can be observed, so even emotional condition of the human can be concluded. Thus, a system based on eye-tracking can be used for studying cognitive processes like learning, problem solving also. In this article the results of an eye-tracking analysis is introduced. The eye-tracking data was registered during a test, problem solving related to an IT-problem. The results shows, that difference was observed in the analysis of eye-tracking, depending on the prior IT knowledge related to the problem. The results of eye-tracking may provide useful information for the effectiveness of problem solving related to the prior knowledge.

15.1 Introduction

The science field which deals with human-computer interactions, includes design of interactive computer systems and studies main relevant phenomena. These tools play the major role in controlling of computer systems. The control of these systems traditionally can be achieved by standard input peripherals, such as the keyboard and mouse. Thanks to the technical development, several controller alternatives were developed. Using eye-tracking measurement, several parameters can be examined and processed to achieve Human-Computer Interactions. The relevant data calcu-

T. Ujbanyi (✉) · G. Sziladi · J. Katona · A. Kovari
University of Dunaujvaros, Tancsics M. u. 1/A, Dunaujvaros 2401, Hungary
e-mail: ujbanyit@uniduna.hu

G. Sziladi
e-mail: sziladig@uniduna.hu

J. Katona
e-mail: katonaj@uniduna.hu

A. Kovari
e-mail: kovari@uniduna.hu

© Springer International Publishing AG, part of Springer Nature 2019
R. Klempous et al. (eds.), *Cognitive Infocommunications, Theory and Applications*,
Topics in Intelligent Engineering and Informatics 13,
https://doi.org/10.1007/978-3-319-95996-2_15

lated from the gaze of a human is much more informative, than the motion of the mouse or pressing a button on the keyboard. In human-computer connection, human behaviour and interactions between the two parties plays a major role. These interactions can be examined in several forms, whether we are talking about brain activity, gestures or eye motions. With modern engineering tools, we can examine these factors and the results can be used for further research. The motivation of this research based on the observation of cognitive processes related to learning, which can be analysed and processed by engineering tools. Measurements are performed without the research methods common in pedagogy, but by the support of a target hardware and software based on eye tracking and manual gesture control. This way, we can get more accurate image about the correlation of eye- and hand moving processes, and we may reveal connections between them. By tracking human eye, visual attention can be defined, thus some hidden cognitive processes can be revealed and examined. Basics of eye-tracking researches are defined by observations. Examining the eye motion, geometric location of visual focal point can be defined during watching an object. By measuring eye motion, observed areas can be defined, as well as which details stay focused on a bit longer. Based on these parameters, we can examine the correlation of visual stimuli with the performance of the test, based on the level of previous knowledge, which cannot be analysed with other methods. Processes introduced above, are influenced by visual information collected by the human eye, and connections between the cognitive process and eye-movement have a great importance. Important element of cognitive processes is human learning, where the most important information sources are the visually observed and detected data. In the observation of learning process, eye-tracking systems give such new opportunities, which enable much wider range of examinations for current and future researches, compared to already available methods.

15.1.1 *Role of Eye Motion in the Observation of Cognitive Processes*

As we have already referred, the human behaviour and interaction play a major role in human-computer relation. This can be examined in different ways, whether through brain activity [1, 2], gesture control [3], or even eye-tracking [4]. Human-computer interaction itself deals with such a science field, which covers design of interactive computers made for humans, and studies main relevant phenomena [5]. Importance of eye-tracking examination is that computer users receive most information via visual stimuli like information channels [6]. Examination of human eye motion is also useful to define the visual attention, and give the opportunity to observe the hidden cognitive processes.

15.2 Parameters of Eye-Tracking, Methods and Tools

The goal of eye-tracking is to define the geometric location of users visual focal point, so it shows, what and how long he watches, observes his environment. By measuring eye motion, we can define what pattern, path does the look follow, and on what details does it stay for longer time with a specific frequency. These data may provide detailed information, how the users perform a task or an activity by using visual stimuli, like searching, reading, viewing images, walking, or driving, which could be less observed by other methods.

15.2.1 Eye-Tracking

In photography, it can be observed in many cases, that some photographers take much better pictures with the same devices and equipment than others. What could be the reason of that? The technology is the same, but the quality of the images is different. Then where should we search the reason of difference? Dealing with the question more thoroughly, process of image taking shall be examined. Before the photographer takes a picture, he looks around, what region of the field of view should he capture. When the location of this zone is ready, he holds the camera in front of him, and takes the picture after directs the focus on the selected area. So, the key of the solution is the human factor, in this case, the photographer himself. When looking for the reason of difference, we have to examine, on what points are the photographers attention aimed at, when taking pictures. The so-called eye-tracking deals with this phenomenon, which is the tracking of eye motion itself.

15.2.2 Motoric Operation of Eye Motion, as a Vision System

The positioning of gazes' direction and focus is performed by human eye muscles, then the image of the observation shall be projected on the location of sharp view, the fovea centralis. Motion of eyeballs may happen consciously by directing them on the selected area of interest, then keeping in the field of view is performed primary by spontaneous eye motions. If we fix a motionless object, then eye performs spontaneous eye motions. This motion may be an approximately 3 kHz frequency tremor, slow floating motion, when the image of fixed object slowly passes in front of the fovea, or quick reflex motion, after the image of the examined object is projected on the fovea again. When the gaze follows a fast-moving object, during reading, recurring observation of image details, the eye fixes the image again and again. Between fixation points, the eye performs quick, spontaneous eye motions, the so-called saccades [7].

Fig. 15.1 Spectacleframe recorded oculography based on infrared light reflection

15.2.3 Methods of Eye-Tracking

Several methods have been worked out to track the human eye and define the location of visual focal point. The electro-oculography (EOG) is based on that the eyeball behaves like an electric dipole: its positive pole is situated on the surface of the cornea, while its negative pole behind the retina [7–9].

Eye location can be examined by applying infrared light beam. Infrared light is advantageous from many aspect, because it is invisible, thus not disturbing. Detecting light reflected from the eye, the position can be revealed [10] (Fig. 15.1).

Nowadays image process based eye-tracking solutions are getting more-and-more widespread. We may choose a target hardware suitable for eye-tracking from several manufacturer, where such tools can be found, like the Tobii, the Eye Tribe, Mirametrics, GazePoint or Smart Eye. The basics of camera systems is the definition of pupils position through image process algorithms. Camera may be placed a bit farther from the eye, for instance fixed to the computers monitor, or close to the eye on an eyeglass [11] (Fig. 15.2).

15.2.4 Eye Tribe Tracker

In this research the Eye Tribe tracker (Fig. 15.3) has been applied. The Eye Tribe tracker is a camera based, portable eye-tracker tool, which enables the observation

Fig. 15.2   Tobii EyeX and Tobii Pro eye tracking systems

Fig. 15.3   The Eye Tribe tracker

of the direction of the eye, and measurements may be performed regarding the eye-tracking parameters. A camera see the users eye motion, even the tiniest motions on the pupil. The device can calculate coordinates, which become visible thanks to a utility software. Camera sensors of the tool work in different light conditions as well, further enhancing users experience. The device works in 30 and 60 Hz sample rate operational mode, with an average spatial accuracy of 0.5°. The Eye Tribe tracker, contrary to its cost-efficiency, can be well-used in psychological researches [12, 13].

15.2.5   Eye-Tracking Parameters

As mentioned before, reception of different visual stimuli, and their process by the brain become available. In the control of eye motion, complex nerve background

Fig. 15.4 Fixations and
Saccades

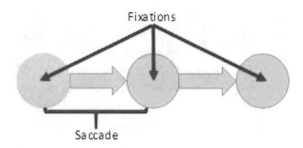

cooperation is required. Motoric operation of vision system is achieved by eye muscles. The eye movement can be divided into two types: fixation and saccadic motion
[13] (Fig. 15.4).

Fixation: It represents focus on a specific point, where the gaze spends more time.
Average fixation length is between 100 and 1000 ms, in most cases this period is
between 200 and 500 ms. The period greatly depends on the quantity of information
to be processed, and on the current cognitive load. Fixation designate that condition
of the eye, when eyeball stands still, and input or decoding of information is in
progress [4, 13].

Saccade: Also known as regression, designates the short, quick and leaping motion
between two fixations. Average length of a saccade is between 20 and 35 ms. It
designates the next looking position for the eye. If a motion above 30° is performed,
then the head moves together with the eye. Index numbers may be derived from
saccades as well the length of gaze [13].

Length of gaze: During the length of look, the length of all fixations and saccades
are understood. It may be also used for a kind of forecast: when look spends a long
time on a particular point, a relevant event caused by it is likely to emerge [13].

15.2.6 Eye-Tracking Analysis

The goal of eye-tracking analysis is to evaluate, how the parameters specific for
eye motion change, while the observed test subject performs a particular activity.
Continuous observation of eye motion is available by eye-tracking systems, but process of data provided by these systems and definition of parameters describing eye
motion require further analysis and evaluation. In order to understand and display
data related to eye motion, a software or software bundle performing such task is
required. For devices listed above, there are analysing software solutions available,
either based on open source, or created by the manufacturer. The free, open source
OGAMA software is being briefly introduced below.

15.2.7 Eye Motion Analysis by OGAMA Software

Manufacturers, producing eye-tracking systems, developed such target software based on the analysis of eye motion, but these software are usually very expensive. However, a freeware, open source software, the OGAMA (Open Gaze And Mouse Analyser) is available, which enables recording, process and analysis of eye motion data. The program can cooperate with most eye-tracking systems, whether we are talking about a target hardware or a web camera. Its main properties are database controlled pre-processing, as well as filtering eye motion. Measured data may be exported in text, or depending on the module, in video format. Data may be displayed and evaluated in several formats, the program contains more (particularly 10) modules accordingly [14].

15.3 Role of Eye-Tracking in the Observation of Cognitive Processes

In case of a research defined in the introduction, aiming to examine students cognitive processes based on eye tracking, knowing the correlation of eye motion and cognitive processes is crucial. During solving IT problems, for example in case of system and network engineering, or software development field, very complex tasks need to be solved. Virtually in these tasks, high-level preliminary knowledge, specific way of thinking of design, description and overview of complex systems, and solution of problems in a logical, sequential way is required. Students must be prepared for solving these kind of tasks, which, beside learning the theoretical principles, can be achieved most efficiently through practical tasks solution. Education of IT-specific subjects has been analysed by different professional fields [15, 16], in the professional literature regarding several software developer related [17–24], and computer network related [25–33] competences. Analysis of eye motion helps to understand the operation of cognitive processes. Thanks to an eye-tracking system, by monitoring eye motion data, human attention can be examined [34, 35]. Eye motion is one of the main elements of cognitive processes, while we can percept visual stimuli by our eyes, which processed by our brain. So visual attention launches, (activates) cognitive processes, which are inevitable during performing a task [36]. Application of systems based on eye-tracking helps the researches to understand and examine the cognitive processes more efficient.

15.3.1 Eye-Tracking in the Observation of Learning Processes

Using the parameters of eye-tracking, human activity, the visual attention can be observed, so even emotional condition of the human can be concluded. Therefore an eye-tracking system can be used to examine the effectiveness of learning processes as well. Lately more have been performing such researches [13]. In the e5learning project Calvi and others studied users behaviour during processing different e-learning contents [37]. Eye-tracking is used in several areas, like studies related to learning. In ADELE project Pivec developed a monitoring framework, whose purpose was to track students gaze during dynamic process of e-learning contents [38]. In iDict project Hyskykari evaluated students needs by observing the look [39]. Similar examination was introduced by Wei and his team for virtual learning environments [40]. A software agent was developed by Wang, which maintained students attention by utilizing emotional reactions [41]. Porta also developed a system, which helped to observe changes of students' emotional state changes based on eye tracking parameters [42]. Porta has taken students eyetracking data into consideration in an experiment aiming to examine learning experience, where examination of learning experience was achieved by an adaptive system [43]. Drewes introduced a system which helps to track in what sequence and what extent do users read text contents in the web browser. The study revealed that eyetracking data may be used to enhance learning efficiency however, authors also note that currently applied adaptive systems are still limited in their features because only a small part of learning conditions can be detected by them [44]. The relation between text and image contents in a multimedia educational material was examined by Schmidt-Weigand. The study show that image contents defined students visual attention, when text contents was replaced by voice [45]. The sudy of Al Wabil revealed that students with different learning styles (visual/verbal) show different visual behaviour during learning. Have been lerned from the studies that verbal students focused mainly on text content while visual typed noticed only some parts of the text (spending a short period of time on the selected parts) and focused rather on multimedia contents [46]. Similar study was performed by Tsianos too, who compared eye-tracking data of students with different learning styles in hypermedia based learning environment [47].

15.4   Examination of Eye-Tracking, Regarding the Solution of an IT Task

The goal of the research aimed to decide, whether there were differences between parameters describing eye-tracking parameters between groups with different prior knowledges when solving an IT task. Main parameters describing the eye-tracking aimed the observed area were evaluated upon eye-tracking parameters, then their comparison were performed in case of testing subjects with different prior knowledges. Conditions, tools of the research, and the received results are introduced below.

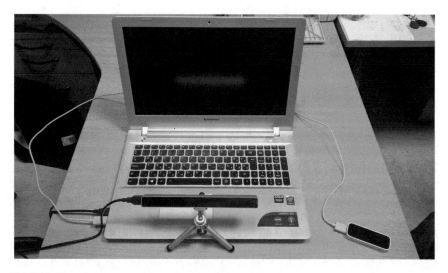

Fig. 15.5 Test environment with eye tracking and gesture control

15.4.1   Test Environment for Eye-Tracking

The examination defined as the goal of the research, if there were differences in parameters describing the eye movement trajectory and eye movement parameters during an IT test in case of groups with different prior IT knowledge in the field. The previously introduced Eye Tribe tracker was used for eye-tracking purposes which connected via USB 3.0 port to a notebook. The notebook was equipped with Intel Core i5 CPU, 4 GB RAM, dedicated graphic card and 1 TB hard drive, which provided appropriate resources for running image processing method requiring more computing capabilities for eye-tracking (Fig. 15.5). The computer was running Windows 10 operation system, as well as the Eye Tribe Server service required for the use of eye-tracking system, and, and OGAMA software performing the evaluation of eye-tracking parameters. Some analysis modules of the software can be used to evaluate the results [14].

15.4.2   Testing Subjects

In the experiment, 32 graduated or prospective IT specialists took part, who had to solve a task related to computer networks. The group of participating testing subjects consisted of students from professional acquaintances, current or former students of the University of Dunaújváros, and students participating in OKJ (Hungarian National Training Register) advanced studies on a voluntary basis, who held themselves healthy. Participants mainly represented the younger generation, and there was a student from abroad. Gender distribution of testing subjects: 28 men and 4 women. Their ages varied between 20 and 30.

15.4.3 Examination of the Test

Prior to the start of the examination tests, testing subjects had to solve a test related to computer networks, which were then evaluated. The results received from the knowledge test, provided the extent of testing subjects prior knowledge. Testing subjects were categorized into three groups, according their answers on the test (max. 10 points): Lower Than Average (LTA, 0–3 points), Average (A, 4–6 points), Better Than Average (BTA, 7–10 points). Based on the test results, 13 persons were put into the group of Average knowledge, 11 into the Lower Than Average, and 8 into the Better Than Average group. Prior to the start of the first test, testing subjects had to sit down in front of the computer performing eye-tracking. Then, thanks to the OGAMA software, calibration was performed as a first step. After the calibration was successful the test was performed. During the test, a computer network related task appeared on the screen, which had to be selected from three options by moving the mouse pointer to the correct one.

15.4.4 Measuring Tool and Process

Recording, process and evaluation were achieved by the testing environment providing eye-tracking and gesture control, which were based on Eye Tribe tracker eye-tracker and Leap Motion gesture controller devices. The recorded data in the test environment was processed, evaluated and analyzed using OGAMA software using the Scanpath, Saliency and Fixations modules.

15.4.5 Steps of Performing Eye-Motion Analysis Test

Prior to the start of the examination using the Recording module of the program, a calibration process was performed in all cases, which helped to check the proper detecting capability of the eye-tracker system (Fig. 15.6).

Calibration process consists of four steps: first, the device needs to be connected by clicking on the Connect buttonThe Eye Tribe Server application also has to run in the background for the connection which monitoring the activity of the system and the connection of the computer via 6555 TCP port. Following the successful connection to the program, providing the testing subjects data is necessary. Data typed in here (name, category, gender, age) are important in a possible later statistic. Then Calibrate feature becomes available, so the 9-point calibration process can be

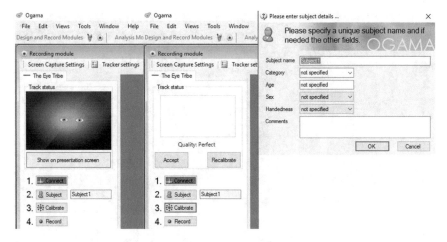

Fig. 15.6 Recording module of OGAMA

initiated. About the accuracy of the calibration process, track status preview screen may provide information. Depending on the accuracy, calibration quality may vary between Perfect Quality, Good Quality, Moderate Quality and Poor Quality. On the preview screen, it is also visible, that during the calibration, watching the whole screen, on which part was the process inaccurate, fullness and colour of calibration points (circles) vary depending on the quality: circles may be semi- or fully filled, colour of filling may be red, yellow or green. Following the performance of calibration, by clicking on the Record button, recording of eye-tracking data can be launched. The question appears on the computers monitor, on the slide displayed by the Recording module of OGAMA software, which were able to be defined in the Stimulus Design module. From the moment of the show-up of slide containing the question to be answered, eyetracking data recording starts until the testing subject selects the answer supposed to be the correct one. Post-process of data was performed following the tests.

15.4.6 Evaluation of Eye-Tracking Results

For data process and display in table and graphic form, built-in modules of OGAMA program can be applied. The program enables the evaluation of qualitative and quantitative results of performed eye-tracking based examinations. Analysation of qualitative and quantitative results of eye-tracking parameters recorded during the solution of computer network related task are introduced in the sub-chapters below.

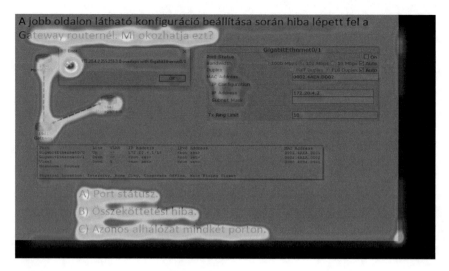

Fig. 15.7 Attention map of Gaze I (Hungarian)

15.4.7   Illustration of Test Results

Here the illustration of test results of eye-tracking are introduced: the relief, fixation and attention map of the observed area. In the saliency module of OGAMA software a salient image can be displayed about the especially observed areas. Different colours may vary depending on the value of salients during solving the task. The map helps to examine, which are those areas, whose observation differs from other parts of the image. On Figs. 15.7 and 15.8, salient map is shown about the observed areas. On the figure the question possible answers and observation of computer network topology and network parameters are well-visible.

Replay module of the program helps to track fixations and connections between them, this way the way of look becomes displayable. This provides information, what way do each subjects look follow, while the task is being read and interpreted. On Figs. 15.9 and 15.10 the way of look is shown during the task, circles refer to fixations, lines between them refer to saccades, which together define the motion of the gaze.

Thanks to the Attention map module, an attention map similar to heat map can be displayed, where areas more or less observed by the subjects can be separated. Figure 15.11 shows that the answer (b), the network topology, and the network parameters of each computers were more observed. Figure 15.12 shows that the answers, the network topology, and the network parameters of each computers were more observed.

Maps introduced above, show each features describing eyetracking is a useful, spectacular way, making the way of gaze and otherwise observed areas identifiable.

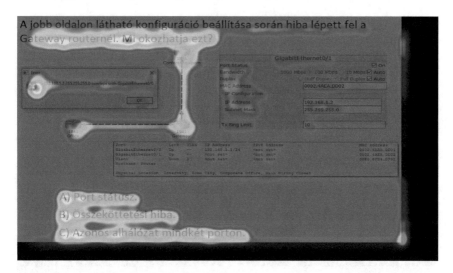

Fig. 15.8 Attention map of Gaze II (Hungarian)

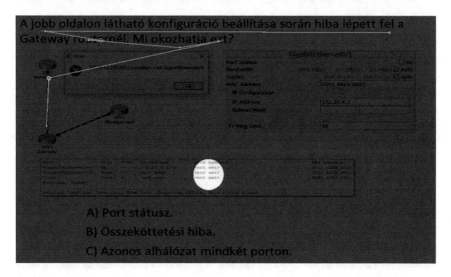

Fig. 15.9 Gaze path, fixations and connections made by one of the users when reading task 1

15.4.8 Quantitative Test Results

Using the statistic module of OGAMA software, some main parameters of eye motion have been defined, whose results are shown in Table.15.1.

As it is shown in the table, the eye-tracking parameters of the three groups showed difference. Data in the table calculated from the results of each groups members. In the LTA group, number of fixations are bigger (296), than in case of A (252) and

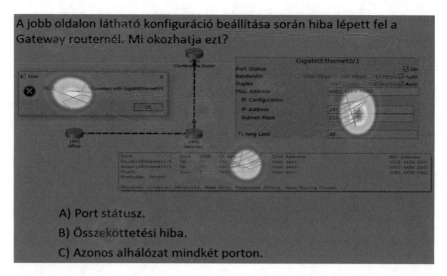

Fig. 15.10 Gaze path, fixations and connections made by one of the users when reading task 2

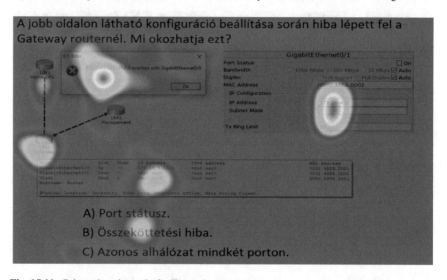

Fig. 15.11 Orientation channel of saliency 1

BTA (218) groups. In contrary, at the mean value of fixation lengths, reverse proportionality is experienced: members of BTA group has more lengths of fixations (303), than A (288) and LTA (283) groups. The number of fixations measured during the particular time period also higher and the proportion of fixations/saccades similarly. In case of the LTA group, the fixations count 1.92 and fixation/saccade ratio 543 are the lowest, while in A group, they are 2.21 and 636, while in BTA group, the values are 2.43 and 736. Proportion of average saccade lengths are similar to those, which

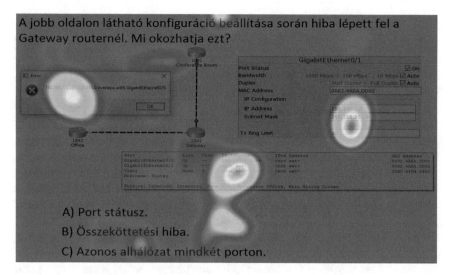

Fig. 15.12   Orientation channel of saliency 2

Table 15.1   Eye metrics result of exam test

Groups	Fixations (count)	Fixations (count/s)	Fixation duration mean (ms)	Fixation/Saccade ratio	Avg. Saccade length (px)
LTA	296	1.92	283	543	228
A	252	2.21	288	636	219
BTA	218	2.43	303	736	198

LTA Lower than average, *A* Average, *BTA* Better than average

were experienced in case of fixations. Average of LTA group is 228, whilst it is 219 in group A, and 198 in group BTA.

The results show that during performing the task, in case of students with better backgrounds, solution of the task was performed in a better organized way, than in case of students with less knowledge. In case of students with less knowledge, watching of each network items in a repeatedly recurring, way in a not well-thought sequence refers to their uncertainty of the solution which might be due to their lacking prior knowledge. In case of students with more knowledge, a well-organized checking of network items to systematically assemble a network system, and parameters was observed in the way of their look, in the location of more observed areas. As a summary the results shows, that during solving the IT test, in the parameters of eye-tracking, difference was observed depending on the prior knowledge.

15.5 Summary

Result of examination of eye-tracking introduced in the article, and the results showed, that among student groups, with different prior knowledge, differences were detected in parameters describing eye motion during solving a test related to an IT-problem. Results got regarding eyetracking give the opportunity in the future to determine, what correlation may be noticed between the success or the lack of success of solving a test with the subjects relevant prior knowledge or learning styles. The results may provide useful information in checking learning process in its examination, so these tools and methods may become organic parts of futures educational infrastructure. In our research we attempted to act thoroughly, but the experiment mentioned above shall be performed again involving more participants, in order to get an even more complete image of the revealed correlations. The fact, that not every IT specialist have the same knowledge in this field of informatics, makes the performance of the examination more complicated, so in the future it is worth to make this test done by other types of tasks too. Calibration process was performed prior to the examination of each testing subjects, and we attempted to provide optimal environment for the calibration process, however, calibration quality is not the same in all cases. Calibration quality might be influenced by several factors, such as light conditions, testing subjects eye motion coordination, posture or seat position. Profitability of the research can be further enhanced by involving other age groups. This research was based on the age group between 20 and 30, in the future, by widening age groups, we may get even more representative data to be evaluated later. Further examinations may be performed to figure out, that during solving practical tasks (programming, electric measurement etc.), in what extent do arm or eye motion parameters differ in case of subjects with more or less prior knowledge; as well as that during the performance of different type practical tasks (primary practical tasks, which can be performed at the same speed), in what extent do observed parameters differ. The results may provide useful information for controlling and examination of learning process, so these devices and methods may become an organic part of futures educational infrastructure. In the future a possible goal could be the creation of such a device system, which contains more measuring tools (suitable for observing cognitive processes), such as a brain wave and eye-tracking interface. This way a complex human-computer interface could be established, which helps through the observation of students cognitive processes, we may conclude students learning capabilities. Educators could utilize these features at the selection of material and formal items of the educational material when compiling it, and during the application of knowledge transfer methods. Educators, during educational process, based on real-time feedback, may receive information about students attention level, direction, and even to determine, how they are able to follow the educational material. It provides the option that through indirect observation of eye motion, at the determination of the forms and method of exams, these results offer new opportunities. Although in this thesis application on the field of pedagogy has been introduced, there are several other possibilities available in human-computer interfaces.

Acknowledgements The project is sponsored by EFOP-3.6.1-16-2016-00003 founds, Consolidate long-term R and D and I processes at the University of Dunaujvaros.

References

1. Katona J (2014) Examination and comparison of the EEG based attention test with CPT and T.O.V.A. In: Proceedings of 15th IEEE international symposium on computational intelligence and informatics. IEEE, pp 117–120
2. Katona J (2015) The examination of the application possibilities of brain wave-based control. DUF Press, pp 167–176
3. Sziládi G Kvári A (2016) Gesztusvezérlésen alapuló ember-számítógép interfészek. Dunakavics, pp 63–73
4. Hari S, Jaswinder S (2012) Human eye tracking and related issues: a review. Int J Sci Res Publ 2(9):1–9
5. Strong G, Hewett TT, Baecker R, Card S, Carey T, Verplank W (1992) ACM SIGCHI curricula for human-computer interaction. ACM, New York, USA
6. Bárdos G, Dètári L, Hajnik T, Kiss J, Schlett K, Tárnok K, Világi I (2012) Élettani gyakorlatok. ELTE, Budapest
7. Yarbus AL (1967) Eye movements and vision. Academy of Sciences of the USSR, New York
8. Chronos Vison (2016) Scleral Search Coils 2D/3D. CHRONOS VISION GmbH, Berlin, pp 1–4
9. Metrovision (2016) Electro-oculography by Metrovision. Retrieved from Electro-oculography by Metrovision: http://www.metrovision.fr/mv-eo-notice-us.html
10. Andrew Murray WJ, Robert T, Kate C, Natalie CM (2007) Monitoring eye and eyelid movements by infrared reflectance oculography to measure drowsiness in drivers. Somnologie 11:234–242
11. Tobii AB website (2016) Tobii. http://www.tobii.com/
12. Dalmaijer ES (2014) Is the low-cost EyeTribe eye tracker any good for research? PeerJ PrePrints 2:e585v1:1–35
13. Ujbanyi T, Katona J, Sziladi G, Kovari A (2016) Eye-tracking analysis of computer networks exam question besides different skilled groups. In: Proceedings of 7th IEEE conference on cognitive infocommunications, pp 277–281
14. Vosskühler A (2016) Description of the main features of Ogama. http://ogama.net/node/1
15. Hadi A, Ping JS, Hantao X, Emmanuel HE, Bart KJ, Guy A, Dominique EM (2013) Implementation of a project-based telecommunications engineering design course. IEEE Trans Educ 57(1):25–33
16. Kori K, Pedaste M, Altin H, Tõõnisson E, Palts T (2016) Factors that influence students' motivation to start and to continue studying information technology in Estonia. IEEE Trans Educ 59(4):255–262
17. Cadenas JO, Sherratt RS, Howlett D, Guy CG, Lundqvist KO (2015) Virtualization for cost-effective teaching of assembly language programming. IEEE Trans Educ 58(4):282–288
18. Chrysafiadi K, Virvou M (2013) Dynamically personalized e-training in computer programming and the language C. IEEE Trans Educ 56(4):385–392
19. Guo Y, Zhang S, Ritter A, Man H (2013) A case study on a capsule robot in the gastrointestinal tract to teach robot programming and navigation. IEEE Trans Educ 57(2):112–121
20. Jorge A, Nadya A, Camilo H (2009) Developing programming skills by using interactive learning objects. In: Proceedings of the 14th annual ACM SIGCSE conference on innovation and technology in computer science education. ACM, New York, USA, pp 151–155
21. Mok HN (2011) Student usage patterns and perceptions for differentiated lab exercises in an undergraduate programming course. IEEE Trans Educ 55(2):213–217

22. Perkins D, Hancock C, Hobbs R, Martin F, Simmons R (1986) Conditions of learning in novice programmers. J Educ Comput Res 2(1):37–55
23. Putnam RT (1986) A summary of misconceptions of high school basic programmers. J Educ Comput Res 2(4):459–472
24. Silva-Maceda G, Arjona-Villicaña PD (2016) More time or better tools? A large-scale retrospective comparison of pedagogical approaches to teach programming. IEEE Trans Educ 99:1–8
25. Bodnarova A, Olsevicova K, Sobeslav V (2011) Collaborative resource sharing for computer networks education using learning objects. IEEE, pp 25–28
26. Dobrilovic D, Jevtic V, Stojanov Z, Odadzic B (2012) Usability of virtual network laboratory in engineering education and computer network course. IEEE, Villach, pp 1–6
27. Lu H-K, Lin P-C (2012) Effects of interactivity on students' intention to use simulation-based learning tool in computer networking education. In: IEEE 14th international conference on advanced communication technology (ICACT), pp 573–576
28. Ma X-M (2010) Probe of bilingual education in computer network course for undergraduates. In: 2nd international conference on education technology and computer (ICETC). IEEE, pp 22–24
29. Nunes FB, Stieler S, Voss GB, Medina RD (2013) Virtual worlds and education: a case of study in the teaching of computer networks using the Sloodle. In: 2013 XV symposium on virtual and augmented reality (SVR). IEEE, pp 248–251
30. Papadopoulos PM, Lagkas TD, Demetriadis SN (2012) How to implement a technology supported free-selection peer review protocol: design implications from two studies on computer network education. In: 2012 IEEE 12th international conference on advanced learning technologies (ICALT). IEEE, pp 364–366
31. Skoric MM (2013) Amateur radio communications, software and computer networks in education. In: 2013 IEEE international conference on signal processing, computing and control (ISPCC). IEEE, pp 1–8
32. Qun ZA, Jun W (2008) Application of NS2 in education of computer networks. In: CACTE '08. International conference on advanced computer theory and engineering. IEEE, pp 368–372
33. Yan C (2011) Bulid a laboratory cloud for computer network education. In: 2011 6th international conference on computer science and education (ICCSE). IEEE, pp 1013–1018
34. Duchowski A (2007) Eye tracking methodology: theory and practice. Springer London, Springer, New York Inc, London, p 334
35. Rayner K (1998) Eye movements in reading and information processing: 20 years of research. Psychol Bull 124(3):372–422
36. Just AM, Carpenter PA (1980) A theory of reading: from eye fixations to comprehension. Psychol Rev 87(4):329–354
37. Calvi C, Porta M, Sacchi D (2008) e5learning, an e-learning environment based on eye tracking. In: International conference on advanced learning technologies. IEEE, Spain, pp 1–5
38. Maja P, Trummer C, Pripfl J (2006) Eye-tracking adaptable e-learning and content authoring support. Informatica 83–86
39. Hyrskykari A, Majaranta P, Aaltonen A, Räihä K-J (2000) Design issues Of iDICT: a gaze-assisted translation aid. In: Proceedings of the 2000 symposium on eye tracking research and applications. ACM, New York, NY, USA, pp 9–14
40. Wei H, Moldovan A, Muntean C (2009) Sensing learner interest through eye tracking. In: 9th IT and T conference. Dublin Institute of Technology, pp 1–8
41. Wang H, Chignell M, Ishizuka M (2006) Empathic tutoring software agents using real time eye tracking. In: Proceedings of the 2006 symposium on eye tracking research and applications. ACM, pp 73–78
42. Porta M, Ricotti S, Perez C (2012) Emotional e-learning through eye tracking. In: Global engineering educalion conference (EDUCON). IEEE, pp 1–6
43. Porta M (2008) Implementing eye-based user-aware e-learning. In: CHI '08 extended abstracts on human factors in computing systems. ACM, pp 3087–3092

44. Drewes H, Atterer R, Schmidt A (2007) Detailed monitoring of user's gaze and interaction to improve future e-learning. In: Universal access in human-computer interaction, ambient interaction, vol 4555, pp 802–811
45. Florian SW, Alfred K, Ulrich G (2010) A closer look at split Visual attention in system-and self-paced instruction in multimedia learning. Learn Instr 2(20):100–110
46. Al-Wabil A, El-Gibreen H, George RP, Al-Dosary B (2010) Exploring the validity of learning styles as personalization parameters in eLearning environments: an eyetracking study. In: 2nd international conference on computer technology and development. IEEE, Cairo, pp 174–178
47. Tsianos N, Germanakos P, Lekkas Z, Mourlas C, Samaras G (2009) Eye-tracking users' behavior in relation to cognitive style within an e-learning environment. In: Proceedings of the ninth ieee international advanced learning technologies. IEEE, pp 329–333

Chapter 16
The Edu-coaching Method in the Service of Efficient Teaching of Disruptive Technologies

Ildikó Horváth

Abstract In the 21st century, an increasing number of studies are concerned with forecasting the informatics and technological development created by the dynamic evolution of the information and knowledge society, which forecasts the acceleration of the scientific and technological progress. In the globalizing economy, governmental leaders, market actors and employers hungry for innovation—to enhance efficiency and productivity—increasingly require a global integration of technological innovations in the educational material, and the training of innovative, creative, problem-solving production and developing engineers. There is a growing demand for shaping high quality higher education, and in accordance with this, for supervising learning material contents and applied methods in the higher education for engineering, for a much wider application of up-to-date technologies and for joining in the scientific developments more dynamically.

16.1 Introduction

In the 21st century, an increasing number of studies are concerned with forecasting the informatics and technological development created by the dynamic evolution of the information and knowledge society, which forecasts the acceleration of the scientific and technological progress.

In the globalizing economy, governmental leaders, market actors and employers hungry for innovation—to enhance efficiency and productivity—increasingly require a global integration of technological innovations in the educational material, and the training of innovative, creative, problem-solving production and developing engineers. There is a growing demand for shaping high quality higher education, and in

I. Horváth (✉)
Faculty of Engineering and Information Technology, University of Pécs, Pécs, Hungary
e-mail: horvath.ildiko@mik.pte.hu; ildiko.horvath@vrlc.hu

© Springer International Publishing AG, part of Springer Nature 2019
R. Klempous et al. (eds.), *Cognitive Infocommunications, Theory and Applications*,
Topics in Intelligent Engineering and Informatics 13,
https://doi.org/10.1007/978-3-319-95996-2_16

accordance with this, for supervising learning material contents and applied methods in the higher education for engineering, for a much wider application of up-to-date technologies and for joining in the scientific developments more dynamically.

It has become inevitable to create new solutions to the new market challenges in the field of the higher education for engineering. More and more disruptive technologies appear every year on the Gartners Hype Curve, so the competitiveness of the higher education for engineering can be maintained if it can provide this extra knowledge for students necessary for innovation in this fast changing world. As long as new technologies are considered cognitive and/or metacognitive tools, it can be stated that they strongly support the learning process. The earlier learning theoretical approaches give the basis of the model defined as the digital taxonomy of learning whose starting point is Benjamin Blooms 1956 model. Blooms taxonomy ranks the developmental levels of knowledge in three categories. These are the cognitive-intellectual, the affective-emotional, or volitional and the psychomotor-motional areas. From the methodological developmental perspective of the on-line learning, mobile-learning which increasingly spread due to the technological boom, the cognitive model is relevant [2, 3].

16.2   Challenges Related to the Integration of Disruptive Technologies in the Educational Material

It can observed that the early introduction of disruptive technologies launches an information hunger in the system of the higher education for engineering. If the education of the new technology is started soon after its appearance it is clearly seen that the learning material presenting the techniques, its features, characteristics and feasibility are not available yet. Internet gives the practical scene for following research results, publications appearing in several scientific fields, for having access to the up-to-date information, which will also become a new learning environment approximating to the digital life of students—for the CE generation students [2, 9]. The above statement is justified as the following: the great amount of multidisciplinary learning material and the Internet related learning environment demand the approximation of the digital life of lecturers to the digital living space to which the CE generation is accustomed to [13]. On the other hand, the extensive amount of knowledge material demands the existence, development of the hyper attention specific to the CE generation which is highly useful in case of the dominance of a multiple information stream in situations demanding rapid reactions, which at the same time facilitates efficient time management. In the light of the above it is clear that the learning by doing pedagogical method should be chosen for the efficient education of disruptive technologies.

16.3 Overview of the Learning by Doing Pedagogical Methods for Teaching Disruptive Technologies

Investigating the present practice of the higher education it can be stated that the educational methodology, the pedagogical views currently being applied and the specific student motivation, attitude and thinking patterns only incidentally ensure the training of engineers with innovative views. The greatest changing effect is in the educational methodology and in the pedagogical practice applying it because the appearance of disruptive technologies—still being formed in the education stimulate to acclimatize a motivating method or methods in the higher education of engineering making CE generation students more involved and requiring a much more active research and development activity from students. Teaching disruptive technologies soon after their appearance obviously differ in several aspects from the routines of the traditional educational frames already when choosing the method in respect of the innovative contents of the education, of the multidisciplinary educational material spanning through several subjects, of the change in the roles of lecturers and the students or of expanding the educational opportunities to a 3D virtual reality space, which facilitates the expansion of the boundaries of the education technological opportunities. Choosing the appropriate method is not a simple task even for the experienced teachers when teaching traditional educational materials. It is important for the lecturer to know the complete scale of the applicable methods and to choose the most effective to teach the given learning material. Naturally the combined application of several methods during the education is expedient. The tools to choose the optimal method group and to elaborate the edu-coaching method were the learning, teaching and training methods applied in the educational and market environment from which the ones that consider the aspects specified by the Futures Wheel method suitable for the integration of disruptive technologies in the educational material of higher education were selected.

First the relevant educational methods were surveyed which were created as a result of the international and domestic research of the 21st century, especially the learning by doing learning-centered methods brought to the fore by education history, particularly the constructivist pedagogy. It is investigated which methods suit the assumptions specified in the education of the disruptive technologies in the future educational practice. Following that, the edu-coaching method applied in the higher education for engineering which connects the teaching and developing methods applied in the educational and market environment is demonstrated. The constructivist pedagogical method was especially highlighted from the pedagogical trends because according to the constructivist pedagogical approach the aim of teaching is not simply knowledge transfer but ensuring the conditions for creating students knowledge through personal constructions [12]. During the education of natural sciences and engineering, knowledge must be incorporated in the base of critical thinking by the student being motivated in applying this knowledge. I followed the deductive way as a learning approach when elaborating the edu-coaching method because the recognition of new technologies derives from the students existing knowledge, and then with the help of generalizations and abstractions new, innovative solutions are created by the emergence of more and more complex knowledge systems. The common feature of the learning by doing methodological trends is that the emphasis is put on facilitating self-reflection expanding to the students activity, independent thinking and each element of the activity, and the learning process is based on the students motivational state besides the internalization of the experiences gained in reality. A knowledge building suiting students character and knowledge level is intended in which the student is an active participant: looking for answers and solutions to a specific situation and phenomenon. This solution suits the pedagogical methods well applicable in teaching disruptive technologies. At the same time, atypical learning methods gain more and more ground mainly in terms of the requirements of lifelong learning and to increase adults learning efficiency following the reform pedagogical endeavors in modern educational practices. We must also consider the opportunities lying in atypical learning methods when creating the educational method of disruptive technologies because getting to know new techniques, technologies extends beyond the frames of formal education. Atypical learning is a form of lifelong learning which aims at developing key competences necessary in the informatics society. The opportunity for adults learning besides working does not necessarily facilitate their participation in formal education. Beyond the learning forms besides formal education,—informal learning, atypical learning includes all the methods which have an effect on the individuals intellectual and physical development by cognitive recognition. During teaching and learning disruptive technologies, both formal and informal learning have a role in the process of information finding, filtering, processing and producing new information. The working forms of atypical teaching-learning are group and team work, peer learning, or cooperative working form where the individual working forms can be experimental learning, individual learning, practical learning, self-experienced, external learning, open learning, distance learning, media learning, flexible learning, digital learning, e-learning, blended learning, alternative education, or coaching widespread in sport and in the market sector. These offer the opportunity of applying methods demanding a much greater independence than common in the traditional higher education for the actors of a wide variety of pedagogical situations, thus their overview and the adaptation of their best practice capital is justified to create the efficient educational method of disruptive technologies.

In the following, the education-organizational, strategic and methodological opportunities suiting the education of disruptive technologies are surveyed. The common characteristics of the following methods are being student-centered, the change in the roles of lecturers and students, the similarity of activities of the educational-learning process and team work, each of which includes facilitating the integration of disruptive technologies in the learning material of higher education, in the phase right after the appearance of technologies.

16.4 Cooperative Learning

The more and more increasing need for the application of cooperative learning forms nowadays can be traced back to the insufficiencies of traditional pedagogy in formal learning [12]. Cooperative learning includes methods which can be both applied in learning and in out-of-class activities (Fig. 16.1). We mean cooperative team work if the team members carry out tasks by contacting each other and to achieve a better common result sharing the work also appears in the team. Its main characteristic is the order, the meaningfulness and the expediency of connections. Cooperative learning-organization can be considered rather an education-organization, whose principles are:

- simultaneous and comprehensive parallel interaction;
- constructive and encouraging interdependence;

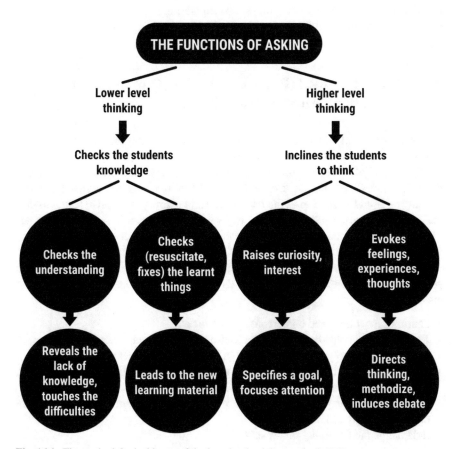

Fig. 16.1 The methodological bases of the learning by doing methods [11]

- equal participation and access;
- individual responsibility and assessment.

Each member has an equal role in the cooperative team. The relationship between he roles is not hierarchical but cooperative in partnership, which shows a significant similarity to the coaching method applied in the market sector, thus it is worth overviewing it in more details to create the new method of higher education. In cooperative learning the members carry out both individual and team tasks. The classical teachers role changes and aims at three main activities:

- preparedness;
- observation;
- intervention.

Nowadays, cooperative learning—with the help of the project method, the less well-known design method and the closely-related and much more widely known research-based learning strategy, and the problem-based learning method—is more and more popular in implementing the student-centered education.

16.5 Project Method (Project Education, Project Based Education/Project Based Learning)

The project method was created at the beginning of the 20th century in the United States as the critique, alternative of the traditional school. It was experienced that the application of knowledge in the traditional school is separated from the knowledge itself, it is not clear what the knowledge learnt through the particular subjects can actually be used for. As during the recognition and learning disruptive technologies, the new knowledge is closely related to its practical application, thus the essence of the project method clearly confirms that this is an educational method which facilitates the rapid teaching of the latest technologies, so it must be a part of the methodology to be developed. The evolving of the project method was based on the principles of John Dewey (1859–1952), which primarily emphasized the following correlations:

- Learning must be based on personal experiences.
- Teaching must consider the students' developing needs and interests.
- The student has to participate actively in forming his own learning processes.
- The student has to be educated to actively take part in the affairs of the community, to become a citizen feeling responsible for the community.
- Shaping the whole personality, the close connection between the curriculum and the social reality are important.
- It demands the flexibility out of school.

According to the following characteristics, the project method adjusts to the characteristics specified in the early phase of the education of disruptive technologies:

- The project can rarely be adjusted to the tight university schedule (2×45), to the system of the classes.
- Due to its interdisciplinary character, it breaks through the frames of the subjects.
- It often steps over the age frames set by the year.
- It demands a new type of teacher-student relationship (the teacher does not control from above, but is a member of the community accomplishing the project, helps the successful accomplishment of the project from inside).
- The project can hardly or not at all be assessed with the traditional grading.

16.6 Design Based Learning

The methodological logics of the design-based learning is that gaining knowledge is the most successful if students do not learn about ready things (objects, environmental elements, processes), but when they have to design and create a specific thing (for example a model object or info-graphics). In case of technological innovations in today's modern education, model creation and the application of simulation pro grammes are ordinary tools in the higher education for engineering, their application has a role during the recognition of disruptive technologies.

16.7 Research-Based Learning–Research-Based Strategy

The base of the research-based learning, which is actually a version of the design-based learning [1], is a specific problem for which students design an investigation, experiment to solve having their preliminary knowledge. Not what they learn or think is important but how they think, that is, the emphasis is on the process of learning things. During learning students understand the concepts and the processes, their knowledge deepens by synthesizing the knowledge elements, their attitudes related to these become enriched and they understand the essence of scientific recognition. Lane [10] interprets the research-based learning as a strategy which actively involves students in the investigation of the content and the results. With this specific characteristic it clearly confirms the education, the recognition, the further research and revealing the fields of the utilization of the still developing technological innovations. Research-based learning can lead to a usable knowledge, to the ability of sensible and lifelong learning, which closely suits the present international and Hungarian educational policy objectives.

16.8 Problem-Based Learning/Teaching (PBL)

The academic literature often mentions Problem-based learning, -teaching besides Project-based education. The first version of the PBL method was elaborated by Barrows and Tamblyn [5].

It was successfully applied at the McMaster university in Canada in the 1960s (http://www.edb.u-texas.edu/mmresearch/Students97/Hemstreet/pbl2.htm). PBL is a teaching method although according to certain researchers [16] it is rather a general educational strategy—where the students work in a small group. They try to understand, solve and explain real-life problems. Their work is helped by the teacher, as a tutor, whose task is to stimulate discussions and arguments. PBL, due to its work form helps students to evolve self-controlling learning, and to develop competences which are not realized or are kept back during the traditional education. Meanwhile, PBL improves adaptability to changes, acquirement of processes facilitating making decisions in unfamiliar situations, assessment and acceptance abilities, problem-solving ability, thus it improves the development of critical and creative thinking abilities.

According to the study titled The future of higher education: How technology will shape learning? published in the Economist Intelligence Unit in October 2008, the technological innovations will have a significant influence on the methodology of education. According to the research, significant changes in the future education: the realization of learning in diverse places and time, the need for personalized learning, the strengthening of students opportunity for a free choice, the intensification of the short-term, project-based learning attitude, the integration of personal experiences facilitated by modern technologies in learning, the ability to successful learning, the change in the teachers role. With keeping the characteristics listed here in view, the benefits of the above-mentioned methods in the elaborated edu-coaching method were integrated because the formation of the teaching material, the planning, the development get into the focus of the learning process during teaching disruptive technologies, where the changing process appearing in the lecturers and students roles is well outlined.

16.9 Edu-coaching Method

The emergence and spread of the coaching method as the introduction of the elaborated edu-coaching method well applicable in the higher education for engineering, which shows the reason for existence of this training method in the higher education of engineering. Tim Gallwey, a lecturer at Harvard and tennis expert, wrote in his books published since 1975 about the factors underlying the athletes performance enhancement and the change in the coaches roles. His Inner Game theory later spread to many fields of business life, as it is an advantage in the market competition to be rapidly adaptable, to have inner motivation, the ability to follow fast changes and the commitment to conscious goals, so the organizations started to investigate methods which are able to achieve really efficient changes and results at the level of the individuals. They gradually recognized the supporting and helping power which the coach provides in sports for his athlete, competitor.

The types of coaching are diverse, with many categorizations, such as life coaching, executive coaching. According to applied methodology: action-oriented coaching, NLP-based coaching, system-approach coaching. Categorization according to orientation: recognition-oriented coaching, achievement-oriented coaching, relation-oriented coaching, and besides these: preventive, strategic, stress, project, conflict, inter-cultural, team coaching, team case-processing coaching etc.

16.10   Presenting the Edu-coaching Method

In most cases the clients contact coaches when they bump into a problem, a situation or a task which they cannot solve on their own. This basic situation is completely identical with the situation when students have to learn a topic for which they do not receive the pre-compiled, logically correlated learning material which is easy to learn, but they have to search for the necessary information and then get from the information to the knowledge on their own (Fig. 16.2). In order to integrate the information in the knowledge system, students have to go through the process of information processing which cannot eliminate one or the other phase.

The developed method is based on the short, solution-centered coaching technique, supplemented with the peculiarities of the higher education system, the science of engineering and modern educational techniques. The solution-centered procedure was formed at the beginning of the 1980s following the research of Insoo im Berg and Steve de Sharez. With their research, they wanted to find the answer to what questions and methods lead to a useful result in a client-counselor relation. Peter Szabo started to integrate the results in coaching in 1997, so the solution-centered coaching model developed which creates a simple and clear thinking frame for the client who draws up his goals, solutions and the steps leading to these in this framework. The coach edits the frame from targeted questions, confirming feedbacks and useful summaries. The client this way receives space and time to arrange his thoughts, to specify his goals, to make his resources real and to plan the next step. The main tool of the coach

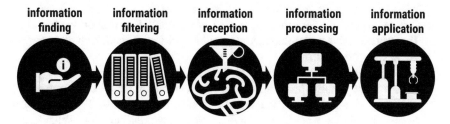

Fig. 16.2 Processing the information

is asking. The questions give the corner stones of the learning process. Children also always ask their peers and adults. They get to know the world based on the answers they receive. In the present public educational system and higher education it is experienced that year after year students ask less and less. The conversations are initiated not by them, but by their parents and teachers, that is the adults, but they also rather speak to children than talk to them. Following the questions, we get answers to our problems, get to know each others thoughts, recognize our limits and ourselves. The better questions the participants that is, both lecturers and students—pose to each other, the more successful the teaching-learning process is. Composing the questions properly in terms of familiarization, deepening, resuscitation and application, and the acquirement of appropriately treating them is not an easy task, as the basic role of the questions is to actuate and to keep thinking in an active state during the learning process. However, in reality often it is just the opposite that happens. The questions posed in the wrong way or at the wrong time can hinder students independent train of thought, and can cause content disturbance, can make them unsure. Lecturers might as well obstruct the thinking process by asking too much, they try to keep the learning process of the teaching material under control with many small questions and meanwhile they control students thinking. By this, not consciously, they absolve students from the efforts of thinking, thus they slowly give up asking, and what is even a bigger problem, individual thinking. The phenomenon—that teachers, lecturers do not welcome students questions, are less helpful to answer them thoroughly with sufficient details, encourage them less to participate in dialogs in connection with the learning material is often experienced. Furthermore, it is important to emphasize that the edu-coaching method strongly relies on modern ICT devices and applications. The 3D virtual educational space completely serves the IT tool demand of the edu-coaching method. The platforms—suiting the digital life of the students, which allow the fast finding, overviewing and processing of the information based on the adequate questions—are available in the 3D VR spaces. Based on these, it is particularly useful to introduce this method in the system of the higher education of engineering which is based on questions (Fig. 16.3). The work hypotheses of the edu-coaching method based on questions are the following:

1. Inventing the solutions
   The question WHY to reveal the source of the problem—common in the classical education is replaced with targeted questions and tools which focus the attention to the solution, so conversation is raised from the level of the problem to the level of solution, thus the solution gets into the focus of the conversation. A question directed to the solution, for instance, is: What would the solution facilitate? or If you could solve the problem, what would be different, what kind of operation could we accomplish?
2. The experience of the solution already exists in the students
   Which is the method/process/solution that worked earlier? With this question we lead students to what resources help them to solve the task. What they have to do for success, what brought a solution in a similar situation. What is the knowledge which they can use to find the solution?

3. TRUST towards the students' resources, competences

Students specify what should be improved to reach the goal. They can work on what they consider important. This trust is a strong motivating factor on the way to reach the common goal as students have to justify their own truth during the solution. Although the lecturer, as a coach can delicately coordinate the direction of further progress with questions and summaries, but is primarily the observer of the process. If necessary, he gives a safety net, helps students if they have chosen the wrong direction, but he always has to keep in mind the freedom of choice.

4. The benefit of being an outsider, not knowing

The lecturer, in the role of the coach, can enforce the unpretentiousness deriving from not knowing against the students hypotheses, at the same time this being an outsider gives the freedom to pose surprising questions, with this he can delicately lead the course of learning in the right direction. Point out why you have chosen this way/method/procedure to find the solution?

5. Extensive IT device usage

According to the Bloom Model (1976), learning is influenced by three factors: preliminary knowledge (cognitive variable), preliminary motivation (affective variable) and the quality of teaching. The existence of all three for the CE generation are supported by the usage of Internet and by ensuring modern IT background. According to the Bruner theory (1974), acquiring knowledge is based on image, symbolic interactive activity. At the edu-coaching method which optimally supports to find the solution, gaining and acquiring knowledge in 3D virtual reality spaces are preferred.

In terms of the working form, team work is considered the most efficient to teach disruptive technologies where team members do both individual and common work. According to this, the steps related to the cooperative education organization, the project-based educational method and the problem-based learning strategy were elaborated. The actual tasks of the edu-coaching method:

1. Choosing the topic: Specifying the content of education primarily 3D virtual reality learning environment
2. Preliminary planning

    a. Planning the optimal educational environment,
    b. Specifying the student target groups to be involve in the education (a target group spanning through courses, years or even training levels is also possible),
    c. Identifying special fields, project parts,
    d. Identifying the sources of information finding

3. Objective: Dual objective specification

    a. Specifying the external target which aims at a new product
    b. Specifying the internal target which aims at learning targets

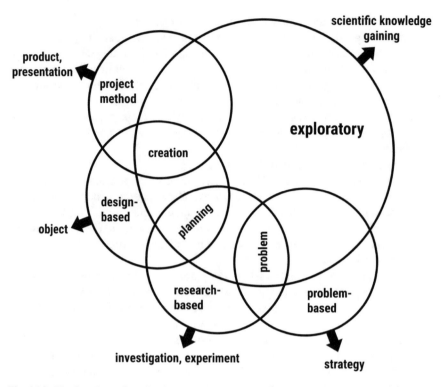

Fig. 16.3 The functions of teacher interrogation in the teaching-learning process [11]

4. Organizing and planning tasks

    a. Making a schedule which includes the compilation of the progress plan and the schedule of the discussions

    b. Specifying the number of people in the group

    c. Specifying the composition of the groups

    d. Considering the initial knowledge

    e. Specifying the initial tools necessary for the work in the 3D VR space

    f. Compiling thought-provoking presentations, videos and other materials

    g. Forming the students self-assessment system

    h. Compiling the question groups helping work suiting the coaching method

5. Education with the edu-coaching method

    a. Coaching agreement fixing the goal

    b. Fixing the time of meetings

    c. Making a list: What do we already know?

    d. Taking resources into account

    e. Specifying the steps of progress

    f. Making a mutual schedule

    g. Dividing the roles and the tasks, assigning deadlines to the tasks

6. Administering a project diary
7. Presenting the results
8. Closing, assessment.

During learning carried out with the edu-coaching method, students experience the satisfaction deriving from creation and innovative activity, thus as the result of the edu-coaching method they can provide a solution to the raised problem by the end of the learning process. The edu-coaching method presented above was successfully applied in the electrical engineering and computer science engineering training at the Faculty of Engineering and Information Technology at the University of Pécs. In our pilot programme teaching the memristor, as a newly appearing disruptive technology was implemented with the application of the edu-coaching method. The target of our pilot project was the extensive recognition of the memristor, as a new disruptive technology in the upward curve of the Hype Graph, revealing its developing, applicational opportunities, developing individual ideas. Students participating in the pilot project offered creative suggestions as solutions for the application of the memristor, as a new circuit element applicable in case of over-voltage and flash protection where they utilized the dissipative feature of the memristor when creating the theoretical models. These suggestions naturally cannot be considered as an implemented engineering development, prototypes were not made, but they served the process of acquiring the secure theoretical material. Last but not least, the common joy of well-done work serves as a positive confirmation for the cooperation in the further projects. The pilot project showed that the edu-coaching method can be well combined with the virtual reality educational environments in favor of the structural development necessary for the efficient education of disruptive technologies.

16.11   Conclusion

As a summary, the presented edu-coaching method gives guidelines for creating a work form which facilitates the rapid introduction of disruptive technologies in the higher education for engineering. If we use the learning method supported by solution-centered questions, then students go through each cognitive level specified in the Bloom taxonomy in the learning process, moreover, the outcome can be a product, a model, a new item or idea for further utilization of the technology.

The traditional lecturer role is replaced with a helping, supporting role, providing a safety net for students paying attention to the fact that students keep in mind the original target during their problem solving attempts (Fig. 16.4). Due to its solution-centered educational approach, the application of the edu-coaching method greatly contributes to the successful teaching of disruptive technologies in the early phase, to the development of engineering innovation. It can be stated that the presented educational methods supplemented with the edu-coaching method include educational tools which facilitate the rapid introduction of disruptive technologies. In conclusion,

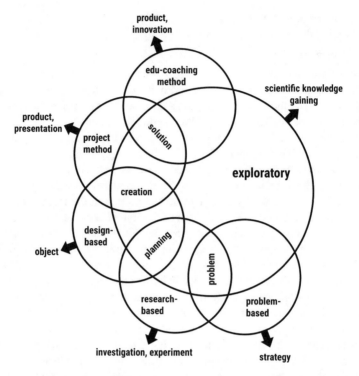

Fig. 16.4 The position of edu-coach in the methodological bases of the learning by doing methods [8]

the scientific topic of this article belongs to the scientific discipline of Cognitive Infocommunications [2, 3]. Related to this, this article presents the application of the new method of edu-coaching.

References

1. Anderson RD (2006) Handbook on research on science education, chapter inquiry as an organising theme for science curricula, pp 807–830. Erbaum
2. Baranyi P, Csapó Á (2012) Denition and synergies of cognitive infocommunications. Acta Polytech Hung 9:67–83
3. Baranyi P, Csapó Á, Sallai G (2015) Cognitive infocommunications (coginfocom). Springer International Publishing
4. Baranyi P, Persa G, Csapó Á (2011) Definition of cognitive infocommunications and an architectural implementation of congnitive infocommunication systems. World Acad Sci Eng Technol Int J Sci Eng Technol 58:501–505
5. Barrows HS, Tamblyn RM (1980) Problem based learning an approach to medical education. Springer Publishing Company, InC.536 Broadway New York N.Y. 10012
6. Csapó Á, Baranyi P (2012) The spiral discovery method: and interpretable tuning model for CogInfoCom channels. J Adv Comput Intell Intell Inf 16:358–367

7. Csapó Á, Baranyi P (2012) A unified terminology for structure and semantics of CogInfoCom channels. Acta Polytech Hung 9:85–105
8. Horváth I, Kvasznicza Z (2016) Innovative engineering training todays answer to the challenges of the future. In: Proceedings of 2016 international science education conference, pp 466–472
9. Komlósi I, Waldbuesser P (2015) The cognitive entity generation: emergent properties in social cognition. In: Cognitive infocommunications (CogInfoCom), pp 439–442
10. Lane JL (2007) Inquiry-based learning. Technical report, Schreyer Institute for Teaching Excellence
11. Makádi M (2005) Tanuljunk, de hogyan!? chapter A termszetismeret tanuálása. A fldrajz tanulsa, pp 170–222. Nemzeti Tanknyvkiad
12. Nagy J (2005) A hagyományos pedagógiai kultúra csõdje. Iskolakultúra
13. Nahalka I (1998) A magyar iskolarendszer átalakulása befejezõdött. Pedagógiai Szemle 48(5)
14. Persa G, Csapó Á, Baranyi P (2012) CogInfoCom systems from an interaction perspective a pilot application for EtoCom. J Adv Comput Int Int Inf 16:297–304
15. Waldbuesser P, Komlósi LI (2015) Empirical findings in cognitive entity management: a challenge in digital era. In: 2015 6th IEEE international conference on cognitive infocommunications (CogInfoCom), pp 433–437
16. Walton HJ, Matthews MB (1989) Essentials of problem-based learning. Med Educ 23(6):542–558

Chapter 17
3D Modeling and Printing Interpreted in Terms of Cognitive Infocommunication

Ildikó Papp and Marianna Zichar

Abstract Digital technologies tend to appear in more and more fields in our life. But what is obvious in the real life, may be rather complicated to represent and reproduce in a digital environment. Nowadays, a growing number of technologies and devices have an associated property called 3D, referring to their ability of approximating reality. This study aims to investigate rather popular 3D-related technologies (3D modeling and printing) in connection of the importance to bring them into the education. Term of cognitive infocommunication is also introduced providing the opportunity to comment the sections in point of its view. Overviewing the main functionalities and properties of four 3D modeling software products gives the reader guidance in this special world, while the case study highlights the additional benefits of acquiring a 3D printer by an institution.

17.1 Introduction

The labels 3D or smart attract the attention of people intensively while these or similar adjectives are assigned to more and more products and services. Counting and denoting of dimensions have a longer history than that of the property smart. This last one seems to have much more flexible definition than determination of dimensions may have. It is a simple and rather clear true statement that we live in a three-dimensional world. But the need to represent, record or reproduce some features of this world in certain forms belongs to characteristics of human beings. The first drawings on the wall of caves made by a caveman can be also considered as

I. Papp (✉) · M. Zichar
Faculty of Informatics, University of Debrecen, Kassai u. 26, Debrecen H-4028, Hungary
e-mail: papp.ildiko@inf.unideb.hu

M. Zichar
e-mail: zichar.marianna@inf.unideb.hu

© Springer International Publishing AG, part of Springer Nature 2019
R. Klempous et al. (eds.), *Cognitive Infocommunications, Theory and Applications*,
Topics in Intelligent Engineering and Informatics 13,
https://doi.org/10.1007/978-3-319-95996-2_17

a two-dimensional representation of his local environment. After a big jump in the history we can mention several further 2D examples such as illustrations from books, hand drawings, paintings, maps etc. Nevertheless, in art also 3D appeared very early, because sculptures, reliefs are definitely of 3D. The appearance and spreading of digital technologies involved 2D contents till the natural demand has arisen again to develop 3D contents. Typical fields of it: games, movies, virtual reality applications and 3D modeling software products.

No doubt, that 3D movies had and have an inevitable impact on accelerating research activities and intensive development of engineering in the field of visualization. As a result, different 3D display devices have become available to facilitate 3D moving pictures, animations even on a desktop computer which means a risk of becoming addictive. Several scientific papers explore and try to identify the (potentially negative) side effects of watching too many 3D movies [8, 15, 28]. Software products used to create or edit 3D contents [17] brought innovation into the education as well in the form of enhancing learning materials by adding 3D presentations and introducing basics of these special software products into the classrooms. Nevertheless, the attribute 3D has been, and also remains to be, an attractive adjective for the masses, especially for youngsters.

Nowadays feasibility of 3D printing and consequently 3D modeling are hot topics in many fields, that is why all educational institutions should introduce the fundamentals of 3D printing and modeling into their educational program taking into account the age of students they teach. The objective of this paper is to analyze the state of 3D modeling and printing in the context of education and cognitive perception. After introducing the cognitive infocommunication, the forms of applying 3D-related products in the education are overviewed that is followed by short introduction to some 3D modeling software. Finally, a case study is described to demonstrate what changes can be implied by the appearance of a 3D printer in the life of an educational institution. Throughout the paper cognitive aspects of the currently discussed topics are also highlighted.

17.1.1 Basics of Cognitive Infocommunication

Cognitive science is a complex, interdisciplinary field devoted to exploring the nature of cognitive processes such as perception, reasoning, memory, attention, language, imagery, motor control, and problemsolving. The goal of cognitive science is to understand the representations and processes in our minds that underwrite these capacities, how they are acquired, how they develop, and how they are implemented in underlying hardware (biological or otherwise). Nowadays, it is a well-known phenomenon that traditionally distinct research areas have started to convergence to each other. The erosion of boundaries between previously individual fields is finally resulted in new forms of collaboration and leads to define new fields to research. This process can be observed in case of informatics, media and communications as well where, beside the original disciplines, new research areas emerged [18]. The fields of

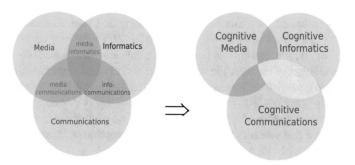

Fig. 17.1 Mutual relationships [7]

infocommunications, media informatics and media communications appeared in the latter half of the 20th century and the Fig. 17.1 describes their mutual relationships.

In the past few years, similar processes run in the field of cognitive sciences and as the Fig. 17.1 represents cognitive media [8, 22], cognitive informatics [2, 5, 31, 32] and cognitive communications [9, 13, 23, 30] seemed to be appropriate to take part in such a synergistic relation.

The first attempt to define the functionality of cognitive infocommunication dates back to 2010 when the First International Workshop on Cognitive Infocommunications was held in Tokyo [6]. From this year IEEE organizes a conference (CogInfoCom) on this topic every year. In 2012 the first definition was refined and a finalized version was released in [7], where the authors say that it "explores the link between the research areas of infocommunications and cognitive sciences, as well as the various engineering applications which have emerged as a synergistic combination of these sciences." The key factor is to study how cognitive processes co-evolve with infocommunications devices meanwhile the human brain is extended through these devices and also interact with any artificially cognitive system. The basic components determining the main characteristics are the medium and the information. According to Baranyi et al. two dimensions of cognitive infocommunication can be specified based on the actors and on the type of information. The properties of cognitive capabilities, the actors have at the endpoints of the communication, can determine the mode of communication:

- Intra-cognitive communication,
  cognitive beings with equivalent cognitive capabilities For example: communication is realized between two humans.
- Inter-cognitive communication,
  cognitive beings with non-equivalent cognitive capabilities For example: communication is realized between a human and an artificially cognitive system.

The type of information transferred between the two entities and the way of its realization can be used to define the types of communications such as sensor-sharing communication, sensor-bridging communication, representation-sharing communication and representation-bridging communication. Several relationships can be explored between CogInfoCom and other fields in computing and cognitive sciences. These synergistic relationships are discussed in detail in [7]. In the case of 3D printing a special type of communication can be observed: data are transformed into a physical object by printing, and transmitting information to humans. Cognitive interpretation of the data can be enhanced by making possible to touch a surface, a graph, a solid, instead of only watching it on the screen (even if the display device is of 3D). Without precise understanding of concepts, the motivation of further study can be hard. It is well-known that visual experience helps the perception of information, and 3D printing can be a tool in the process of visualization.

17.2 Emergence of 3D in the Education

Watching 3D movies at the cinema or even at home is not an exceptional event any more. This statement is especially true for youngsters who have been surrounded by digital devices since their birth. One of the most important factors for dynamic develop of any technology is to introduce its fundamentals into the education. The age of students has to be taken account of course, when the textbooks, exercises or tutorials are produced according to the demands of educational institutions. This section provides a general overview on the different forms of 3D related applications appearing in the education while some comments about the cognitive aspects are mentioned too.

17.2.1 3D Tutorials

In the 21st century, nobody debates why the different 3D technologies have to be involved into the education. Several researchers studied its impacts on learning curves. Between October 2010 and May 2011, a team led by Professor Dr. Anne Bamford, Director of the International Research Agency, carried out an extensive research on the implementation of 3D technology in classrooms and the way it affects students understanding and learning of science-related material. The subjects were 740 students between the age of 10 and 13 from 7 European countries, with a 3D and control class pair set up at each school involved.

In the paper summarizing the results, Bamford stresses the need for the incorporation of 3D into science education by referring to today's elementary school students as "digital natives", who have already been exposed to technological devices and 3D since their early childhood; she also notes that they tend to have a positive attitude towards the idea of 3D exposure at school.

As for the benefits of being taught via 3D projection, Bamford notes that most students prefer visual learning to auditory learning, as visualizations make it easier for them to understand complex structures, the part-whole relations within them, and their overall functionality. The acquisition of the material through 3D is also claimed to facilitate retention, with 3D classes having a quantitative and qualitative edge over control classes; their recalls of the material tended to be more systematic while being more elaborate. In addition, according to feedback from teachers and students, there were major developments in classroom interactions as well. Students tended to ask more questions and questions with greater complexity; they appeared to be more motivated to engage in classroom activities, and their attention span notably increased. Teachers' methods also underwent changes in 3D classes: they encouraged more conversation and student participation, which made them more popular and learning more fun based on student feedback.

Towards the end of the paper, Bamford gives some guidelines for 3D implementation in classrooms, listing the necessary equipment as well as some tips for the effective use of this technology when working with students [5].

17.2.2   3D Software Products

Current and future university students will enter a job market that highly values up-to-date skills. Being familiar with new technical solutions also entails that the future employee will be able to adopt further innovation in his or her profession. The term *lifelong learning* explicitly refers to one of the largest challenge that will be met by the students during their professional life [3]. In the 21st century lifelong learning means more than simply to acquire new knowledge. People have to be able to get to know brand new methods, procedures that were unknown for example a decade before. In the past years, the education has been tending to open new horizons in technology-related areas not only in science, but also in technology, engineering, arts and mathematics education (STEAM). These disciplines can interact continuously and their combination results in experimental learning. Worldwide, there are initiatives to prefer and promote project-based trainings focusing on STEAM disciplines. These new teaching strategies require more financial support, just think about purchasing computers, software products or even 3D printers. Fortunately, several very good free 3D software packages are available [2].

So the first challenge is to decide which software to select that probably meets our requirements and current proficiency.

Most of us would play with wooden blocks in our childhood, which shows some similarity to additive manufacturing. This old-fashioned game is based on additive compilation; that is, blocks can be placed on the top of each other including optional holes, protrusions and ledges as well; but one cannot build something that connects to the side faces of the edifice and cannot subtract or intersect the bricks.

Digital building or modeling eliminates these deficiencies. Basically, Constructive Solid Geometry (CSG) allows one to create a complex solid by using Boolean operators to combine simpler objects (Fig. 17.2).

The simplest solid objects used for representation are called primitives. Typically, they are objects of simple shapes (cuboid, cylinder, prism, pyramid, sphere, cone or torus). The set of allowable primitives is limited by each software package. The CSG

Fig. 17.2 CSG objects can be represented by binary trees. In this figure, leaves represent primitives and nodes represent operations

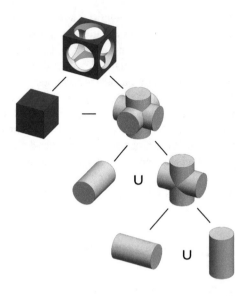

technique is built into the geometric modeling kernels of nearly all engineering CAD packages.

Most modeling systems give you the ability to create new basic models received as a result of applying various operations: extrusion, revolution, or a new primitive can also be constructed by generating its boundary surfaces (using swept or loft).

In general, computer-aided modeling provides a higher degree of freedom to design solids, but 3D printing has its own limitations: extra thin veneers or shells without real volume are not printable. As for which software packages can be used for design, one can use any CAD software that supports the creation of solids.

17.2.3 Expectations for the Model

The standard data transmission file format between a CAD software and a 3D printer is the STL. It is widely used, although few agree on what it stands for. The most frequent suggestions include Standard Triangle Language, STereoLithography, and Standard Tessellation Language. Chuck Hull, the inventor of stereo-lithography and founder of company *3D Systems* reports that file extension originated from the word stereo-lithography. An STL file is nothing else than a series of x, y and z coordinate triplets describing connecting triangular facets. Each facet is determined by three points and the surface normal. The boundary surface of a solid is transformed to this format, but sometimes there can be some mistakes in the mesh: gaps, holes or overlapping of some facets. A well-functioning solid modeler will automatically produce closed surface of a solid (sometimes called a "closed manifold surface") which is watertight; in simple terms triangular mesh without holes. Furthermore, the

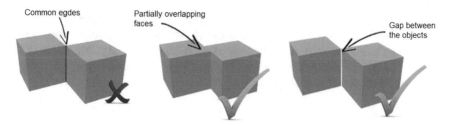

Fig. 17.3 Requirements for the model: instead of common edges use small gaps between the objects or move them to get partially overlapping faces

boundary surface cannot be self-intersecting and contain shared edges or corners. Meeting these requirements for the model ensures successful printing although most slicing software can deal with minor issues (Fig. 17.3).

During slicing a special software divides a 3D design into printable layers and helps to plan the path of the print head. The primary property of the printing is the layer height, on which quality of print is partially dependent. Other essential supplements are chosen or defined in this step such as the rate (or shape) of internal structure of the object (infill), number of shells, supports for any overhanging sections or the need of a raft serving as a base. The user is also informed about the printing time and the amount of material needed in advance.

17.2.4   3D Printing

3D printing is a term used to identify technologies, where the basic principles of creation of three dimensional products are: build your object layer-by-layer, store the information in digital files, use a 3D printer of the appropriate additive manufacturing technology for the realization. 3D printing will radically change the dynamics of consumer culture, affect all aspects of our lives and usher in the next industrial revolution just like the appearance of web 2.0 did some years ago.

17.2.4.1   Milestones from the History

Although 3D printing started to become a known procedure only some years ago, the technology was invented much earlier. The term '3D printing' was first used at MIT in 1995, where two PhD students Tim Anderson and Jim Bredt hacked a traditional ink-jet printer. Their idea was to use a binding solution onto powder to create 3D shapes. It is Chuck Hull, the American engineer that invented the first, really 3D printing process called 'stereo-lithography' in 1983. In his patent, he defined stereo-lithography as 'a method and apparatus for making solid objects by successively "printing" thin layers of the ultraviolet curable material one on top of the other' [12].

Chuck Hull is considered the "father" of 3D printing. The first website appeared in 2008 and has been offering for customer to design their own 3D products by uploading a computer aided design (CAD) file. Nowadays, there are several technologies to 3D print a model and the used materials vary in a large scale from different types of filament till the metal powder [14].

17.2.4.2 Some Benefits and Drawbacks of 3D Printing

First let us consider some thoughts why all educational institutions should adopt this innovative technology at every level from elementary school till university:

- Sometimes even the best 3D rendering cannot help to really capture a concept or geometric shape, while its 3D solid representation that can be touched physically, definitely gives another opportunity to grasp its characterization. Artistic sculptural forms can be designed and printed; geographers can generate 3D terrain models; mathematical shapes, architectural and historical buildings or structures can be studied as 3D models; molds for food products; cells, atoms, DNA and other scientific concepts can be modeled.
- Students prefer hands-on learning, when they can simply learn by doing and the result of their work can be evaluated by themselves. 3D printing makes it possible to go through all the phases of producing: the digital plan can be turned into a physical object, a prototype that can be tested to have immediate feedback about the model properties such as quality, usability, accuracy. The study programs for students of engineering, architecture and multi-media arts should include courses dealing with this special knowledge.
- The final goal that is to touch and use the designed and printed solids inspires the students to work harder, deal with more mathematics thereby to increase their proficiency at designing.
- The joy of creation makes the students more self-confident, which can be increased by experiencing the practical usage of the 3D printed solid.

The option of 3D printing entails introducing of new terms into the design process that have to be considered before starting the printing itself. The placement and position of the model on the build platform influence, for example the necessity of supports and raft, while the angle of inclination also counts when decision on support is made.

It seems to be a challenge to mention some drawbacks of 3DP. The relatively high price of the device, its maintaining cost, including the printing material, the need for gaining extra knowledge and proficiency that were not taught for current educators could be issues to be solved for educational institutions. Fortunately, a number of companies realized how they can help with acquisition of 3D printers. Calls to have a free device was a great success among the institutions, although not every applicant could be awarded. For them, and for the public 3D printing as a service could be an option.

17.2.4.3   Fused Deposition Modeling

This modeling technique (shortly FDM) is more user-friendly than the industrial methods. FDM is a filament-based technology where a temperature-controlled head extrudes a thermoplastic material layer by layer onto a build platform. It is also more affordable, making it the most popular method for desktop 3D printers. In addition to carrying a lower price tag, the thermoplastic filaments are also environmentally and mechanically stable. However, printing an object with FDM generally takes longer than printing the same object using SLS (Selective Laser Sintering) or SLA (Stere-olithography), and the final product probably will need some retouching. Discussion of all 3D printing technologies is beyond the scope of this work, but a comprehensive description on them is available in [12]. FDM often produces objects with rougher surfaces due to the non-sufficient resolution. Smoothness of the solid is limited by layer height in the direction of Z axis and by nozzle size parallelly to coordinate plane XY. Furthermore, printed models are weaker along the Z axis, forces of such direction may cause delamination of layers (Fig. 17.4). It is also a rather important aspect in designing everyday objects that how to cut and how to position them in order to have physically much resistant fighters to tolerate the loads they need to do [27]. Formation of large curved overhangs is especially difficult, parts hanging in the air require supports.

Basically, it is worth considering to print rather large or too complicated shapes in multiple pieces and then to assemble (glue) the segments. This procedure definitely reduces the amount of support or even can eliminate the need for it (Fig. 17.5). There are some optimization techniques and frameworks for dividing objects into a set of smaller segments and preparing them for the best positioned printing. The location of splits can be controlled based on several parameters, including the size of the connection areas or volume of each segment. For example, the overall printing time of the 'Stanford bunny' can be reduced in a 30% and the supports material can be saved in a 35% using automatic segmentation of PackMerger framework [29].

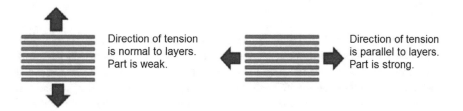

Fig. 17.4 The printed models are weaker in the way of Z axis, such directed forces may cause delamination of layers

Fig. 17.5 Left part of this figure illustrates which rate of overhangs requires supports. The borderline case depends on technical implementation of an FDM 3D printer, which is usually 45° for obliquity. On the right side we can see an example how to leave support. The original model must be divided into multiple pieces

17.2.4.4 Toy or Device for STEAM Idea?

The 3D printing is a great opportunity to join different disciplines of STEAM idea exactly by shapes, inner mechanism and structure of the things. Printing may be game activity as well. A good example of this is the *da Vinci miniMaker*, which is similar to the children's toys in appearance, while suitable for adult's work [20]. Collections of 3D educational projects are often shared by the manufacturers in order that instructors download and incorporate them into their classes. Most of the FDM printers in school environment use biodegradable, non-toxic PLA (Polylactic Acid) filament. This is a thermoplastic material made from dextrose by fermentation, it has gone through various tests to ensure that it is DEHP-free and heavy-metal-free for the sake of the user safety. Young students prefer to use tools that were at least partially manufactured and assembled by themselves. Such tools are prepared consciously and out of necessity, and thus their production and usage are expected to influence future consumer habits.

17.2.5 3D Scanning

3D printing requires a digital model of a physical or an imagined object. 3D modeling produces precise digital description for geometric shapes, but our world is mainly free-form. 3D scanning is invented to create a digital model of an arbitrary-shaped physical object by collecting information about its surface. The resulting model can also be edited as needed or the scanning process can be repeated with other settings.

A wide range of devices can be called as a 3D scanner. Notwithstanding the precise description of the operating principles, a 3D scanner provides information about the spatial extent of the physical objects using lasers, lights or X-rays and can detect hundreds or millions of points from its surface. During the scanning process, a point cloud is being generated from surface data in almost real time, which finally can be stored in polygon meshes as well. Special software enables us to process such point clouds or polygon meshes; among others a point cloud can be cleaned by removing certain points, new polygon mesh can be generated over multiple conditions.

While 3D printer prices have continued to drop, the same cannot be experienced in the case of 3D scanners. A "good" scanner will still set you back over $1000 although there are kits for makers that help to bring the cost down. Fortunately, there are many mobile applications geared towards 3D scanning. Following the trends several applications use cloud processing during calculation but there are some (faster) applications where all the computation is done locally on the device as well. That allows users to scan almost any object with their smart phone. The resulting 3D models can be stored, shared, and edited by third party applications, and can be used in augmented or virtual reality applications [1].

One of the principal benefits of 3D scanning is that its maintenance cost is low, because its operation does not require any physical material. This type of data capturing is preferred to support design processes, robotic control, quality control, documentation of historical sites, making maps for GIS, etc. What is more, customization for virtual reality applications, such as virtual avatars, can be enhanced with usage of a 3D scanner [21].

17.3   Selected 3D Modeling Software Products from Top 25

Nowadays, unimaginable amount of software products are available for solid modeling from beginner to professional levels. Therefore, it is not easy to decide which one to choose. Different communities tend to make recommendations, but the educational institutions generally prefer freely available ones. One of the most popular innovative site dealing with 3D technologies ranked the 3D modeling software according to several factors: the general popularity of the software as well as its usage rate within the 3D printing community were counted [4]. The scores are summarized in Fig. 17.6. The list of Top 25 contains freeware and commercial software products as well. Next sections introduce four modeling software from the list: two ones (a free and a commercial) from the beginning of the list, another one that runs in web browser and can be recommended for young children too, and one more where solids are described by code lines. These descriptions focus only on their main properties; for further detailed information their websites have to be visited.

17.3.1   SketchUP

SketchUp is considered to be one of the simplest modeling software to use although experts say that it facilitates design of complicated models as well. It is often recommended for hobbyist, but seasoned 3D designers, experienced users, or artists prefer it too. Trimble Inc. offers two versions: SketchUp Make and SketchUp Pro from

		General		3D Printing Community				Total
		Social	Website	Forums	YouTube	Databases	Google	Score
1	Blender	61	91	100	100	27	100	80
2	SketchUP	87	82	79	49	80	74	75
3	SolidWorks	95	81	42	52	25	75	62
4	AutoCAD	100	78	46	43	4	85	59
5	Maya	91	80	35	50	3	93	59
6	3DS Max	90	83	24	53	2	78	55
7	Inventor	98	80	29	31	15	75	55
8	Tinkercad	78	57	38	5	100	31	51
9	ZBrush	83	69	45	42	4	50	49
10	Cinema 4D	84	76	6	28	1	62	43
11	123D Design	85	67	21	14	18	50	42
12	OpenSCAD	1	65	33	2	100	29	38
13	Rhinoceros	17	75	50	21	6	49	36
14	Modo	82	63	10	9	1	45	35
15	Fusion 360	93	81	10	3	2	4	32
16	Meshmixer	1	62	18	7	9	28	21
17	LightWave	23	52	1	8	0	32	19
18	Sculptris	0	67	7	6	4	26	19
19	Grasshopper	9	60	4	5	1	32	18
20	FreeCAD	4	59	15	8	11	5	17
21	Mol3D	0	53	3	1	0	28	14
22	3Dtin	4	57	0	0	11	1	12
23	Wings3D	0	66	1	1	0	2	12
24	K-3D	0	62	1	1	0	2	11
25	BRL-CAD	0	60	1	0	0	1	11

Fig. 17.6 Top 25: most popular 3D modeling and design software for 3D printing. Social media score and website authority factors show the overall popularity of the software. The last four factors (how often the software is mentioned in forums, YouTube, databases and Google) are related to 3D printing community [4]

which the first one is free of charge for personal or educational use. Both versions are available for Windows and Mac as well.

Two reasons for using SketchUp [25]:

- It has an easy learning curve that allows even beginners to create models effortlessly.
- It offers the user the ability to draw highly complex models, while several other ones (e.g. AutoCAD, SolidWorks, Blender) have steep learning curve and are intimidating compared to the ease of using SketchUp.

Its freeware version considerably increases the popularity of SketchUp, that is why the version *Make* is in the focus of this section.

When our final target is to print the model a special template called 3D Printings is recommended to use, which contains a dynamic component (3D Printer Build Volume) showing the build area of the machine that will host the model. The user interface gives tools to draw lines (to form edges and faces) and planar figures:

Fig. 17.7 Not printable
solids because of the lack of
wall width [25]

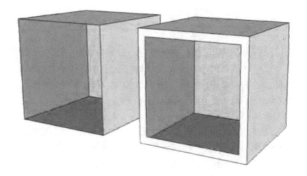

rectangles, circles, regular polygons, and arcs. As soon as lines enclose an area, SketchUp will consider it automatically as a face. The shaped 2D surfaces can be converted into 3D models by the Push/Pull tool that is based on a patented technology of SketchUp. Additional useful modeling tools:

- *Offset*, copies lines and surfaces in line with the original model
- *Follow me*, extrudes a face along a path. Creating curved surfaces is based on applying this tool.

3D models can be shared using 3D Warehouse that is an open source library storing more than 2.2 million 3D models for download and insert them right into the modeling window. When encountering a problem, an outstanding community of passionate experts can provide support in form of on-line tutorials, forums and blogs. The core features of SketchUp does not contain any tools for dealing with STL files, although it supports the export of your work as a 3D model (for example Collada model). To export your model as an STL file, the SketchUp STL extension from the SketchUp Team has to be installed, which can be also downloaded and installed within SketchUp. This extension, which is also available via the 3D Warehouse, enables you to export any models you create as an STL file. You can also import STL files that were created in other modeling programs into SketchUp. Additional useful extensions for 3D printing: Solid Inspector (with its required component: TT_Lib2), and Cleaning. Since SketchUp is not restricted to create 3D models to print you should check the model for any errors before going on printing. The most important criterion that a SketchUp model must meet to be printable is to be a solid. In a solid model every edge is bounded by exactly two faces. Figure 17.7 represents examples for non-solid SketchUp models, where missing wall thickness prevents them to be printed.

SketchUp works best on rectilinear and geometric-type models, but can create curvy, organic models of lower quality. Nevertheless, it can be recommended for people of any age as a first modeler to get to know.

17.3.2 SolidWorks

SolidWorks is a professional, commercial solid modeling computer-aided design (CAD) and computer-aided engineering (CAE) software that utilizes a parametric feature-based approach to create models and assemblies. Although it was primarily worked out for mechanical engineering, several tools support the user during the different kinds of design steps.

There are no predefined primitives in SolidWorks, each block can be easily constructed. Building a model usually starts with a sketch (2D or 3D) that consists of geometry elements such as points, lines, arcs, conics (except the hyperbola), and splines. Dimensions are added to the sketch to define the size and location of the geometry. Relations are used to define attributes such as tangency, parallelism, perpendicularity, and concentricity.

The parametric nature of SolidWorks means that the dimensions and relations drive the geometry itself. Features are the shapes and operations that construct the part. Shape-based features typically begin with a 2D or 3D sketch of shapes such as bosses, holes, slots, etc. This shape is then extruded or cut to add or remove material from the given part. Operation-based features are not sketch-based and include features such as fillets, chamfers, shells, etc. In an assembly, the mates define equivalent relations with respect to the individual parts or components, allowing the easy construction of assemblies [10, 24].

The FeatureManager Design Tree (DT) automatically stores the settings of shape-based and operation-based features, and it can be considered as outline of design reflecting the user thoughts. The parameters in DT can be altered at any time, determining different versions of a single design component. DT provides some folders (such as Sensors, Annotation, etc.) and hosts error and warning messages as well (Fig. 17.8).

Quality of 3D printing is greatly affected by the resolution of STL output. Solid-Works supports its both encoding (binary and ASCII) and allows to control the STL file with deviation and angle of triangle facets (Fig. 17.9).

SolidWorks is a professional engineering software with high performance and accuracy. It is the best choice as a 3D CAD software by providing an integrated 3D design environment that covers all aspects of product development. Nevertheless, becoming an expert requires significant efforts, but gives the feeling of being able to design everything your mind gives birth to.

17.3.3 TinkerCAD

TinkerCAD, provided by Autodesk Inc., is the first browser-based free 3D design platform (Fig. 17.10). It runs in any web browser that supports HTML5/WebGL on Windows, Mac or Linux. Usually Chrome and Firefox perform the best.

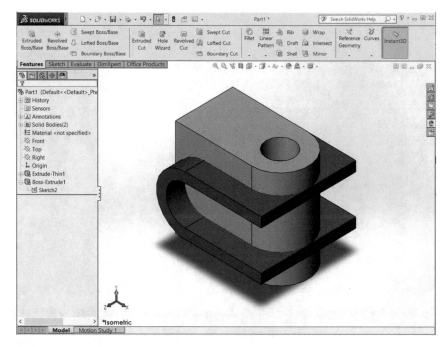

Fig. 17.8 User interface of SolidWorks. The left panel displays the feature manager design tree of actual work containing two solid bodies with their names Extrude-Thin1 and Boss Extrude1. These solids as design components can be illustrated in isometric axonometry, but other views are available as well. These components can be combined by applying Boolean operators

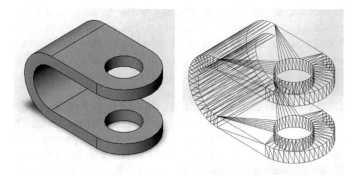

Fig. 17.9 The left side shows the common part of the components which was prepared in Fig. 17.8. The right side displays the STL triangle facets of the boundary surface. It contains 676 triangles with standard resolution of SolidWorks output (Deviation Tolerance 0.0667906 mm, Angle Tolerance 10.00000°)

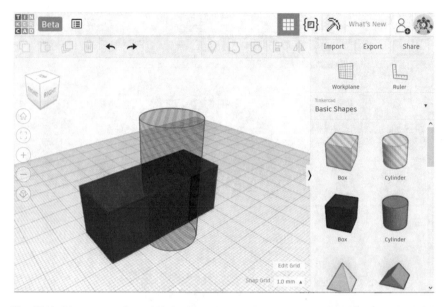

Fig. 17.10 The user interface of TinkerCAD. A resized box and the cylindrical hole define a well known shape

TinkerCAD is used by designers, hobbyists, teachers, and kids to make toys, prototypes, home decor, Minecraft models, jewelry, etc. The list is truly endless. After registration, on-line tutorials, and also sample models become available for free and the user is connected with an on-line community as well. The members can support each other by giving advice, sharing models, writing posts and comments. Every user can work alone or as a member of a team while shared models can be saved to your own account to improve them or to be inspired by them. 3D designs are stored in the cloud, so you can easily access them from anywhere in case of Internet connection.

TinkerCAD is easy to start using and has both simplistic beginner tools and some powerful features once you become more proficient in 3D modeling. The base of simple operation is ability to easily add all sorts of pre-made or parametric shapes into your models (Fig. 17.11). The application uses Constructive Solid Geometry (CSG) that allows one to create a complex solid by using Boolean operators to combine simpler objects. Novices can find a number of great textbooks even on Internet [9, 16, 26].

Additional useful and time-saving feature of TinkerCAD is that it can import hand drawings as SVG graphics, while building a sophisticated model from multiple pieces is also available for the designer.

There is possibility to browse the collection of Customizable models (Community Shape Generator), which are published with or without their Java code. The Autodesk

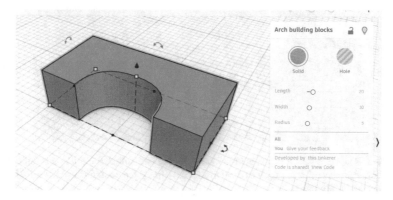

Fig. 17.11 A customizable model on workplane of TinkerCAD. A user can change the role of the model (solid or hole), in case of solid color and size can be changed in right-hand side dialog box

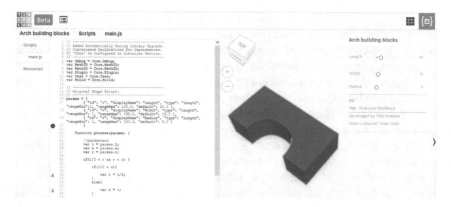

Fig. 17.12 Autodesk creative platform: arch building block and a part of the JavaScript describing this customizable model. The script starts with the definition of parameters that are changeable on slider bars during the usage

Creative Platform provides all the users to create their own shapes, to alive them with/by JavaScripts and to share for the community (Fig. 17.12).

It is surprising, but TinkerCAD does not have any explicit tool to intersect two shapes. A shape can be used in two roles during the design: solid or hole. The solid role gives material to our design, while removing material from our design is linked to hole role. The user must combine these roles of shapes perfectly to create common part of them (Fig. 17.13). Only a downside can be highlighted that there is no possibility to influence the density of STL output. It affects the quality of printed objects, especially for curved surfaces will be conspicuous. The user interface of TinkerCAD application is clear, simple, the features are easily accessible, so it can be recommended to beginner young designers from the age 10.

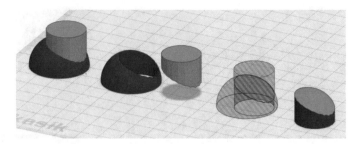

Fig. 17.13 The steps how to construct the intersection of two solids. From left to right: consider their union, then their differences, turn them to be holes, finally the desired intersection is the group of the union and differences

17.3.4 OpenSCAD

While it seems fitting for engineers to use AutoCAD, SolidWorks or other CAD software products, programmers tend to prefer graphical systems such as OpenGL or OpenCSG, since, as IT professionals, they have a better understanding of software development kits and data structures (e.g. binary tree). This means they have different schemes to apply during solving problems. Choosing a solid modeling system based on your knowledge (such as programming skills, experience with data structures) can speed up the acquisition process of becoming familiar with the design system. Beside the specificity of the model, personal preferences, preliminary knowledge and additional skills of the user also determine which type of 3D modeling software to select. While all of the modeling software products discussed in the previous sections are based on the visuality, OpenSCAD breaks this rule and uses its own script language to describe a solid. The language has parametrized commands for 2D and 3D primitives, strings, basic transformations, essential mathematical functions, Boolean operators and contains also conditional and iterator functions. Basically, it can be considered as the programmers solid 3D CAD modeler that is based on a functional programming language.

Although OpenSCAD is a software for creating solid 3D CAD objects based on program codes instead of visual interaction, it represents the object in a separate window (Fig. 17.14). Design of the models is described within a single text file, that it parses to generate and show 3D shapes. The mouse is only used for navigation, such as to view, zoom and rotate around the generated shape. As its name suggests it is an open-source and also free software, and is available for several platforms such as Linux/UNIX, MS Windows and Mac OS X. Furthermore, people who do not want to (or cannot) install new software on their computer can use OpenJSCAD (www.openjscad.org) that is an on-line version running in a web browser.

OpenSCAD has two main operating modes to represent visually the model described with the code lines. Preview is relatively fast using 3D graphics (OpenCSG and OpenGL) and the computer's GPU, but is only an approximation of the model. On the contrary, Render generates exact geometry and a fully tessellated mesh, using

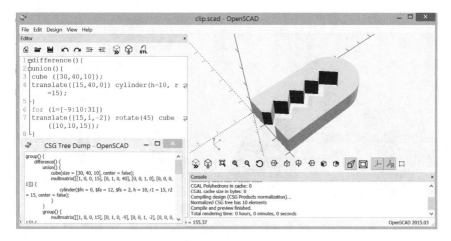

Fig. 17.14   The user interface of OpenSCAD

CGAL as its geometry engine. In the case of larger models with a number of curved faces, rendering is often a lengthy process taking minutes or even hours.

Concerning the modeling itself both principles of CSG and extrusion of 2D primitives can be applied. The set of CSG primitives contains surprisingly few shapes: square, circle, polygon as 2D primitives, and spheres, cylinders, cubes and polyhedron as 3D primitives. Setting parameter values of the primitives and application of CSG operations (union, difference, intersection) and geometric transformations (translation, rotation, mirror, and scaling) provide great freedom to create complex models as well. OpenSCAD provides two parameterized commands to create 3D solids from a 2D shape located on the XY plane: linear_extrude() and rotate_extrude(). Linear extrusion with its (currently six) parameters provides us extremely flexibility how to extend a 2D shape (Fig. 17.15).

Rotational extrusion spins a 2D shape around the Z-axis to form a solid which has rotational symmetry. Idea of this modeling operation may come from the way like potter's wheel works. Conditional and iterator functions of OpenSCAD help building the CSG tree of objects in a convenient way sparing lines of code. To set the resolution or "smoothness" of the mesh, there are two dedicated variables: "$fa" and "$fs". They respectively define the "minimum angle" and "minimum size" of the facets. The minimum angle makes OpenSCAD break a facet in two pieces when its angle with the neighbors exceeds a certain level. The minimum size results in subdivision of the facet when it becomes too large [11].

Beyond the above mentioned tools, models with surfaces determined by functions like $z = f(x, y)$ can be constructed as well. Creating surfaces entirely in OpenSCAD is based on module polyhedron [11]. If the computation cannot be performed within OpenSCAD, then a DAT file has to be created for example by some Python code lines. In this case another module called *surface* is used to reconstruct the faces of the model (Fig. 17.16).

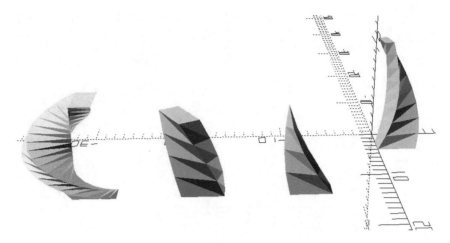

Fig. 17.15 Different linear extrusions of a rectangle

Fig. 17.16 Printed model
generated form DAT file

Most important arguments pro OpenSCAD:

- Models of great precision.
- Beyond the 2D and 3D primitives, even the installed True Type fonts can be utilized as building blocks.
- User defined functions and modules allow grouping portions of script for easy reuse with different values.
- Support of parametric design.
- Models can be exported as STL files.

To sum up, OpenSCAD is best suited to mechanical designs, parametric modeling and not an appropriate choice if you are interested in designing and printing organic shapes or artistic models.

	Free	Desktop	Online	Need for preliminary knowledge	Quality of the model	Support of multiply platform	Recommended form the age of
SketchUp	yes	yes	no	no	high	Windows Mac	10
SolidWorks	no	yes	no	recommended	high	Windows Mac	15
TinkerCAD	yes	no	yes	no	limited	Windows Mac Linux	10
OpenSCAD	yes	yes	yes	recommended	high	Windows Mac Linux	12

Fig. 17.17 Comparison chart of the presented software products

17.3.5 Comparison

Appropriate selection of 3D modeling software is always of great importance, although most of the users probably try another modeler sooner or later. There are properties which usually hold for every product, such as support for exporting the model into STL, but there can be some which are worth to consider. Figure 17.17 summarizes some basic properties of the above presented modelers.

17.4 Our Case Study

Experience shows that 3D printing technology changes the dynamics of consumer culture by turning users into active creators from passive consumers. Although most of the educational institutions are interested in adopting this innovative technology, the lack of a 3D printer presents a barrier. Fortunately, industrial companies are committed in introducing 3D printing into the education which sounds promising.

17.4.1 3D Printers to Every School

At the beginning of 2016, FreeDee Ltd., as the authorized distributor of MakerBot products in Hungary, made a call with title "MakerBot in the Classroom." In the first round, 11 applying educational institutions were selected to get a free printer, while additional 10 were awarded after another two Hungarian companies, CraftUnique Ltd. and Leopoly Ltd., joined the innovative initiative "3DTECH at Schools."

As one of the winners, The Faculty of Informatics has become the owner of a 3D printer of the kind MakerBot Replicator 5th Generation (Fig. 17.18). It uses the most common technology, fused deposition modeling (FDM), which works by laying

Fig. 17.18 The MakerBot replicator

down material in layers; a plastic filament is unwound from a coil and serve as the material to produce objects.

The printer made it possible for us to combine different disciplines also in practice while fulfilling the challenges of STEAM idea. Instead of dealing with individual disciplines such as Science, Technology, Engineering, Art and Mathematics, main goal of STEAM movement is to integrate them into a cohesive learning paradigm based on real-world applications. Basically, it is a curriculum-based idea of educating students in an interdisciplinary and applied fashion.

17.4.2 Events Where Our 3D Printer Was in Focus

A single desktop printer, like our MakerBot Replicator, can offer a number of opportunities for introducing people to the world of 3D printing. Events where the device was in focus: "Girls in ICT" Day, where young high school girls are invited to the faculty to draw their attention to carrier possibility in the field of ICT; Campus Festival, which is a three-day-long open-air music festival accompanied by cultural activities; and it will presumably be a regular participant at Researchers' Night. The institutions that won a printer at the tender form a Facebook group, where experience, results, success, failures and issues can be shared and discussed. This group uses the social media to exchange custom digital models as well.

17.4.3 Influence of the 3D Printer on Our Degree Programs

Summer of 2016 had a special role in life of the undergraduate degree programs at the Faculty of Informatics. It was this summer, when our team developed the curriculum of a new course to ensure the appearance of 3D printing and additive manufacturing in the study programs of bachelor students. Our own training material was compiled based on excipients provided by the FreeDee Ltd. and Stratasys Ltd. companies supplemented with new practical tasks nearby IT or Engineering [2]. The course titled "Introduction to 3D Printing and Design" is open in all semester to all students of IT education starting from Software Engineering to IT teachers.

In the following years, we have intention to create a hand-on 3D Printing Classroom combined with a Lab for Visual Studies. The mission is to serve students of all majors and backgrounds in their advancement connecting the real and virtual, visual world in interactive work in groups. The students and faculty could design and print mathematical objects, spare parts or prototypes to support research projects and education.

17.4.4   Additional Collaborations

The presence of the printer has already implied collaboration with researchers or committed students. One student needed, for example a special tool to his biological experiment that was printed based on the model designed by himself. A mathematician professor also contacted us to print him the boundary of the closure set of a contractive polynomial. First his desire seemed to be a real challenge, but a bit Python programming followed by an OpenSCAD script created the point cloud which made it possible to generate the solid and finally we could print this special solid. Similar ad-hoc projects are typical when a new technology starts to spread widely [19]. To touch and watch the solid was an exceptional experience for the colleague proofing that cognitive perception is of great importance.

17.5   Conclusions

It is a natural life-cycle of technological inventions that first they are considered either odd or brilliant while only few have access to them. The development of technology goes on his way including spreading of the former versions mainly due to the decreasing pricing. The more people use and know a novelty, the faster it will become more perfect. Not so long ago, for example, people dreamed of having a computer at home, and then PC's came along. Some people used to dream of "on-the-move" communication for the masses, and now almost everyone has a mobile phone. What is more, some appliances have computing capacity rivaling desktop computers.

This process could be observed in the case of 3D modeling and printing too, during the last decade. Nowadays, a key factor of the further development is to make them know with youngsters. It was this idea, that motivated us to outline the role of 3D-related techniques, which belong now to the next step of the evolution of the manufacturing. As additional benefit we have to mention that artwork can be considered to be a form of cognitive infocommunication. It is envisaged that 3D printing will be more accessible to the masses to make spare parts for broken appliances, create bespoke components, objects, etc. just download the design and print it at home. According to these ideas, 3D printing should belong to digital literacy in the near future.

References

1. Adam F (2016) SCANN3D App Review, Phones review. http://www.phonesreview.co.uk/2016/04/20/175648/. Accessed 20 Feb 2017
2. Amundsen M, Arden E, Lentz D, Lyttle P, Taalman L (2015) MakerBot in the classroom, an introduction to 3D printing and design. MakerBot Publishing, Brooklyn, NY. ISBN 978-1-49516-175-9
3. Aspin DN, Chapman JD (2000) Lifelong learning: concepts and conceptions. Int J Lifelong Educ 19(1)
4. Backer F (2015) Top 25: most popular 3D modeling and design software for 3D printing. https://i.materialise.com/blog/top-25-most-popular-3d-modeling-design-software-for-3d-printing/. Accessed 20 Feb 2017
5. Bamford A (2011) The 3D in education. White paper. http://www.eonreality.com/download/The%203D%20in%20Education%20White.pdf. Accessed 15 July 2016
6. Baranyi P, Csapo A (2010) Cognitive infocommunications: CogInfoCom. In: 11th international symposium computational intelligence and informatics (CINTI), pp 141–146
7. Baranyi P, Csapo A (2012) Definition and synergies of cognitive infocommunications. Acta Polytech Hung 9(1):67–83
8. Bombeke K, Van Looy J, Szmalec A, Duyck W (2013) Leaving the third dimension: no measurable evidence for cognitive aftereffects of stereoscopic 3D movies. J Soc Inf Display 21:159–166
9. Bouchard F (2016) Introduction to 3D design with TinkerCAD. University of Ottawa, Faculty of Engineering, 3D Modelling Project. http://engineering.uottawa.ca/sites/default/files/tinkercad_guide.pdf. Accessed 15 July 2016
10. Castro-Cedeno MH (2016) Introduction to SolidWorks, 3rd edn. CADeducators.com, 2016, ISBN: 1534699910
11. Francois J (2014) How to use OpenSCAD (1), tricks and tips to design a parametric 3D object. http://www.tridimake.com/2014/09/how-to-use-openscad-tricks-and-tips-to.htm. Accessed 15 Feb 2017
12. Grimm T (2004) Users guide to rapid prototyping. Dearborn, Michigan. ISBN 0-87263-697-6
13. Hewes D (1995) The cognitive bases of interpersonal communication. Routledge
14. Horvath J (2014) Mastering 3D printing. Apress, Berkeley. ISBN-13: 978-1-48420-026-1
15. Ji Q, Lee YS (2014) Genre matters: a comparative study on the entertainment effects of 3D in cinematic contexts. 3D Res 5(3)
16. Kelly JF (2014) 3D modelling and printing with TinkerCAD: create and print your own 3D models. Pearson Education
17. Koppal S, Zitnick CL, Cohen M, Kang SB, Ressler B, Colburn A (2010) A viewer-centric editor for stereoscopic cinema. IEEE Comput Graph Appl 31(1):20–35
18. Kozma R (1991) Learning with media. Rev Educ Res 61(2):179–212
19. Novac M, Novac O, Indrie L (2005) Using CAD programs for computation of the 3D transient temperature distribution. In: Proceedings of 5th international conference on electromechanical and power systems, pp 904–906. ISBN: 973716-230-7

20. Official website of da Vinci miniMaker. http://us.xyzprinting.com/us_en/Product/da-Vinci-Mini-Maker. Accessed 20 Feb 2017

21. Rácz R, Tóth Á, Papp I, Kunkli R (2015) Full-body animations and new faces for a WebGL based MPEG-4 avatar. In: Proceedings of 6th IEEE conference on cognitive infocommunications, pp 419–420

22. Recker MM, Ram A, Shikano T, Li G, Stasko J (1995) Cognitive media types for multimedia information access. J Educ Multimed Hypermed 4(2–3):183–210

23. Roschelle J (1996) Designing for cognitive communication: epistemic fidelity or mediating collaborative inquiry?, Computers, communication and mental models, pp 15–27. Taylor & Francis

24. Reyes A (2014) Beginner's guide to SolidWorks 2015—Level I: Parts, assemblies, drawings, photoview 360 and SimulationXpress, SDC Publications, 2014, ISBN-10: 15-8503-918-7, ISBN-13: 978-1-58503-918-0

25. Ritland M (2014) 3D printing with SketchUp. PACKT Publishing. ISBN: 978-1-78328-457-3

26. Roskes B (2016) Getting started in TinkerCAD. http://www.3dvinci.net/PDFs/GettingStartedInTinkercad.pdf. Accessed 15 July 2016

27. Smyth C (2014) Anisotropy, or how i learned to stop worrying about making everything in one piece and love my 3D printer. http://3dprintingforbeginners.com/stop-3d-printing-everything-in-one-piece/. Accessed 20 Feb 2017

28. Solimini AG (2013) Are there side effects to watching 3D movies? A prospective crossover observational study on visually induced motion sickness. PLoS ONE 8(2):e56160. https://doi.org/10.1371/journal.pone.0056160

29. Vanek J, Garcia Galicia JA, Benes B, Mech R, Carr N, Stava O, Miller GS (2014) PackMerger: a 3D print volume optimizer. Comput Graph Forum 33(6):322–332. https://doi.org/10.1111/cgf.12353

30. Vernon D, Metta G, Sandini G (2007) A survey of artificial cognitive systems: implications for the autonomous development of mental capabilities in computational agents. IEEE Trans Evoluti Comput 11(2):151–179

31. Wang Y (2007) The theoretical framework of cognitive informatics. Int J Cognit Inf Nat Intell 1(1):1–27

32. Wang Y, Kinsner W (2006) Recent advances in cognitive informatics. IEEE Trans Syst Man Cybern 36(2):121–123

Chapter 18
Constraints Programming Driven Decision Support System for Rapid Production Flow Planning

Grzegorz Bocewicz, Ryszard Klempous and Zbigniew Banaszak

Abstract The proposed attempt should be considered as a structure for creating a task-oriented Decision Support System (DSS). This attempt is addressed at interactive investigative for reactions to queries expressed in both ways (direct or reverse), while assisting a decision maker in finding answers to such routine questions as: Does the given premise imply a desired conclusion? Is there a premise which implies the given conclusion? The main goal consists of a declarative representation allowing a decision maker to formulate direct and reverse scheduling problems regarding projects portfolio under constraints imposed by a company's multi-project environment. It is easy to observe that our contribution can be treated as an alternative approach to Decision Support System (DSS) project that can handle distinct and imprecise decision variables and also multi-criteria questions problems. Instances, are used to illustrate the proposed methodology.

G. Bocewicz (✉) · Z. Banaszak
Faculty of Electronics and Computer Science, Koszalin University of Technology,
Śniadeckich 2, 75-453 Koszalin, Poland
e-mail: bocewicz@ie.tu.koszalin.pl

Z. Banaszak
e-mail: zbigniew.banaszak@ie.tu.koszalin.pl

R. Klempous
Department of Control Systems and Mechatronics, Wrocław University of Science
and Technology, 27 Wybrzeze Wyspianskiego, Wrocław, Poland
e-mail: ryszard.klempous@pwr.edu.pl

© Springer International Publishing AG, part of Springer Nature 2019   391
R. Klempous et al. (eds.), *Cognitive Infocommunications, Theory and Applications*,
Topics in Intelligent Engineering and Informatics 13,
https://doi.org/10.1007/978-3-319-95996-2_18

18.1 Introduction

The ever shorter product life cycles and the increasing competition mean that, to maintain their market position, companies have to make accurate innovation decisions, which are, by definition, both expensive and risky. This statement seems quite obvious when applied to competing companies which have equal possibilities (the same access to raw materials and energy, human resources and "know-how", bank loans and industrial infrastructure, IT technologies, etc.) such firms can only compete in the area of decision-making.

Various methods developed so far, in such domains as including operations research, mechanical engineering, production engineering and so on, become employed for making rational choices. Aided by various techniques grounded in information science, artificial intelligence and cognitive psychology, and implemented as computer software, these methods can be put into practice as person-computer systems with specialized problem-solving expertise. The term expertise is understood here as knowledge and understanding of a particular area and its problems as well as the skill necessary to solve these problems. In that context such systems called Decision Support Systems (DSS) [14, 17] provide integrated computing environments for complex decision making. For the sake of further discussion, in which DSSs are considered as instruments supporting scrutiny of possible alternatives, they are defined as interactive computer-based systems designed to help users in judgment and choice making [12].

The multi-criteria nature of production-planning problems (in particular, production flow scheduling), the complexity of these problems, and the need to make decisions on-line, spur the development of techniques and methods for building dedicated DSSs. Such systems should be designed to allow integrated on-line analysis of alternative scenarios for completing production orders and early detection of errors in the order execution method used. In other words, because the main objective of operational planning concentrates on decision making considering the constrains generated by a company's multi-project environment, the appropriately dedicated DSSs are of crucial importance.

An important role in the corresponding decision support process is played by the time horizon within which the effects of the decisions are assessed. The effects of the same decisions can be evaluated differently in different time horizons. The relativity of such assessments, as well as the individual predispositions of the operator, related, for example, to the possibility of simultaneously evaluating several or a dozen parameters, suggest what form the already available or newly designed human-computer interfaces should have and how they should be used.

In a majority of systems operating in practice, many decisions are made on a individual basis (regarding batching, allocation, scheduling and routing) basis. In the last half-century, various decision-support methods and techniques have been developed and explored, for instance, [11, 15], B&B (Branch and Bound) attempts [7] or, more lately, Artificial Intelligence algorithms, to mention just a few. The last type of approaches (AI-based methods) are predominantly fuzzy-set-theory-driven techniques and constraint programming frameworks. Constraint Programming (CP) and Constraint Logic Programming (CLP) [4, 9, 16] are languages of choice when it comes to the designation and realization of DSO (Decision-Support-Oriented) programs forasmuch they employ the tools of declarative modelling, which provides reference models that take into consideration both a company's powers and its portfolio of manufacturing orders. What is more, Constraint Programming (CP)/ Constraint Logic Programming (CLP)-driven reference models can be used to deal with decision variables (imprecise as well as crisp) adopted in Decision Support System (DSSs).

The computationally hard character of production flow scheduling problems, as well as the Diophantine nature of numerous planning problems associated with the fact that the decision variables that describe them have integer values, implicates the need for developing methods based on the paradigm of abductive reasoning. Abductive methods take into account the constraints associated with the indeterminacy of numerous Diophaninte problems, particularly those related to the attempt to achieve the desired behavior of a system under arbitrarily given structural constraints. One example of such problems are the relatively frequent timetabling problems associated with the search for a desired solution, e.g. a timetable without empty slots that is feasible with respect to constraints related to the availability of classrooms, teachers, the structure of the courses taught and the number of contact hours. In the absence of an appropriate solution, the paradigm of abductive inference tells us to seek methods oriented towards the synthesis of structures that guarantee the performance of expected behaviors. In other words, the desired values of object functions are generally easier to obtain by synthesizing an appropriate system structure (which can guarantee an expected level of performance), than by analyzing the (usually exponentially growing) variants of behavior supported by the arbitrarily given structure of the system.

From this angle, the approach advanced in this Chapter should be viewed as a a structure for creating a task-oriented Decision Support System (DSS) [1, 2] the purpose of which would be to interactively search for answers to forward (direct) and reverse queries, and, at the same time, to help a decision maker find answers to such routine questions as: *Is it possible to finish the production order before an arbitrarily given deadline? Can a new production order be executed using the available resources (when resource availability is constrained in time) without disrupting the execution of the orders that are already being processed?*

To put it differently, the ability to quickly answer questions such as: *Does the given premise imply a desired conclusion? Is there a premise which implies the given conclusion?* allows one to weigh different scenarios related to the decision-making situations under consideration. Playing out such scenarios resembles running a classical computer simulation. For example, working in this mode, the decision maker is capable of assessing the development of a situation determined by, say, order execution deadlines and the availability of resources, the effects of changing the values of parameters such as technological operation times, running times, transfer times, etc., and the consequences of changing resource allocation, order priorities or the resource availability calendar.

The remainder of the present article is constructed in the following way: Sect. 18.2 describes a declarative modelling structure for constraint satisfaction problem driven formulation of an operational planning model. Then, Sect. 18.3 devotes the notion of operational planning of production orders. Section 18.4 provides an illustrative example, and Sect. 18.5 offers some concluding remarks.

18.2 The Declarative Modeling Framework

Environments that are well suited to solving operational problems typical for production orders prototyping are constraint logic languages, which represent the structure of problem constraints in a natural manner. The important shortcomings which, unfortunately, limit the practical use of these environments include the time- and memory-consuming nature of the process of variables distribution and the need to verify the consistency of the solutions obtained using these languages. It is easy to notice that the practical utility of a solution to a given Constraint Satisfaction Problem (CSP) implemented in these environments and understood as a vector of the values of the decision variables which satisfy all the constraints of the problem, is determined by the relationships between the different subsets of variables. These relationships usually have an IF...THEN... rule structure. Therefore, if, in a set of feasible solutions, there are two vectors such that the same values of premise variables correspond to different values of conclusion decision variables, then this set of feasible solutions must be regarded as inconsistent. The problem of analyzing the consistency of a CSP is also a computationally hard problem.

18.2.1 The Reference Model

Let us take into account the reference model for a decision problem concerning multi-resource task allocations in a multi-product job shop, assuming that the decision variables have an imprecise character. The model specifies both the job shop capacity and the production order requirements in a unified way, by describing the sets of variables which determine them and the sets of constraints restricting the domains of discrete variables. Let us consider that sets of non-continuous (discrete) alpha-cuts (i.e. accurate (crisp) values) defining all fuzzy decision variables in the investigated model. Let us also assume that relations sets merging the fuzzy variables represented the fuzzy constraints. A FCSP (Fuzzy Constraint Satisfaction Problem), rather than a standard CSP, has to be used to handle the fuzzy variables and their constraints. It is evident that fuzzy decisions are not acceptable by standard CP platforms [13].

18.2.2 The Constraint Satisfaction Problem

Each decision problem can be formulated in terms of decision variables (values of which are usually belong to discrete domains) and linking them constraints. Since values of decision variables usually belong to discrete domains, the question whether there exists a set of values following all constraints or not can be directly formulated in the CP framework. To put it more formally, CP is a framework for solving combinatorial problems specified by tuples: *(a set of variables and associated domains, and a set of constraints restricting the possible combinations of the values of the variables)*. With this in mind, the CSP [4] can be formulated as follows:

$$CS = ((Y, D), K), \qquad (18.1)$$

where:

$Y = \{y_1, y_2, \ldots, y_n\}$ a finite set of discrete decision variables,
$D = \{D_a | D_a = \{d_{a,1}, d_{a,2}, \ldots, d_{a,b}, , d_{a,ld}\}, a = 1, \ldots, n\}$ a family of finite domains of the variables,
$K = \{k_a | \alpha = 1, \ldots, L\}$ a constraints set (finite) encompassing relations merging the decision variables and bounding the variables realm. Every k_a constraint can be understood as a relation determined over an appropriate subset of the variables $Y_a \subset Y = \{y_1, y_2, \ldots, y_n\}$.

A vector $V = (v_1, v_2, \ldots, v_n) \in D_1 \times D_2 \times \cdots \times D_n$ is the CS solution and the entry assignments fulfill all the constraints K. Therefore, the vector V is fully considered, as the possible solution of the CS.

Decision variables

Let be given number lz of renewable resources bo_a (for example workers and/or industrial robot workers) precised by the sequences of resources $Bo = (bo_1, \ldots, bo_a, \ldots, bo_{lz})$ and their availabilities $Wo = (wo_1, \ldots, wo_a, \ldots, wo_{lz})$, where wo_a—is the availability of the a-th resource, i.e. the value of wo_a denotes the available amount of the a-th resource within the discrete time horizon $H = \{0, 1, \ldots, h\}$. Given a set of production routes $P = \{P_1, \ldots, P_a, \ldots, P_{lp}\}$, where each particular route $P_a, (a = 1, \ldots, lp)$ where the set of operations is specified by $P_a = \{A_{a,1}, \ldots, A_{a,lo_a}\}$, where:

$$A_{a,b} = (y_{a,b}, c_{a,b}, Cp_{a,b}, Cz_{a,b}, Dp_{a,b}), \qquad (18.2)$$

where:

$A_{a,b}$ the operation starting time i.e. the time counted from the beginning of time horizon H,
$c_{a,b}$ the standing of operation $A_{a,b}$,

$Cp_{a,b} = (cp_{a,b,1}, cp_{a,b,2}, \ldots, cp_{a,b,lz})$ the sequence of time moments of operation $A_{a,b}$ requires new units/amounts of renewable resources: $cp_{a,b,k}$ the time counted from moment $y_{a,b}$. This means a resource is allotted to an operation during its execution period: $0 \leq cp_{a,b,k} < t_{a,b}; k = 1, \ldots, lz$,

$Cz_{a,b} = (cz_{a,b,1}, cz_{a,b,2}, \ldots, cz_{a,b,lz})$ the sequence of moments of operation $A_{a,b}$ releases the subsequent resources: $cz_{a,b,k}$ the time counted from moment $y_{a,b}$, $dp_{a,b,k}$ units of the k-th renewable resource released by operation $A_{a,b}$. It is assumed that a resource is released by an operation during its execution: $0 < cz_{a,b,k} \leq c_{a,b}; k = 1, \ldots, lz$, and $cp_{a,b,k} < cz_{a,b,k}; k = 1, \ldots, lz$,

$Dp_{a,b} = (dp_{a,b,1}, dp_{a,b,2}, \ldots, dp_{a,b,lz})$ the sequence of the k-th resource units $dp_{a,b,k}$ are allocated to the operation $A_{a,b}$, i.e., $dp_{a,b,k}$ the number of units of the k-th resource allocated to operation $A_{a,b}$. That assumes: $0 \leq dp_{a,b,k} \leq wo_k$; $k = 1, 2, \ldots, lz$.

Therefore, each particular route $P_a, (a = 1, \ldots, lp)$ is defined by the following sequences composed of:

- start times of operations executed along route $P_a : Y_a = (y_{a,1}, y_{a,2}, \ldots, y_{a,lo_a})$, $0 \leq y_{a,b} < h; a = (1, \ldots, lp)$,
- duration of operations on route $P_a : C_a = (c_{a,1}, c_{a,2}, \ldots, c_{a,lo_a})$,
- start times for allocation of the b-th resource to operation $A_{a,k}$ on route $P_a :$ $CP_{a,b} = (cp_{a,1,b}, \ldots, cp_{a,k,b}, \ldots, cp_{a,lo_a,b}))$,
- release times of the b-th resource by operation $A_{a,k}$ executed along route $P_a :$ $CZ_{a,b} = (cz_{a,1,b}, \ldots, cz_{a,k,b}, \ldots, cz_{a,lo_a,b})$,
- numbers of resource units assigned to the b-th resource allotted to operation $A_{a,k}$ executed along route $P_a : DP_{a,b} = (dp_{a,1,b}, \ldots, dp_{a,k,b}, \ldots, dp_{a,lo_a,b})$.

In general, precise (crisp) values define the operation times, and those for which only rough estimates are known are treated as fuzzy variables. Accordingly, operation $A_{a,b} = (\widehat{y}_{a,b}, \widehat{c}_{a,b}, Cp_{a,b}, Cz_{a,b}, Dp_{a,b})$ is defined by the successive sequences of:

- start times of operations on route P_a:

$$\widehat{Y}_a = (\widehat{y}_{a,1}, \widehat{y}_{a,2}, \ldots, \widehat{y}_{a,lo_a}) , \tag{18.3}$$

- duration of operations on route P_a:

$$\widehat{C}_a = (\widehat{c}_{a,1}, \widehat{c}_{a,2}, \ldots, \widehat{c}_{a,lo_a}) , \tag{18.4}$$

where:
$\widehat{Y}_a$ is a fuzzy set indicating the start times of operations $A_{a,b}$, where $\widehat{y}_{a,b}$ indicates the start time of operation $A_{a,b}$,
$\widehat{C}_a$ is a fuzzy set specifying operation execution times, where $\widehat{c}_{a,b}$ marks the execution time for operation $A_{a,b}$,
$Cp_{a,b}, Cz_{a,b}, Dp_{a,b}$ sequences defined as in (18.1).

The fuzzy variables under consideration are defined by fuzzy sets described by the convex membership function [18]. Since the distinct variables can be seen as special cases of imprecise variables, further discussion is focused on the fuzzy kind of variables.

Constraints on the order of operations

Let P_a be production routes set composing of lo_a precedence- and resource-constrained, non-preemptible operations executed on renewable resources. Let lz be amount of renewable resources, and sequences units denoted by $r_i = (bo_1, bo_2, \ldots, bo_f), i = 1, \ldots, lo_a$, which govern the fixed demand for resources required by the a-th operation. The available units total number of the b-th resource, $b = 1, \ldots, lz$, is limited by wo_b. It is possible to arbitrarily assign resources to operations from set $\{1, \ldots, wo_b\}$ but the resources which have already been assigned to the i-th operation must be achievable at determined times $Cp_{a,b}, Ts_{a,b}$.

Let us take into account routes P_a in an operation-on-node network. Set of arcs define the order in which production operations are executed. Let us presume that there exist constraints which determine the operations order:

- the a-th action is preceded by the k-th operation:

$$\widehat{y}_{a,b} \widehat{+} \widehat{c}_{a,b} \widehat{\leq} \widehat{y}_{a,k} , \tag{18.5}$$

- the k-th action proceeds other operations:

$$\widehat{y}_{a,b} \widehat{+} \widehat{c}_{a,b} \widehat{\leq} \widehat{y}_{a,k}, \ldots, \widehat{y}_{a,b+n} \widehat{+} \widehat{c}_{a,b+n} \widehat{\leq} \widehat{y}_{a,k} , \tag{18.6}$$

- the k-th action is proceeded by other operations:

$$\widehat{y}_{a,b} \widehat{+} \widehat{c}_{a,b} \widehat{\leq} \widehat{y}_{a,k+1}, \ldots, \widehat{y}_{a,b} \widehat{+} \widehat{c}_{a,b} \widehat{\leq} \widehat{y}_{a,k+n} . \tag{18.7}$$

The fuzzy arithmetic-like operators $\widehat{+}, \widehat{\leq}$ used above are discussed in [1].

Let us theorize that each a-th fuzzy constraint K_a (e.g. $\widehat{v}_a \widehat{<} \widehat{v}_l$) can be expressed by the logic value $E(K_a)$, $E(K_a) \in [0, 1]$. Values $E(K_a)$ are used to establish the level of uncertainty DE of the reference model, i.e. the kind of uncertainty threshold (18.8):

$$DE = \min_{i=1,2,\ldots,lo_c} \{E(K_a)\} , \tag{18.8}$$

where: lo_c is the number of constraints on the reference model.

Resource conflict constraints

Let us consider imprecise operation execution times $\widehat{c}_{a,b}$ and operation start times $\widehat{y}_{a,b}$. Let us also introduce functions f_k^* and g_k^* restricting the number of available resource units allotted to the k-th resource at time $\widehat{v}$ [1]:

- $f_k^*(\widehat{v}, \widehat{Y}, DEf_k)$ is the function that gives the number of k-th resource units required at fuzzy moment $\widehat{v}$, which is contingent upon the assumed fuzzy operation start times $\widehat{Y} = (\widehat{Y}_1, \widehat{Y}_2, \ldots, \widehat{Y}_{lp})$. It is assumed that set H is the domain of the membership function $\mu(v)$ of the variable $\widehat{v}$ which determines $DEf_k \in [0, 1]$. Because function f_k^* is calculated for a given level of uncertainty, a $DEf_k = 0.8$ should be understood as meaning that the number of k-th resource units required at time $\widehat{v}$ is not larger than the number of units f_k^* with an uncertainty level of 0.8.
- $g_k^*(\widehat{v}, DEg_k)$ is the function that gives the number of units of the k-th resource available at time $\widehat{v}$, with an uncertainty level $DEg \in [0, 1]$. Moreover, the number of available resource units assigned to the k-th resource is constant over the assumed time horizon H, i.e. $g_k^*(\widehat{v}, DEg_k) = gv_k$, where $gv_k = const, \forall v \in H$.

The following inequality holds for cases of a closed loop of resource requests: $f_k(v_b, Y) > g_k(v_b)$ [3]. If variables are assumed to be imprecise, the inequality $f_k(v_b, Y) > g_k(v_b)$ can be viewed as an effect of a closed loop of resource requests, with an uncertainty level $f_k^*(\widehat{v}, \widehat{Y}, DEf_k) > gv_k$. This generalization implies that the inequality given below is a necessary condition for the occurrence of a closed loop of resource requests, with the uncertainty level DEf_k:

$$f_k^*(\widehat{v}, \widehat{Y}, DEf_k) > gv_k , \tag{18.9}$$

Additionally, if it is assumed that execution of operations cannot be interrupted, the following Lemma holds.

Lemma *[1] If allocation of resources to operations in the projects portfolio P at the given moment follows the condition* $f_k^*(\widehat{v}, \widehat{Y}, DEf_k) \leq gv_k, \forall k \in \{1, \ldots, lz\}$, *for assumed* $\widehat{Y}, \widehat{T}, CP_{a,b}, CZ_{a,b}, DP_{a,b}, H$, *then the execution of the operations does not lead to deadlocks, with uncertainty level* $DEf = \min_{k \in \{1,2,\ldots,lz\}}\{DEf_k\}$.   □

Consequently, if at any fuzzy point in time $\widehat{v}$ over the time horizon H, the following condition holds $f_k^*(\widehat{v}, \widehat{Y}, DEf_k) \leq gv_k, \forall k \in \{1, 2, \ldots, lz\}$, then operation execution is deadlock-free, with the uncertainty level $DEf = \min_{k \in \{1,2,\ldots,lz\}}\{DEf_k\}$.

Due to the above introduced assumptions the functions f_k^* and g_k^* have the following form:

$$f_k^*(\widehat{v}, \widehat{Y}, DEf_k) = \sum_{i=1}^{lp} \sum_{j=1}^{lo_a} [dp_{a,b,k} \widehat{1}(\widehat{v}, \widehat{y}_{a,b} \widehat{+} cp_{a,b,k}, \widehat{y}_{a,b} \widehat{+} cz_{a,b,k}, DEf_k] ,$$

$$\tag{18.10}$$

where:

$cp_{a,b,k} < cz_{a,b,k}, lp$—the number of projects,
lo_a—the number of operations in the i-th project,
$dp_{a,b,k}$—the number of resources of the k-th resource in use by operation $A_{a,b}$,

$\widehat{1}(\widehat{v}, \widehat{a}, \widehat{b}, DEf_k) = \widehat{1}(\widehat{v}, \widehat{a}, DEf_k) - \widehat{1}(\widehat{v}, \widehat{b}, DEf_k)$— is a unary fuzzy function determining the time of recourse occupation, where: $\widehat{1}(\widehat{v}, \widehat{a}, DEf_k)$ is the unary fuzzy function defined as follows [1]:

$$\widehat{1}(\widehat{v}, \widehat{a}, DEf_k) = 1 - \frac{DEf_k - E(\widehat{v} \ge \widehat{a})}{1 - 2E(\widehat{v} \ge \widehat{a})} , \qquad (18.11)$$

where: $\widehat{1}(\widehat{v}, \widehat{a}, DEf_k) \in \{0, 1\}$, $DEf_k \in [0, 1]$.

Function $g_k^*(\widehat{v}, DEg_k)$:

$$g_k^*(\widehat{v}, DEg_k) = gv_k = wo_{k,1} , \qquad (18.12)$$

where: $wo_{k,1}$—the available number of the k-th renewable resource.

18.3 Operational Planning of Production Orders

What interests the decision-maker most is finding answers to the following questions: *Can the order be successfully executed? (i.e. Can it be executed under specific conditions, achieving a specific profit?), What are the alternative scenarios for the execution of the order? (i.e. What are the consequences of the decisions taken?)* And How well prepared must the company be to execute the order at a pre-established efficiency level? The answers to these questions should be supplied to the decision-maker on-line; at the same time, the process of searching for those answers with the help of a decision support system should be intuitive and user-friendly, based on the mechanisms of forward inference, e.g. an answer to the question: What production effects can be achieved under the given constraints (state of preparation) of the company? And reverse inference, e.g. an answer to the question: What state of preparation of the company will guarantee the achievement of the expected production effects? Implementations of these expectations in a task-oriented DSS should, therefore, take into account the production flow models used.

The novelty of this contribution lies in that it employs a declarative (i.e. constraint logic programming based) approach that allows for questions to be asked both directly from cause to effect and in reverse, from effect to cause. The possibility of alternately formulating these types of questions allows one to determine, in an interactive mode, solutions that satisfy the constraints imposed both by the capability of the production system and the expectations associated with the execution of the incoming production order. The interactive character of the method adopted for seeking admissible solutions is an important property from the point of view of CogInfoCom research, particularly in terms of the mode and type of communication [5, 6]. Since information transfer occurs between an operator and an artificial cognitive system (e.g. the proposed DSS), the mode of communication can be classified as an inter-cognitive and sensor-bridging one [6]. In that context, the main idea of our approach is that the DSS should play a role of a cognitive actor, i.e. a system that can change its behavior based on reasoning, using observed evidence and domain knowledge [19]. In particular, this means that when the functions of a decision-making system are extended to include a function that enables recognition and classification of the operators orders (queries), the generation of DSS obtained in this way will acquire a new feature allowing these systems to learn cognitively. Cognitive learning, in turn, will enable us to take into account the influence of the particular task context on decision-making by both actors the operator and the DSS. When this level of progress is achieved, it will be possible to consider problems of co-evolution of cognitive entities with different, complementary cognitive capabilities.

Within the reference model considered in Sect. 18.2.1, many different CSPs aimed at different types of production flow operational planning can be formulated.

18.3.1 Direct and Reverse Problem Formulation

Given the time horizon $H = \{0, \ldots, h\}$, the set of production orders P, the set of resources and their availabilities Wo over H as well as the distinct and imprecise variables expressed as fuzzy numbers, i.e. sequences $\widehat{C}_a$, $\widehat{CP}_{a,b}$, $\widehat{CZ}_{a,b}$ one should be able to answer the following two questions:

Will the given resource allocation result in a schedule with a makespan that will allow the producer to meet deadline H? This question can be answered by determining the following sequences $\widehat{Y}_1, \widehat{Y}_2, \ldots, \widehat{Y}_{lp}$ have to be determined.

Does there exist a resource allocation which can ensure that the production orders are completed within deadline H? The expected answer to this question is associated with the sequences: $\widehat{C}_1, \ldots, \widehat{C}_{lp}$.

These questions express, respectively, a forward and a reverse problem of multi-product scheduling. Of course, many other questions of this type could be asked:

- **forward problems** (analytic problems in which the reasoning goes from premises to results)

  How long will a portfolio of production orders take to complete, given the routings, the operation times and the number of available resource units? Does there exist a schedule for a portfolio of production orders that can be completed within the given deadline H while satisfying the resource availability constraints at $NPV > 0$?

- **reverse problems** (synthetic problems in which the reasoning goes from results to the causes that produced them)

  Do there exist specific values of operation time and number of available resource units for which a given portfolio of production orders can be completed within the assumed deadline? What values of what variables $C_1, \ldots, C_4$ will result in a makespan of the project portfolio that will meet the given deadline, subject to limits on the available quantity of resources?

The questions related to those problems can be formulated in the same model, which can be viewed as a combination of variables and constraints. All the needs and concerns discussed above have incited the present authors to investigate the possibilities of developing a new-generation interactive decision support system based on the concept of a user-friendly, self-organizing multi-screen interface (a "touch screen"). Such a solution could be successfully applied in small and medium-sized production and/or services companies which have:

- multi assortment production or provide multi-assortment services (associated with the simultaneous execution of various projects) or which have/provide,
- one-off or short-run production/services,

and whose owners want to make direct and rational decisions (i.e. ones that are made without a chain of intermediary administrators, programmers, etc. and which guarantee feasibility) about:

- whether (or not) to accept further orders,
- how to solve random events associated with the formation of bottlenecks.

The DSS can be used as a tool to support the owner/decision maker in the process of pre-implementation analysis conducted prior to the acceptance of a new production order, which makes possible:

- evaluation of alternative execution routes for production/service orders in terms of time and cost,
- sufficiently early detection of bottlenecks and/or collisions with other projects/ orders already in progress; it can also be used in situations of random noise associated with the execution of production/service processes, caused by factors such as:

  - employee absenteeism, breakdowns of machinery and equipment,
  - changes in technology and/or order execution conditions.

In the present context, the conception proposed harks back to the structure of the so-called interactive system for navigation-like project prototyping.

18.3.2   A Dedicated Operational Planning Decision Support System

Consider a DSS that can be used to define the decision variables characterizing the production orders to be executed and the company's overall capacity. A system like this helps one to establish how to organize work flows and to determine important production parameters such as batch sizes, production cycles, work-in-progress, and so on.

When approached from this angle, the problem of supporting operational planning of production orders is seen as a problem of prototyping a production orders portfolio.

Such a portfolio is driven by an iterative schema, in which solutions to forward and reverse decision problems are formulated and assessed in an alternating sequence. A schematic diagram of navigation-like driven searching is shown in Fig. 18.1.

The available set of "folders" and "tabs" allows one to establish the values of selected elements which determine the structure and behavior of the Production Orders Operational Planning System (POPS) under consideration. The set of tabs at the user's disposal (reflecting the specific character of the adopted model) may include: resources (resource types and resource limits), their assignment to operations, execution times for the individual operations (nominal, fuzzy), order execution costs, revenue from the execution of certain order tasks, as well as system performance evaluation criteria, such as order completion times (deadlines), timely execution of orders, usage of resources, and profitability of an order or an order package, etc.

Figure 18.1 shows an example of a system interface layout which makes possible flexible selection and arrangement of tabs. The screen menu is composed by selecting tabs from the available set to suit the way the decision-making problem under consideration has been formulated. The tabs that have been selected are grouped on the screen into those related to the elements of the structure of the POPS under consideration and those related to system performance evaluation.

In the example, the structure tabs and the performance evaluation (analysis) tabs have been grouped at the top and bottom of the screen, respectively. The menu contains "folders" which comprise groups of elements that represent components of the company's structure. The user chooses the elements he is interested in on the touch panel and moves them using the drag and drop feature either to a structure tab or to an analysis tab. In this way, he uses the available set of elements to build a structure adapted to the specific needs of his/her company.

The user is only required to change the values of relevant company or project parameters, which will automatically generate changes in the project execution schedule (Fig. 18.1). These changes can be viewed by clicking the analysis tab, which activates the analysis categories pre-selected by the user regarding, for example, the execution schedule or the company's renewable and financial resource capacity. In other words, the solution proposed here allows the user to activate certain fragments of a set of screens corresponding to selected parameters (characteristics) of the company, which gives rise (as a specific reaction) to changes in the parameters assigned to other screen windows. This corresponds to a situation in which each change in the structural parameters of the system is accompanied by a change in the corresponding functional characteristics of the system. The use of the interface is very intuitive, and allows the user to analyze on-line the company's different reactions to the changing work conditions and the decisions being made.

The selecting and arranging of the tabs is a game-like situation. The aim of this game may be to search for those elements of the structure, for example, decision rules, resource limits, or costs, which either make a given set of criteria extreme, or guarantee the achievement of (are sufficient to achieve) given values of the set of criteria under consideration, for example, timeliness, utilization of resources, costs, etc. This means that in the first case one searches for such an organization of the

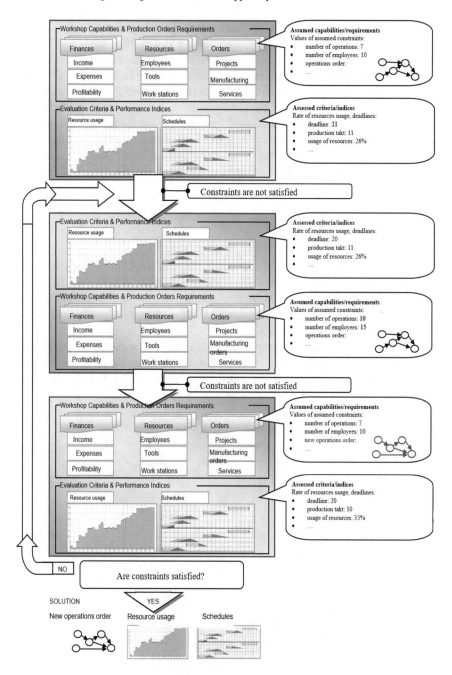

Fig. 18.1 An example of a navigation-like, iterative searching strategy

structure of the POPS, which makes extreme the given criteria of its functioning, while in the second case one searches for such a structure of the system (e.g., a permitted structure), which guarantees the achievement of the expected values of the parameters characterizing its behavior. The action scenario for the first strategy is to arbitrarily determine the values of the selected parameters of the system's structure and evaluate the effect of the changes introduced on the values of the selected system performance evaluation criteria, and the scenario for the second strategy is to determine the values of the selected evaluation criteria and to verify whether, in the given ranges of variation of system structure parameters, there exist values of these parameters which guarantee the fulfilment of the adopted performance criteria.

18.4 An Illustration

Consider the set of production routes $P = \{P_1, P_2\}$ constituted of two routes P_1, P_2, corresponding to products W_1 and W_2, accordingly, see Fig. 18.2. The sets $\{A_{1,1}, \ldots, A_{1,5}\}$ and $\{A_{2,1}, \ldots, A_{2,5}\}$ of production operations are fulfilled by the workers skilled to fulfill the following three roles: role 1—(bo_1, bo_2, bo_3); role 2—(bo_4, bo_5); and role 3—(bo_6, bo_7), see Table 18.1.

When decision variables are imprecise (fuzzy), the performance indices that follow from them also have an imprecise character. Taken under consideration here the realization times for articles W_1 and W_2 are imprecise. Moreover, because the

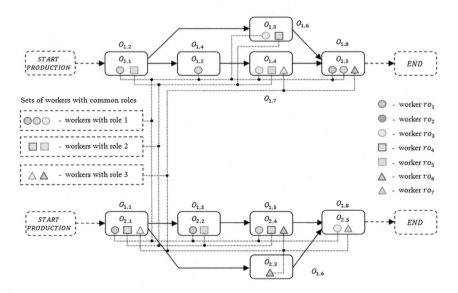

Fig. 18.2 An example of a job shop structure with operations of products W_1 and W_2 [2]

Table 18.1 Allocation of workers to operations executed along the production routes P_1 and P_2

		$A_{1,1}$	$A_{1,2}$	$A_{1,3}$	$A_{1,4}$	$A_{1,5}$	$A_{2,1}$	$A_{2,2}$	$A_{2,3}$	$A_{2,4}$	$A_{2,5}$
Role 1	bo_1	0	0	0	0	1	1	1	0	0	0
	bo_2	1	1	0	0	1	0	0	0	1	0
	bo_3	0	0	1	1	0	0	0	0	0	1
Role 2	bo_4	0	0	1	0	0	1	0	0	1	0
	bo_5	1	0	0	1	0	0	1	0	0	0
Role 3	bo_6	0	0	0	0	1	0	0	1	1	0
	bo_7	0	0	0	1	0	1	0	0	0	1

constraints coupling the decision variables are imprecise, too, the suitable grades of the membership function should be taken into consideration.

Let us assume that two products W_1 and W_2 manufactured along production routes P_1 and P_2 are processed (see Fig. 18.2) using five operations each . The related problem of operational planning of manufacturing flow can be determined by subsequent adjustments of the decision variables and goal function estimation.

A direct problem

The fuzzy variables determining operation times are denoted as z-cuts:

$$\widehat{C_1} = (\widehat{c}_{1,1}, \widehat{c}_{1,2}, \ldots, \widehat{c}_{1,5}), \alpha = \{0; 0.5; 1\} :$$

$$\widehat{c}_{1,1} = (\{[2, 6], [3, 5], [4, 4]\}, \alpha),$$
$$\widehat{c}_{1,2} = (\{[3, 5], [3, 4], [3, 3]\}, \alpha),$$
$$\widehat{c}_{1,3} = (\{[1, 5], [2, 4], [3, 3]\}, \alpha),$$
$$\widehat{c}_{1,4} = (\{[2, 4], [2, 3], [2, 2]\}, \alpha),$$
$$\widehat{c}_{1,5} = (\{[5, 5], [5, 5], [5, 5]\}, \alpha),$$

$$\widehat{C_2} = (\widehat{c}_{2,1}, \widehat{c}_{2,2}, \ldots, \widehat{c}_{2,5}), \alpha = \{0; 0.5; 1\} :$$

$$\widehat{c}_{2,1} = (\{[1, 3], [2, 3], [3, 3]\}, \alpha),$$
$$\widehat{c}_{2,2} = (\{[1, 3], [1, 2], [1, 1]\}, \alpha),$$
$$\widehat{c}_{2,3} = (\{[1, 5], [2, 4], [3, 3]\}, \alpha),$$
$$\widehat{c}_{2,4} = (\{[2, 4], [3, 4], [4, 4]\}, \alpha),$$
$$\widehat{c}_{2,5} = (\{[5, 5], [5, 5], [5, 5]\}, \alpha).$$

Seven different renewable resources $bo_1, bo_2, bo_3, bo_4, bo_5, bo_6, bo_7$, are used. The resource allocation $DP_{a,b} = (dp_{a,1,b}, \ldots, dp_{a,k,b}, \ldots, dp_{a,lo_{a,b}})$ follows Table 18.1.

Resource assignment times follow operation start times and resource release times follow operation completion times.

Consequently, $cp_{1,b,k} = cp_{2,b,k} = 0$, $j = 1, \ldots, 5$, $k = 1, \ldots, 7$. The sequences: $\widehat{cz}_{a,b,k} = \widehat{c}_{1,b}$, $j = 1, \ldots, 5$, $k = 1, \ldots, 8$, where: $\widehat{cz}_{a,b,k}$ is the fuzzy variable, as well as $Wo = (wo_1, \ldots, wo_7)$ such that $wo_1 = \cdots = wo_4 = wo_8 = 1$, are considered.

Given are the time horizon $H = \{0, \ldots, 20\}$ and the threshold $DE \geq 0.8$. $\widehat{cz}_{a,b,k} = \widehat{c}_{1,b}$, $j = 1, \ldots, 5$, $k = 1, \ldots, 8$, where $\widehat{cz}_{a,b,k}$ is the fuzzy variable, as well as $Wo = (wo_1, \ldots, wo_7)$ such that $wo_1 = \cdots = wo_4 = wo_8 = 1$, are considered. Given are the time horizon $H = \{0, \ldots, 20\}$ and the threshold $DE \geq 0.8$.

Consider the following question:

Is it possible to produce the given product within the assumed deadline of 20 time units?

The answer that is being sought regards the makespans: $\widehat{Y}_1 = (\widehat{y}_{1,1}, \widehat{y}_{1,2}, \ldots, \widehat{y}_{1,5})$ and $\widehat{Y}_2 = (\widehat{y}_{2,1}, \widehat{y}_{2,2}, \ldots, \widehat{y}_{2,5})$ where the moments $\widehat{y}_{a,b}$ are fuzzy numbers with a triangular membership function. The makespans $\widehat{Y}_1$ and $\widehat{Y}_2$, which were the most accurate approximations of the required solution, i.e. which followed the time horizon $H = \{0, \ldots, 21\}$, (see Fig. 18.3) were calculated in less than 5 min (Processor Intel i5-3470 3.2 GHz, RAM 8,00 GB).

A reverse problem

Consider production routes P_1 and P_2 consisting of ten operations (see Fig. 18.2) in which two products W_1 and W_2 are manufactured. The operation execution times are fuzzy variables determined by z-cuts $\widehat{C}_1 = (\widehat{c}_{1,1}, \widehat{c}_{1,2}, \ldots, \widehat{c}_{1,5})$ and $\widehat{C}_2 = (\widehat{c}_{2,1}, \widehat{c}_{2,2}, \ldots, \widehat{c}_{2,5})$. Operation times are unknown, but the constraints limiting those times are given [14]:

$$K1 : \widehat{c}_{2,1} \widehat{+} \widehat{c}_{2,2} \widehat{=} \widehat{8},$$
$$K2 : \widehat{c}_{1,1} \widehat{+} \widehat{c}_{1,4} \widehat{=} \widehat{5},$$
$$K3 : \widehat{c}_{1,2} \widehat{+} \widehat{c}_{2,4} \widehat{=} \widehat{7},$$
$$K4 : \widehat{c}_{1,3} \widehat{+} \widehat{c}_{1,5} \widehat{=} \widehat{3},$$
$$K5 : \widehat{c}_{2,3} \widehat{+} \widehat{c}_{2,5} \widehat{=} \widehat{3}, \tag{18.13}$$

where:
Vectors $\widehat{8}, \widehat{5}, \widehat{7}, \widehat{3}$ were determined in similar way as in [10]
In the case considered, the constraint $K1$ determines a fuzzy relationship linking the durations of operations $A_{2,1}$ and $A_{2,2}$.

In general, each constrain forms part of the set $\{a\widehat{t}_{a,b} \widehat{+} b\widehat{c}_{k,l} \widehat{=} \widehat{c} : a, b \in N \wedge \widehat{c}_{a,b}, \widehat{c}_{a,b}, \widehat{c}$ are fuzzy numbers$\}$. Seven different renewable resources $bo_1, bo_2, bo_3, bo_4, bo_5, bo_6, bo_7$ are used. The resource allocation $DP_{a,b} = (dp_{a,1,b}, \ldots, dp_{a,k,b},$

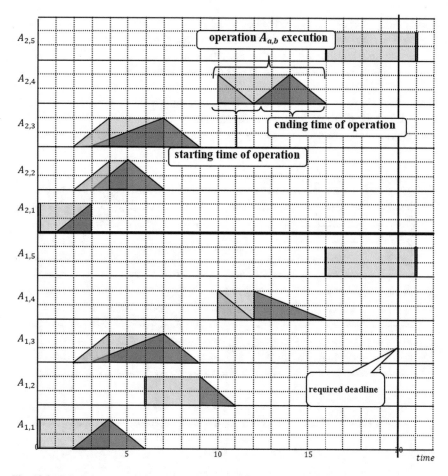

Fig. 18.3 Schedule proposed for the time horizon $H = \{0, 1, \ldots, 21\}$

$\ldots, dp_{a,lo_{a,b}})$ follows Table 18.1. Resource assignment times follow operation start times and resource release times follow operation completion times. Consequently, $cp_{1,b,k} = cp_{2,b,k} = 0$, $b = 1, \ldots, 5$, $k = 1, \ldots, 7$. The sequences: $\widehat{cz}_{a,b,k} = \widehat{cz}_{1,b}$, $b = 1, \ldots, 5$, $k = 1, \ldots, 8$, ($\widehat{cz}_{a,b,k}$ means the fuzzy variable $\widehat{cz}_{a,b,k}$) as well as $Wo = (wo_1, \ldots, wo_7)$ such that $wo_1 = \cdots = wo_4 = wo_8 = 1$, are considered.

Given are the time horizon $H = \{0,1,\ldots,21\}$, and the threshold $DE \geq 0.8$.
How long must an operation last for the given production orders portfolio makespan not to exceed deadline H?

The solution requires determining of the following sequences: $\hat{C}_1 = (\hat{c}_{1,1}, \hat{c}_{1,2}, \ldots, \hat{c}_{1,5})$, $\hat{C}_2 = (\hat{c}_{2,1}, \hat{c}_{2,2}, \ldots, \hat{c}_{2,5})$ and $\hat{Y}_1 = (\hat{y}_{1,1}, \hat{y}_{1,2}, \ldots, \hat{y}_{1,5})$, $\hat{Y}_2 = (\hat{y}_{2,1}, \hat{y}_{2,2}, \ldots, \hat{y}_{2,5})$, where $\hat{c}_{a,b}$ is the duration of the a-th operation and $\hat{y}_{a,b}$ is the start time of the a-

th operation; both of them described by a triangular membership function. OzMozart was used to implement the constraints on the order of operations (18.5), (18.6) and (18.7), resource conflict (21) and operation time limits (23). A first admissible solution $\hat{C}_1$, $\hat{C}_2$ and $\hat{Y}_1$, $\hat{Y}_2$ (see Fig. 18.4) was obtained within 180 s.

The methodology proposed in this study is consistent with the general approach to industrial applications of cognitive infocommunications in areas such as production management and production engineering. The major advantage of this approach is

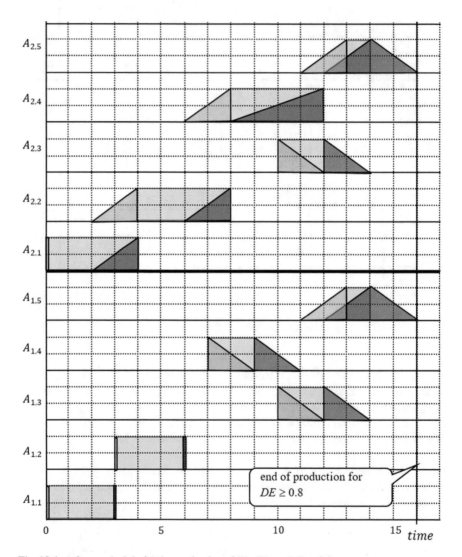

Fig. 18.4 A fuzzy schedule for the production of W_1, W_2, at $DE \geq 0.8$

that the search for the appropriate decision is based on different game-like strategies which allow the decision maker to change his/her viewpoint, e.g. from operated to operating and then from operating to operated, and thus cultivate his/her cognitive functions, ultimately leading to a better decision. In the present study, an Artificial (A) Cognitive (C) Capability (C) (ACC) for Infocommunication (I) Systems (S) is propounded [6].

18.5 Concluding Remarks

Needless to say that expected (not all) and/or system behaviors (required) can be actualized subject to the supposed structural constraints. The same holds true for system structure: not all feasible structures can produce an assumed system behavior. This study advances a method in which both of these perspectives can be considered simultaneously in the decision-making process. To put it differently, this contribution is about two types of decision-making strategies: one that focuses on the conditions which can lead to the required system behavior and one that concentrates on the conditions that have to be met to achieve a system structure capable of producing the desired behaviors. Systems of this type can be applied in online batching and routing of manufacturing orders, task scheduling, and resource allocation, to give just a few examples. They may prove particularly useful in settings involving unused resources and/or system capabilities. In future, we plan to delve into the subject of robust scheduling of work orders which allows to take account of both internal disturbances (e.g. due to workstation malfunctions) and also external ones (e.g. due to changes in manufacturing orders).

References

1. Bach I, Bocewicz G, Banaszak Z, Muszyski W (2010) Knowledge based and CP-driven approach applied to multi product small-size production flow. Control Cybern 39(1):69–95
2. Banaszak Z, Bocewicz G (2016) Declarative modeling driven approach to production orders portfolio prototyping. In: New frontiers in information and production systems modelling and analysis, vol 98. Springer, pp 141–168
3. Banaszak Z, Bocewicz G, Bach I (2008) CP-driven production process planning in multiproject environment. Decis Mak Manuf Serv 2(12):5–32
4. Barták R (2004) Incomplete depth-first search techniques: a short survey. In: Figwer J (ed) Proceedings of the 6th workshop on constraint programming for decision and control. pp 7–14
5. Baranyi P, Csapo A (2012) Definition and synergies of cognitive infocommunications. Acta Polytech Hung 9(1):67–83
6. Baranyi P, Csapo A, Sallai G (2015) Cognitive infocommunications (CogInfoCom). Springer International. ISBN 978-3-319-19607-7. https://doi.org/10.1007/978-3-319-19608-4
7. Beale EML (1979) Branch and bound methods for mathematical programming systems. In: Discrete Optimization II. North Holland Publishing, Amsterdam, pp 201–219
8. Bocewicz G, Banaszak Z (2013) Declarative approach to cyclic steady state space refinement: periodic process scheduling. Int J Adv Manuf Technol 67:137155

 9. Bocewicz G, Banaszak Z, Wójcik R (2007) Design of admissible schedules for AGV systems with constraints: a logic-algebraic approach. LNAI 4496, Springer, pp 578–587
10. Bocewicz G, Klempous R, Banaszak Z (2016) Declarative approach to DSS design for supervisory control of production orders portfolio. In: 7th IEEE international conference on cognitive infocommunications (CogInfoCom). pp 385–390
11. Dang Q-V, Nielsen I, Steger-Jensen K, Madsen O (2014) Scheduling a single mobile robot for part-feeding tasks of production lines. J Intell Manuf 25:1–17
12. Druzdzel MJ, Flynn R (2002) Decision support systems. In: Kent A (ed) Encyclopedia of library and information science, 2nd edn. Marcel Dekker, Inc., New York
13. Dubois D, Fargier H, Fortemps P (2003) Fuzzy scheduling: modeling flexible constraints vs. coping with incomplete knowledge. Eur J Oper Res 147:231252
14. Holtzman S (1989) Intelligent decision systems. Addison-Wesley, Reading, MA
15. Khayat GE, Langevin A, Riope D (2006) Integrated production and material handling scheduling using mathematical programming and constraint programming. EJOR 175(3):1818–1832
16. Krenczyk D, Kalinowski K, Grabowik C (2012) Integration production planning and scheduling systems for determination of transitional phases in repetitive production. Hybrid Artif Intell Syst 7209:274283
17. Mora M, Forgionne GA, Gupta JND (2003) Decision making support systems: achievements. In: Trends and challenges for the new decade. Idea Group Publ, London
18. Zimmermann HJ (1994) Fuzzy sets theory and its applications. Kluwer Academic Publishers, London
19. http://www.vernon.eu/euCognition/definitions.htm

Chapter 19
Improving Adaptive Gameplay in Serious Games Through Interactive Deep Reinforcement Learning

Aline Dobrovsky, Uwe M. Borghoff and Marko Hofmann

Abstract Serious games belong to the most important future e-learning trends. Yet, balancing the transfer of knowledge (the serious part) and the entertainment (the playful part) is a challenging task. Whereas a commercial game can always focus on creating more fun for the player, a serious game has to ensure its didactic purpose. One of the major problems in this context is the rigidity of standard script-driven games. The script may guarantee the educational success of the game, but due to its rigid storyline and, consequently, its fixed gameplay, the game may also be boring or overtaxing. Adaptive gameplay seeks to overcome this problem by individualizing the storyline, the difficulty of the game or the amount of context information given to the player. Our idea is to use interactive deep reinforcement learning (iDRL) to maximize the individualization with respect to the context information. Although DRL has been applied successfully to automated gameplay it succeeded mainly at optimization-like tasks in largely short-horizon games. Our approach is to augment DRL with human player and trainer feedback in order to direct the learning process. Our goal could be described as a synergistic combination of human involvement in games and DRL as an emergent cognitive system that helps to adapt the game to players' preferences and needs. We call this approach interactive deep reinforcement learning for adaptive gameplay.

A. Dobrovsky (✉) · U. M. Borghoff · M. Hofmann
Fakultät für Informatik, Universität der Bundeswehr München, 85577 Neubiberg, Germany
e-mail: aline.dobrovsky@unibw.de

U. M. Borghoff
e-mail: uwe.borghoff@unibw.de

M. Hofmann
e-mail: marko.hofmann@unibw.de

© Springer International Publishing AG, part of Springer Nature 2019

411

R. Klempous et al. (eds.), *Cognitive Infocommunications, Theory and Applications*,
Topics in Intelligent Engineering and Informatics 13,
https://doi.org/10.1007/978-3-319-95996-2_19

19.1   Introduction

The term serious game (SG) origins from [1] and refers to games not exclusively developed for mere entertainment, but primarily for creating educational value. SGs are counted among the current e-learning trends and are gaining more and more acceptance and influence [12]. SG development slightly differs from that of entertainment games: SGs are often individual products for a restricted target audience, industrial branch or company. This results in high expenditure and deficient reusability, although the market shows growing interest in cost-efficient and customized applications [12]. Some of the most relevant, but also most time-consuming tasks are automated content adaptation and the creation of believable non-player character (NPC) behaviour [7, 13, 27]. Thus, simplifying the authoring of AI and supporting automated adaptation in SGs seem desirable and profitable.

Deep Reinforcement Learning (DRL) has already been successfully used for automated NPC behaviour generation. Nevertheless, several issues arise when applying DRL to create believable and diverse behaviour in complex scenarios. Moreover, the higher-level abstract task of game content control holds additional difficulties. We indicate a way to overcome these challenges by combining DRL with human guidance, including effective collaboration between a learning system, human players and domain experts.

In this paper, we give an introduction to the background of reinforcement learning (RL), deep reinforcement learning and interactive reinforcement learning (iRL) and to the ideas and challenges of adaptation in serious games. We show related approaches and depict current issues of applying DRL methods to games and SGs. We briefly outline a simplified strategy game we designated as a suitable test scenario for our method and shortly introduce the SG SanTrain as a future application possibility. Furthermore, we indicate some challenges and ideas for applying adaptive mechanisms to our game and SGs like SanTrain. We introduce our approach of interactive deep reinforcement learning (iDRL), show how it will be integrated into a flexible framework and describe some specific aspects regarding DRL and iRL. Finally, we indicate how our approach could enhance the use of automated behaviour generation and automated adaptation in games and SGs by exemplarily indicating several application opportunities. [1]

---

[1]This paper is an extended version of a conference paper that was presented at CogInfoCom 2016 [9]. In contrast to a further, previously published extended version [10], which concentrated on the different approaches in interactive learning and our future research scenario SanTain, this version focuses on the aspect of adaptivity, introduces our current research scenario and adds several new insights.

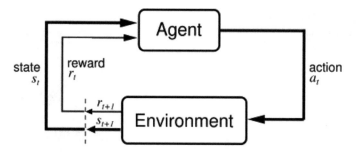

Fig. 19.1   Agent-environment interaction in reinforcement learning [42]

19.1.1   Deep and Interactive Reinforcement Learning

In order to comprehend the presented approach to the fullest extent, it is advised to acquire basic knowledge of deep reinforcement learning (DRL). Hence, this section only briefly mentions the basics of reinforcement learning (RL), DRL and interactive reinforcement learning (iRL).

Figure 19.1 shows an outline of the standard reinforcement learning process over a sequence of discrete time steps. In RL, an agent learns a mapping of situations to actions by interaction with and getting reward from an environment. This mapping is the agent's 'policy' and describes the probability of selecting action 'a' in state 's'. The goal of a RL agent is to maximize the received reward. For further, detailed descriptions the comprehensive RL overview of [42] is recommended.

In deep reinforcement learning, RL is combined with machine learning, particularly neural networks, that are used as function approximators. When applying DRL to a SG, the game provides the environment from which the agent receives a representation of the current game state and a reward value. The action to be executed is selected depending on the current input of the state representation and the learned behaviour policy.

In interactive reinforcement learning, the learning of an RL agent is augmented with additional (or exclusive) feedback provided by a human trainer. Expected benefits are that feedback can be given at arbitrary times during learning, that convergence is accelerated and that, in complex tasks, the learning process can be guided. Different interaction methods are possible, e.g. by providing additional numeric rewards, advising a concrete action or demonstrating a specific desired behaviour. Figure 19.2 shows an outline of a basic iRL process; more information is for example provided in [41].

Fig. 19.2 Example of an
iRL framework with
interactive feedback [41]

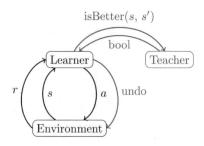

19.1.2   Objectives

Our motivation is to support the development of adaptive gameplay and believable
AI and thereby simplify serious game development, enrich player experience and
support learning. Although machine learning can support automated development of
diverse AI behaviour, the application of ML requires ample availability of machine
learning experts, complex software integration processes into existing games or profound changes to an existing game and long training times on usually powerful
hardware. Furthermore, the common vague objectives of game AI like *fun* in games,
plus *learning* in SGs or the frequently used *staying in flow* are hard to encode as goals
for an optimization ML algorithm. We aim to overcome some of the current problems by making use of expert knowledge and player feedback and thereby strive to
establish a kind of synergistic collaboration between humans and cognitive systems.
Our concepts are to be implemented within a framework that is simply adaptable to
a variety of games.

19.2   Adaptive Gameplay in Serious Games

An adaptive computer game customizes itself to some extent to each individual player.
In most computer games, content (central stories, levels, maps, events) and context
information (storyline decoration, visual and sound effects, background information,
narrative help functions, etc.) are rather static and rigid. A pre-defined script, however,
often leads to predictable and impersonal gameplay, which in turn results in reduced
attention. Adaptive gameplay "has therefore been recently proposed to overcome
these shortcomings and make games more challenging and appealing" [29].

In commercial games, adaptivity is often reduced to the fitting of a few difficulty
parameters of the game to the player's performance or his self-assessment. The fitting
mechanism ensures that the player stays within the *flow channel* (see Fig. 19.3) of
human behaviour, which avoids boredom as well as excessive demands.

More sophisticated approaches for adaptive gameplay are generally based on a
model of the player, which is created on the basis of the player's actions (via mouse or
keyboard), selected preferences (in the pre-setting of the game), style of interacting

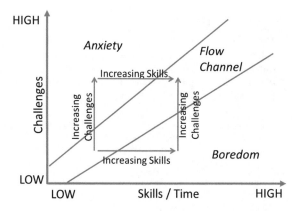

Fig. 19.3 Concept of flow, based on Csikszentmihalyi [8]

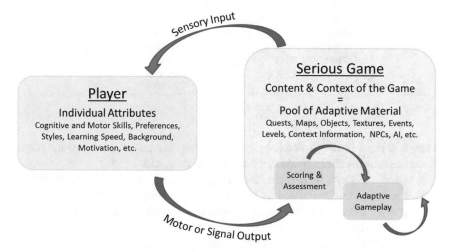

Fig. 19.4 Outline of SG to player adaptation in general

with the game (for example frequency and intensity of play, explorer or achiever type of player) or measurement and analysis of psychophysiological signals. The adaptation mechanism exploits this information in order to create a dynamic and flexible gameplay, potentially enthralling the players (see Fig. 19.4).

The game's difficulty is just the most obvious aspect that can be influenced by the adaptation mechanism. More ambitious is the generation of player-specific *game content* or *context* (see Table 19.1). Even in commercial games it is not trivial to implement such *higher-level adaptations*, because they might change the semantics of the game (in contrast to the purely syntactical change of difficulty). High-level adaptive gameplay might become "standard" in commercial games, too. Yet, it seems more likely that individualized content and context manipulation will first start to

Table 19.1 Overview of types of adaptation in SGs

	Description	Based on	Effected with	Goal
a	Adjusting difficulty to the player's performance	Current scoring or direct manipulation of difficulty by the player	Adjustment of different parameters	Keep player in flow corridor
b	Content and context adaptation to player's deficits	Automated debriefing and selection of levels	Predefined rules	Knowledge transfer
c	Content and context adaptation to player's performance	Current scoring and selection of content	Optimization methods	Increased performance
d	Content and context adaptation to player's attention	Attention measurements	Methods stated above, new approach: interactive deep reinforcement learning?	Increased attention to the game

flourish in the serious gaming field, where the ongoing concentration of the player is as paramount as the individualized adaptation to specific skills.

It is important to note that the last remarks are only adequate for *unsupervised* serious games and simulations. A trainer who constantly controls and steers the learning of a trainee could easily ensure pertinent knowledge transfer. With respect to the knowledge transfer into reality, the human trainer is in general far superior to any machine. The human adviser could, for instance, discuss and emphasize unrealistic events and results of actions taken in the game. Yet, in the applications domains we have in mind, the expert trainer is a scarce resource. All the learning in the serious games addressed in this paper should allow for taking place in unsupervised form. Any solutions that necessitates an adviser foil a central advantage of serious games: the possibility of autonomous learning.

The high-level adaptation mechanisms we know of are all based on rule sets that steer the selection of new levels, maps or events. We are convinced that the inherent limitations of this rule-set-approach known from research in AI will also limit the power of adaptation in gameplay. Our idea is to replace the rules by an interactive deep reinforcement learning approach that individualizes the adaptation mechanism to every single player. Whereas the rule set of adaptation is fixed for all players, the DRL network is intended to develop a user-specific logic of content or context selection. The basis of this approach is a direct and continuous measurement of the attention of the player while varying content or context. The task of optimizing the adaptation process to the individual player is transferred from the game developer to the DRL network.

This idea can be illustrated with a simple example: The attention of the player, which is used as *evaluation function*, is measured via user interactions (mouse clicks) per time unit (10 s, for example). The input to the net are the variations of context information like sound effects or visual presentation of environment objects or behaviour patterns of NPCs. Deep reinforcement learning will optimize these aspects individually for each player.

In our approach for unsupervised serious games, iDRL is hitherto only intended to optimize the attention of the player and not the learning itself. In contrast to some other approaches (e.g. [39, 40]), we are convinced that it is too difficult to improve the effectiveness of the knowledge transfer directly between the game and the player in unsupervised games. The most important reason for this self-restriction is the intricate process of knowledge transfer from the game environment to the real world reference system (without the direct control of an adviser). This transfer can only be assessed by costly evaluations on the basis of series of experiments (in the reference system). The level of attention, in contrast, is a parameter that is attributed to the game. It is obviously only a precondition for learning, not the actual learning itself. However, it is hardly imaginable how a serious game could transfer knowledge without strong attention.

The measurement of attention by mouse clicks or keyboard strokes can only be the very first idea of capturing attention. More sophisticated approaches are for example presented in [21, 43, 44]. In the long run we expect to evolve the iDRL approach in order to capture the different learning preferences and backgrounds of students in the adaptation mechanisms of serious games.

19.3   Related Work

We thoroughly investigated the possibilities of of combining DRL and iRL and found no contiguous research in our research area. In particular, we are not aware of any framework or approach offering interactive DRL to support SG adaptation and supported AI behaviour development. Nonetheless, we were encouraged by different inspiring ideas and current research in the areas of DRL, iRL, General Game Playing and game adaptation. In [10] we already gave an overview of several closely and distantly related approaches. This paper repeats those approaches that present a major influence on our work, adds some recently published works considering concrete applications and covers the aspect of adaptation in particular.

At first, our idea was inspired by the research area of General Game Playing [18] and General Video Game Playing(GVGP) [38]. These approaches prove that the development of algorithms that are able to play previously unknown computer games immediately is possible. However, it is essential that the game itself is modelled in the stipulated game description language, which is a declarative logic programming language in GGP and a relational, object-oriented representation of the game world's features in GVGP. In contrast, we rather envisioned a system that can learn on an customary game, developed with a usual game engine, without the need of profound changes to the game itself or the development of a specific representation language.

This lead us to the investigation of the approach introduced by DeepMind technologies (later Google DeepMind) in 2013. The authors state to present the first successful approach of applying RL for learning control policies directly from high-dimensional sensory input [33, 34]. They use the raw game screens provided by the Arcade Learning Environment (ALE) [5] as input for training a convolutional neural network with their novel method *Deep Q-learning with Experience Replay*. The reward needed for RL is provided by the ALE game score. The output is composed of the estimated values for the game actions represented by possible joystick controller moves. The approach is tested on several games of the ALE and partially outperforms other approaches and even human players.

This approach of Deep-Q-Learning is also taken up in [26] and applied to a simple 3D FPS game. As input, the agent receives the game screen and, during training, additional game information. In the evaluated deathmatch test scenario the trained agent outperforms an average human player.

An influential advancement certainly is the method of asynchronous advantage actor-critic (A3C) learning introduced in [35]. The proposed DRL agent architecture can be trained efficiently on Atari games on customary hardware and yet outperforms previous approaches. Furthermore, this method succeeds on motor control tasks and on 3D games like the TORCS car racing simulator and a 3D random maze. Due to its success, the A3C approach is subject to ongoing enhancement and is more and more extensively examined and applied. Besides, the actor-critic method can easily be extended from discrete, low-dimensional to continuous action spaces [28].

The recently published overview "Deep Learning for Video Game Playing" [20] examines various state of the art uses of several deep learning approaches in different game genres. The authors compare the methods with regard to applicability to and successes in different problem settings in different genres. It is pointed out that strategy games pose a very complex challenge for deep learning and only restricted sub-problems can be reasonably dealt with so far. It can also be concluded that A3C has been successfully applied in several game genres and that it features a good combination of learning quality and speed.

Adaptivity in games is an extensive and diverse research area, hence we can only present a small proportion of basic and specialized literature in this field. The authors of [11] provide a broad overview of game adaptation basics, player and learner modelling, adaptive storytelling and sequencing. The literature review about adaptation in affective video games [6] offers a comprehensive overview of affective adaptation; covering behavioural signal measurement and interpretation, models of emotion recognition and player representation, different adaptation mechanisms and a comparative analysis of affect-based adaptation methods.

A concrete example of the use of psychophysiological signals, often called biofeedback signals, for difficulty adjustment is described in [43, 44]. The authors present their design of a reusable framework for integration of physiological signals (EEG in particular) into educational gaming on mobile devices. Through measurement of a player's signals and game performance, his mental state can be inferred. This is used to hold the player within the flow state and increase learning effectiveness; by adjusting the level of difficulty and game reward automatedly and in combination with an external supervisor.

In the thesis [48], a framework for automated adaptation of collaborative multiplayer SGs is presented. Through tracing different criteria (game variables), game situations can be recognized and classified. Based on this and a group model (considering player characteristics), the best possible adaptation is evaluated and applied via defined adaptation objects with described effects. Another approach described in the thesis [29] uses gameplay semantics for adaptive generation of game worlds. A gameplay semantics model is defined, including gameplay descriptions of meaning and value of a game world and its objects. By using existing player and experience modelling techniques and a mapping between game content and experience, game worlds tailored to specific player models can be generated.

19.4 Problem Statement

Our examinations focus on providing assistance to SG AI developers in game adaptation and supported NPC behaviour development. As we stated in [9, 10] we envision a general approach, to the extent possible, and therefore consider requirements for the use in different games, game representations and game genres. We concluded that our idea requires a method that covers different state-action space representations and possibilities to access in-game information. The classical methods used in the game industry for both of our problem settings, automated adaptation and NPC behaviour generation, seem to be mainly based on scripting and rule-based systems, which implies laborious hand-crafting of AI. This can lead to less dynamic and less adaptive behaviour and thereby makes these hard-coded approaches easily exploitable [37]. By contrast, automated adaptation in games could relieve developers of the need to anticipate reasonable behaviour for every possible situation, especially in complex worlds (cf. [31]).

In the following subsections, we explain the application possibilities of DRL in games and accompanying issues and challenges. We introduce our current research scenario and the SG SanTrain as an example of a modern 3D serious game we anticipate to use as an application scenario in future research. At last, we discuss some aspects concerning the application of adaptivity in these scenarios.

19.4.1 Applicability of DRL in Games

In an expansive survey reviewing the use of AI techniques in SGs [15], the authors examine decision-making and machine learning methods that are applied in released SGs. It can be concluded that classic approaches, e.g. decision trees and finite-state machines, are used more frequently than ML. Regarding the less often applied ML techniques, they are primarily used for supervised learning tasks, e.g. gameplay analysis and user classification. However, ANN and case based reasoning are also used for game adaptation, content customization and NPC behaviour generation.

In the computer games industry, a variety of machine learning methods suitable for utilization in games are known (cf. [32, Chap. 7]). Some state of the art applications are for example shown in [36]. Nonetheless, the corresponding authors also point out the prevalent challenges in applying ML to games. ML algorithms often need long time and vast data for training. Additionally, a meaningful target function as learning objective must be defined. However, most often only restricted time and limited resources are available to game AI.

Furthermore, as learning methods are known to sometimes produce unpredictable outcomes and to offer only limited control over the learning process, improved explainability of algorithms and their results is an important factor to gain the trust of developers in the algorithms' decisions.

On the one hand, learning with a high-level symbolic knowledge representation allows for a more straightforward and understandable AI manipulation for game developers (cf. [16]) than sub-symbolic, connectionist systems. On the other hand, the symbolic approach also requires expense in obtaining this knowledge representation. This includes the explicit specification of relevant game objects, goals, etc. for AI and thus a mandatory, individual modification of the game.

Reinforcement learning can easily be applied to games, because they provide a suitable RL learning environment, including an input signal (current game state), possible game actions and a reward that can be derived from the game score. Nonetheless, specific problems of RL must be handled in computer games [30]: curse of dimensionality (large state-action spaces), partial observability problem (hidden states), generalization and exploration-exploitation dilemma, credit structuring and temporal credit assignment problem (delayed and sparse rewards). This can lead to huge efforts in setting up and running a RL implementation, particularly with regard to slow convergence rates and convergence to undesired behaviour.

In many cases, the size of the state-action space limits the application of RL in games. Considering modern, complex 3D multiplayer games like real-time strategy games, the AI has to act in real-time in environments that contain a great deal of various game objects and many (non-)deterministic events. It is hard to provide a comprehensive definition of the state-action space in such a game.

The application of DRL can at least reduce the issues connected with large state-action spaces and generalization, because it has proven to enable efficient training on large datasets and raw data input. The use of artificial neural networks (ANN) improves generalization, even on unknown states. The approach presented in [33] shows that fundamental changes to an existing game are not always necessary and that efficient application is possible.

The exploration-exploitation dilemma can be faced by fine-tuning of DRL hyperparameters. The partial observability and the credit structuring and credit assignment dilemmas largely remain open issues, which is demonstrated by the game score results of [34]. According to previous research, DRL can handle quick-moving, complex games based on visual input; but severe issues remain with long-horizon games that offer only sparse rewards (e.g. platformers like *Amidar* or games with open, complex worlds and differently visualized information like *Battle Zone*).

19.4.2 Real-Time Strategy Game Research Scenario

Real-time strategy (RTS) games offer various challenges for AI. An implementation to successfully play an RTS would need complex analysis and planning abilities. The tasks involve map exploration including partial observability, resource management and economic planning, setting up buildings and troops, tactical decisions and group movements. Many SGs belong to the strategy games genre or at least contain some of its core characteristics. This covers training games and simulations in military, disaster management or humanitarian aid scenarios and many games that address training of analysis and planning capabilities in general. Developing AI to play a regular RTS end-to-end presents a serious challenge and it appears that this task couldn't be realised so far. Research in this area rather concentrates on solving small subproblems in these games (cf. [20]).

Instead of extracting a small part of a complex game, a complete game with a limited range of tasks should be taken into consideration. Games related to the Tower Defense (TD) subgenre definitely seem suitable. TD games omit or decrease the challenges of a complex economy, resource management, manual troop movement, construction of functional buildings etc. Yet, numerous interesting tactical challenges remain for AI game playing and multiplayer scenarios: the control of units (single and group behaviour, decision making, pathfinding) and choice and positioning of defensive elements (towers). Hence, our first and main research game scenario shall be a TD-like RTS game in 1v1 mode. AI can control type and placement of towers, unit waves, types and behaviour patterns. Possible applications of iDRL in this research scenario are mentioned in Sect. 19.5.4.

19.4.3 Example: SanTrain—A Serious Game for Tactical Combat Casualty Care (TCCC)

SanTrain (Fig. 19.5) is a serious game in development in the domain of military first aid application. It provides a game-based learning and training platform to train TCCC decision making skills in an ego-shooter perspective. TCCC means specialized first aid on the battlefield and is based on a series of simple life saving steps and clear priorities for the first minutes following injuries. It is practised regularly in armed forces by medical personnel and also regular servicemen.

SanTrain is aimed at providing parts of TCCC training by simulating major aspects in a SG (complementary to practical training) and includes, among others, a plau-

Fig. 19.5 Screenshot of SanTrain

sible pathophysiologic model of the human body. The game part of SanTrain has to enhance the learning motivation through appropriate visualization, scenario realism and convincing supportive AI capabilities. Furthermore, there are customary requirements for cost-effectiveness, simple operability and adaptability, especially when considering the limited number of potential users and limited availability of trainer capacities in very specific domains. Further descriptions of SanTrain, its architecture, its development and TCCC in general can be found in [10].

As SanTrain offers a very flexible architecture we endorse the integration of elaborate AI methods. We are convinced that, in the future, our approach can be profitably applied to NPC control and scenario adaptation in a SG like SanTrain.

19.4.4 Aspects of Adaptive Gameplay

Adaptive Gameplay is of vital interest in games and SGs because the actual game could be played by lots of very different students. Taking into account their individual skills, preferences, background and gaming styles is decisive for the game's success. After examining traditional adaptation mechanisms (steering the difficulty level, selecting new levels according to deficits, optimizing map selection for specific skill training) we started to investigate into the possibilities of machine learning approaches. The main advantage of incorporating iDRL into the adaptation process is the flexibility of the concept and the achievable individuality of the steering. An iDRL steering mechanism could adapt itself to each single player without any predefined rules.

As already mentioned in the overview on adaptive gameplay in serious games (Sect. 19.2), we are convinced that steering the learning process directly is a difficult endeavour in all serious games. The reason is that in serious games, the learning should be effective in the real world reference system, which is not assessable in the game. From military simulations it is well-known that players often adapt to the intricacies of the game rather than to the needs of reality. This is called the "gamer mode". In order to avoid the manifold impasses generated via the gamer mode, either advisers have to guide the learning or, with machine learning techniques, the adaptation has to be restricted to attributes of the game itself.

Since SGs like SanTrain are often intended to be played without any trainers we focus our first conceptual test environment for iDRL on player attention to the game. We presume that the level of interaction with the game (motoric output of the player) is a simple yet sufficient indicator of attention, although the exemplary frequency of clicks is a game-specific attribute. As first ideas of adaptation to the player we have identified different areas: general behaviour patterns of enemies, tactical positioning decisions of an adversary player, textures and acoustic context hints (weapon effects, guiding communications, distracting noise), with the latter being ubiquitous in first-person shooter-like games like SanTrain. The approach can be easily extended to other aspects like visual effects or context information.

An open question remains the effect of increased game attention for the learning effects in the real world reference system. This question must be answered in one of the evaluation experiments planned after the realization within the test environment.

Notwithstanding the preliminary results of our research we can already conclude that iDRL seems a promising approach to supplement traditional methods of adaptive gameplay in serious games. We are convinced that the approach will help to increase the individual adaptivity of steering mechanisms in computer games.

19.5 Approach to an Interactive Deep Reinforcement Learning (iDRL) Framework

To overcome the mentioned challenges and support SG developers, we propose to offer relevant functionalities within a new interactive DRL (iDRL) framework. We aim for a system that provides as general a solution as possible for applying our iDRL approach to SGs. This includes that no profound changes should be conducted to an existing game except to provide the necessary input and output values and control options needed for DRL.

In the following, we present an outline of our framework, point out how our chosen concept of DRL is applied and then focus on the communication and its immanent challenges between the framework and human experts or players. At last, we discuss possible application areas of iDRL in our research environment and in general. Our overall prospect is a modular, reusable and easy to use framework, which provides support in developing believable and variable AI for adaptation and NPC control through interactive learning. A concrete realisation is currently being implemented in conjunction with a TD like strategy game as research scenario, as indicated in Sect. 19.4.2. The game is being developed on a usual game engine in order to show that our approach is feasible on common games and to indicate which concrete changes and adaptations are needed during development. Furthermore, we have started to examine several related aspects as DRL architectures, DRL visualization possibilities, game information representation and reward structure of the game and interacting humans.

19.5.1 Outline of the iDRL Framework

The general demands we impose on our framework are modularity, efficiency and scalability (concerning deployment on customary hardware) and parametrized control of structures and processes. Our basic architecture outlined in Fig. 19.6 will offer some default components and functionalities.

Regarding future research, the flexible and generic approach will allow for the possibility to compare different learning and expert knowledge integration methods. The possibility to run multiple learning instances should be integrated to facilitate efficient use and quick convergence, which can be provided by using the A3C approach (cf. [35]). Game instances can be operated by players or be used for autonomous, parallel learning. The connection between several instances is established through a separate, adaptable game interface.

Our default learning process will be a variation of DRL. The learning process and information exchange is controlled by the framework. In order to connect the learning agent to the game, several game aspects should be made known through the inter-

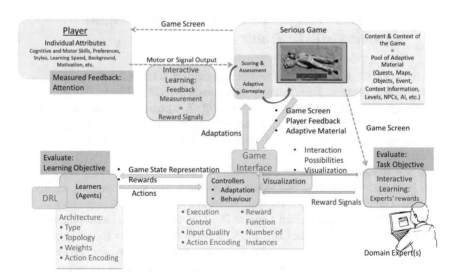

Fig. 19.6 Outline of essential components of and interactions within the iDRL framework regarding game adaptation

face. These include adaptive material, possible adaptations, all actions, game score rewards and possible input information. The latter can include variants of processed screenshots and maybe further game information. Therefore, the necessary changes to the game have to include this information access and execution control, actions and adaptation application. Further configuration possibilities include the specification of input types and quality (e.g. game screen resolution) and the action encoding (e.g as keyboard inputs, basic or composite game actions). We leave the architecture as flexible as possible, thus allowing for individual configuration options, exchange and addition of components etc. Nevertheless, the developer will get support through default parameters and automated configuration options.

Some of the interactive aspects of the framework are marked in red colour. The player's actions or biofeedback signals have to be measured and translated into usable reward values. Additionally, domain expert's rewards have to be integrated. A suitable visualization of the DRL's state and actions could provide an insight in the learner's decision-making. The interaction of the learner with the player or domain expert has to be masked in a way that makes it simple and natural to be used by non programmers and non ML experts.

More specific DRL and iRL aspects and associated design objectives are discussed in the next subsections.

19.5.2   Application of Deep Reinforcement Learning

Deep reinforcement learning is the central aspect in our framework. In this section, we shortly describe which conditions for its application in SGs arise from the issues described in Sect. 19.4 and why the existing approaches of using DRL in games are an appropriate basis for a first realisation in our framework. As already stated, we assume to have no or sparse explicit state information (but access to the raw game screen) and no forward model, a variety of possible actions (depending on game and encoding) and probably no game log data from actual games for (pre-)training of ML algorithms. If high-dimensional information representation should be used as input for the training, huge state spaces must be handled. Utilizing the visual representation of the game screen means the state space is at least composed of the set of all possible screen representations as pixel values. This obligatorily leads to the application of model-free learning methods and approximation functions. Model-free RL methods learn a value function directly instead of a state-transition model and don't presume the functionality of predicting next states and consequences of actions. Widely used model-free methods are Temporal Difference (TD)-learning based approaches, like Q-learning, Sarsa and actor-critic methods (see [42]). Another advantage of these methods is the possibility to update the learning process each time some kind of reward value can be provided. In addition, the use of approximation functions allows handling of large state-spaces. They scale well and are able to generalize over unknown states, whereby retaining a compact representation. At first, we pursued an approach similar to Google's Atari deep Q-network (DQN)

[33]. Deep Learning based on CNNs has proven to work well on high dimensional input like game screens, but still showed drawbacks on problems including time dependencies, sequences and planning, as needed in long horizon games. By now, we have decided to rely on the A3C approach (cf. [35]) because it outperforms DQN in direct comparison on games [20] an can also be trained efficiently on restricted hardware resources.

19.5.3 Application of Interactive Reinforcement Learning

Interactive Reinforcement Learning is the second key aspect of our approach and has been extensively discussed in the previous publications [9] and [10]. Hence, this section only sums up the major aspects. We explain why we examine interactive learning additional to DRL, propose a specific method for involving human feedback and reward and show how interactive learning will be integrated in our framework.

The main reason for investigating possibilities to improve DRL are the mentioned issues regarding complex games including long-horizon planning tasks and the fact that DRL is an optimization method. A pure DRL agent learns to optimize its behaviour in means of game point rewards. Adding interactive learning presents the possibility to shape the learning outcome directly towards a specific, desired behaviour and thus also helps to fill the gaps in sparse reward games. Furthermore, it has been shown that, in complex settings, interactive learning improves learning speed and quality compared to non-interactive learning and that interaction seems to make learning more robust [41].

Our approach should satisfy two preconditions concerning the nature and extent of human integration in the learning process. First, the interaction should not be dependent on programming skills or machine learning knowledge. Second, the basic process of DRL should remain unchanged in order to minimize additional effort. And, above all, the agent should be able to learn autonomously even, temporarily, without any human feedback.

In the context of the area of cognitive infocommunications, this problem falls within the categories of inter-cognitive, sensor-sharing and sensor-bridging communication (cf. [3, 4]). The DRL agent disposes of an ANN and hitherto can be considered as a connectionist emergent cognitive system that is able to adapt and act effectively through interaction with its environment (cf. [47]). Thus, combining DRL and iRL means investigating the infocommunication between the DRL system as cognitive thing and humans, in the role of experts and players, as cognitive beings.

In our approach, we decided to use a combination of DRL and interactive shaping as method. In interactive shaping, an agent receives exclusively human reward in the form of positive and negative values [22]. Thus, the pure interactive shaping approach requires intensive participation and effort on the part of domain experts. However, combining the capabilities of both approaches, DRL and interactive shaping, means that an agent is able to receive rewards and simultaneously learn from both, environment and human reward.

Another interactive method should be noted in this context: learning from human demonstration or inverse reinforcement learning. This method has successfully been applied to games several times, e.g. in TD-learning in backgammon by offline policy learning from experts plays [46] and learning policies for first person shooter games to generate human-like behaviour [45]. Nonetheless, sufficient training data, e.g. in the form of extensive logs or replays created by human players are required because this methods needs ample data for the reconstruction of a reward function from a sample of policies.

By contrast, interactive shaping is advantageous in cases when sufficient training data are not available. Another strength of giving direct reward compared to demonstration is that, by receiving reward, an agent learns the relative values of actions instead of merely when to choose a specific action (cf. [23]). Further advantages are relatively simple realization through an interface and simplicity of use (cf. [24]).

Interactive learning is used at two points in our framework. First, the data gained from assessment of player interactions with the game or, if possible, physiological signals, can be analysed and the values be used for adaptation to the players' needs. Second, domain experts are able to actively provide feedback to the current agent behaviour and thereby share their task-knowledge in a natural way.

Both cases don't require any level of programming skills or machine learning knowledge from the interacting user. The players' feedback is obtained through passive measurement of their current state. The quality of adaptation to the player should only be dependent on the method of feedback and measurement used (e.g. tracking of clicks or actions, measuring biofeedback etc.). This kind of feedback corresponds to a myopic reward, meaning an almost immediate reward for the current state and action. In contrast to this, the rewards provided by trainers should ideally be non-myopic (cf. [24]). Trainers can take into account possible future progresses of a state and in doing so reward a learning agent in a rather goal-oriented way. Several concrete combination techniques of human and environment reward, applicable in action-value functions, are described and compared in [23].

If possible, domain experts should be provided with various feedback and control mechanisms [2], because users seem to favour transparency and are willing to learn how a system works to give nuanced feedback [25]. For example, a visualization of higher level features of convolutional layers in deep networks [14], as was used in [19], could be integrated. We leave it in the hands of pedagogues and trainers to decide if also players should be allowed to get a visualization of the influence of their decisions and feedback to the concrete realization of adaptation. In the long term, we consider it useful to offer a comprehensive repository of visualized information, resulting in system transparency and understandable behaviour.

Considering the time of human reward and feedback integration, the most promising time period of expert involvement, at least for NPC training, seems to be at the beginning of learning [23]. In contrast, a player's biofeedback for adaptation should be used constantly throughout the whole gaming time. Additionally, before adaptation to a player, pre-training of the network on a (simplified) game version should be envisaged in order to reduce training time and volatility later in real application.

19.5.4 Application Scenarios

This section offers a short outline of how SGs in general and our described research environment in particular can profit from iDRL techniques. We cover the fields of NPC behaviour generation and game adaptation and indicate some additional ideas. We think the several possible application scenarios we found to be inspiring to support SG development and to offer some additional value over classical approaches.

Player and NPC Control in Single-Player and Multiplayer Games   This is one of the most obvious possible uses of iDRL and has been one focus of our previous descriptions. Applications of DRL have been concentrated on game playing, meaning to take the player's place. When using DRL in the related application scenario of taking over the position of one of the players in a multiplayer game, this approach can be applied unalteredly. In contrast, if DRL is to be applied to explicit NPC character control, one should consider that notably the graphical state input can differ, unless an explicit NPC game screen representation is available. Then again, entrusting DRL with NPC control offers the chance for less effort in behaviour programming and could lead to heterogeneous and adaptive NPC AI. Using automated players as player replacement in multiplayer games reduces the need for additional real persons in these scenarios. Additionally, we assume that iDRL as machine learning technique, opposed to using scripted opponents, leads to more varying, realistic and entertaining behaviour. In our research scenario one or both of the players in 1 versus 1 can be replaced by our iDRL agent. Also multiple units could be controlled and several techniques exist to enhance existing DRL approaches to allow for multiple agent control.

Scenario and Group Control   Multiple instances of NPCs in a scenario can be controlled as a group (often called 'wave' in TD-like games) in order to manage an entire scenario development. Apart from that, the control objectives or learning objectives of a scenario have to be defined differently than pure success in earning game points. The evaluation criteria should consider the learning purpose, e.g. in a multiplayer scenario in SanTrain, at least one member of the trainees' team should get injured in order to ensure the training of the TCCC procedures. Furthermore, the definition of the scenario evaluation should be an indication of an appropriately challenging task. In games similar to our research strategy game, control possibilities include for example waves of enemies (including their strength, number, types, pathfinding and steering in a group) and strategic decisions as tower choice and placement. We assert that approaches based on DRL are genuinely suitable for such scenario control tasks, because the game screen input, potentially enhanced with additional information, can be read just like a tactical map.

Game Asset Adaptation   As has been the other focus of this paper, we are convinced of the beneficial use of iDRL for game adaptation. The above mentioned mechanisms of scenario and NPC control plus a pool of modifiable context information can be used for adapting the game to individual player capabilities and preferences. As in scenario control, the success depends on an appropriate definition of the objective function in form of rewards for DRL (desired player behaviour, psychophysiological feedback, etc.). In our game, the most interesting specific application seems to be the scenario adaptation (opponent units and towers) but we are also experimenting with textures and sounds.

Game Testing   Automatically created NPCs behaviour could also be used for gameplay testing during development e.g. to support determining flaws and inconsistencies in gameplay and level-design.

We are aware that autonomous learning, especially in serious games, should be used with caution. Sometimes, a strict and scripted scenario is more important for learning than varying behaviour patterns. Furthermore, the success of applying a learning method is also heavily depending on finding and defining an appropriate reward function.

In SGs, it's generally reasonable that algorithms concerning learning content should be validated ideally already in early development phases. Therefore, the application of iDRL for behaviour generation and adaptation should also be integrated in the development process from the very beginning. The learning and calibration process before the real game is finished can be partially but neatly realized by using the concepts of fish tanks and sandboxes (cf. [17]).

Ideally, combining interactivity with DLR will not only relief limited trainer availability by automatically generating behaviour but we also expect faster convergence while biasing the behaviour in the direction of players' needs. We also expect that directed learning counteracts unpredictable outcomes and therefore increases the general acceptance of the generated, resulting behaviour. By supporting this with our proposed framework, we are able to include domain experts and players more directly without the need of translating a desired behaviour into a technical specification.

19.6 Conclusion

Combining deep reinforcement learning methods with interactive human guidance and player feedback may be a promising solution to reduce effort and complexity issues in NPC behaviour generation and game adaptation while at the same time creating believable and diverse scenarios. Since SGs often require convincing human-like NPC behaviour, their development regularly includes laboriously hand-crafted AI design. Furthermore, the increased desire for game adaptation to players, especially in SGs, and the needed effort for and low flexibility of scripted rules lead to the investigation of new approaches.

We discussed that, although the application of machine learning techniques like DRL has already proven successful in different games, there are still enormous issues with more complex scenarios, resulting from incomprehensibly large state-action spaces and RL specific problems. We also mentioned that previous studies have shown that interactive RL methods can improve learning speed and quality, particularly in complex tasks. We therefore proposed to combine DRL with interactive learning.

Overall, we strive for a flexible and easy to use framework, providing the possibility to combine the genuine qualities of these approaches. Our approach is to be examined at first in a simplified strategy game like scenario, as it provides complex but manageable challenges and the problem setting is transferable to many game and SG tasks. In the future, we plan to further test our approach in concrete SGs like SanTrain as a recent serious game in development.

In the long term, we hope our approach aids in expanding possible application areas of DRL and in reducing existing issues e.g. in games that include analysis and planning tasks. Through interactivity, we also strive to facilitate a simple, automated and natural communication between humans and learning agents to overcome some communication gaps in AI development and better adapt to a player's preferences and needs. We envision to contribute to the transformation of the generation of reasonable NPC AI and the increasingly important aspect of game to player adaptation into an efficient, collaborative process with reduced complexity.

References

1. Abt CC (1987) Serious games. University Press of America
2. Amershi S (2012) Designing for effective end-user interaction with machine learning. PhD thesis, University of Washington
3. Baranyi P, Csapo A (2012) Definition and synergies of cognitive infocommunications. Acta Polytech Hung 9(1):67–83
4. Baranyi P, Csapo A, Sallai G (2015) Cognitive infocommunications (CogInfoCom), 1st edn. Springer International
5. Bellemare MG, Naddaf Y, Veness J, Bowling M (2013) The arcade learning environment: an evaluation platform for general agents. J Artif Intell Res 47:253–279
6. Bontchev B (2016) Adaptation in affective video games: a literature review. Cybern Inf Technol 16(3):3–34
7. Brisson A, Pereira G, Prada R, Paiva A, Louchart S, Suttie N, Lim T, Lopes R, Bidarra R, Bellotti F et al (2012) Artificial intelligence and personalization opportunities for serious games. In: Eighth AIIDE Conference, pp 51–57
8. Csikszentmihalyi M (1975) Beyond boredom and anxiety. Jossey-Bass, San Francisco, USA
9. Dobrovsky A, Borghoff UM, Hofmann M (2016) An approach to interactive deep reinforcement learning for serious games. In: Proceedings of 7th IEEE conference on cognitive infocommunications (CogInfoCom 2016). IEEE, Wroclaw, Poland, 16–18 Oct. https://doi.org/10.1109/CogInfoCom.2016.7804530
10. Dobrovsky A, Borghoff UM, Hofmann M (2017) Applying and augmenting deep reinforcement learning in serious games through interaction. Periodica polytechnica electrical engineering and computer science online first: paper 10,313. https://doi.org/10.3311/PPee.10313
11. Dörner R, Göbel S, Effelsberg W, Wiemeyer J (2016) Serious games: foundations, concepts and practice. Springer International Publishing. https://doi.org/10.1007/978-3-319-40612-1
12. Doujak G (2015) Serious games und digital game based learning. GRIN Verlag, Spielebasierte E-Learning Trends der Zukunft
13. Eirik V, Viola I, Hauser H (2009) State of the art report on serious games: blurring the lines between recreation and reality. Universitet i Bergen, Institutt for Informatikk

14. Erhan D, Bengio Y, Courville A, Vincent P (2009) Visualizing higher-layer features of a deep network. Tech. Rep. 1341, University of Montreal
15. Frutos-Pascual M, Zapirain BG (2015) Review of the use of AI techniques in serious games: decision making and machine learning. IEEE Trans Comput Intell AI Games 9. https://doi.org/10.1109/TCIAIG.2015.2512592
16. Galway L, Charles D, Black M (2008) Machine learning in digital games: a survey. Artif Intell Rev 29(2):123–161
17. Gee JP (2005) Learning by design: good video games as learning machines. E-Learn Digit Media 2(1):5–16
18. Genesereth M, Love N, Pell B (2005) General game playing: overview of the aaai competition. AI Mag 26(2):62
19. Guo X, Singh S, Lee H, Lewis RL, Wang X (2014) Deep learning for real-time atari game play using offline monte-carlo tree search planning. In: Advances in neural information processing systems, pp 3338–3346
20. Justesen N, Bontrager P, Togelius J, Risi S (2017) Deep learning for video game playing. arXiv:1708.07902
21. Kivikangas JM, Chanel G, Cowley B, Ekman I, Salminen M, Järvelä S, Ravaja N (2011) A review of the use of psychophysiological methods in game research. J Gaming Virtual Worlds 3(3):181–199
22. Knox WB, Stone P (2010) Combining manual feedback with subsequent MDP reward signals for reinforcement learning. In: Proceedings of the 9th international conference on autonomous agents and multiagent systems: volume 1-volume 1, international foundation for autonomous agents and multiagent systems, pp 5–12
23. Knox WB, Stone P (2012) Reinforcement learning from simultaneous human and MDP reward. In: Proceedings of the 11th international conference on autonomous agents and multiagent systems-volume 1, international foundation for autonomous agents and multiagent systems, pp 475–482
24. Knox WB, Stone P (2015) Framing reinforcement learning from human reward: Reward positivity, temporal discounting, episodicity, and performance. Artif Intell 225:24–50
25. Kulesza T, Burnett M, Wong WK, Stumpf S (2015) Principles of explanatory debugging to personalize interactive machine learning. In: Proceedings of the 20th international conference on intelligent user interfaces. ACM, pp 126–137
26. Lample G, Chaplot DS (2016) Playing FPS games with deep reinforcement learning. arXiv:1609.05521
27. Lara-Cabrera R, Nogueira-Collazo M, Cotta C, Fernández-Leiva AJ (2015) Game artificial intelligence: challenges for the scientific community. In: Proceedings 2nd Congreso de la Sociedad Española para las Ciencias del Videojuego Barcelona, Spain
28. Lillicrap TP, Hunt JJ, Pritzel A, Heess N, Erez T, Tassa Y, Silver D, Wierstra D (2015) Continuous control with deep reinforcement learning. arXiv:150902971
29. Lopes R (2014) Gameplay sematics for the adaptive generation of game worlds. PhD thesis, Delft University of Technology. http://graphics.tudelft.nl/Publications-new/2014/Lop14
30. Mahajan S (2014) Reinforcement learning in complex real world domains: a review. INDIAN J Comput Sci Eng (IJCSE) 5(2)
31. Mehta M, Ontanón S, Ram A (2008) Adaptive computer games: easing the authorial burden. In: Rabin S (ed) AI game programming wisdom 4, Charles River Media, pp 617–632
32. Millington I, Funge J (2009) Artificial intelligence for games, 2nd edn. Morgan Kaufmann, Boston. https://doi.org/10.1016/B978-0-12-374731-0.00007-4
33. Mnih V, Kavukcuoglu K, Silver D, Graves A, Antonoglou I, Wierstra D, Riedmiller M (2013) Playing atari with deep reinforcement learning. arXiv:13125602
34. Mnih V, Kavukcuoglu K, Silver D, Rusu AA, Veness J, Bellemare MG, Graves A, Riedmiller M, Fidjeland AK, Ostrovski G et al (2015) Human-level control through deep reinforcement learning. Nature 518(7540):529–533
35. Mnih V, Badia AP, Mirza M, Graves A, Lillicrap TP, Harley T, Silver D, Kavukcuoglu K (2016) Asynchronous methods for deep reinforcement learning. arXiv:160201783

36. Muñoz-Avila H, Bauckhage C, Bida M, Congdon CB, Kendall G (2013) Learning and game ai. Dagstuhl Follow-Ups 6
37. Ontañón S, Synnaeve G, Uriarte A, Richoux F, Churchill D, Preuss M (2013) A survey of real-time strategy game AI research and competition in starcraft. IEEE Trans Comput Intell AI Games 5(4):293–311. https://doi.org/10.1109/TCIAIG.2013.2286295
38. Perez-Liebana D, Samothrakis S, Togelius J, Lucas SM, Schaul T (2016) General video game ai: competition, challenges and opportunities. In: Thirtieth AAAI conference on artificial intelligence
39. Rowe J, Mott B, McQuiggan S, Robison J, Lee S, Lester J (2009) Crystal island: a narrative-centered learning environment for eighth grade microbiology. workshop on intelligent educational games at the 14th international conference on artificial intelligence in education. Brighton, UK, pp 11–20
40. Rowe JP, Lester JC (2010) Modeling user knowledge with dynamic Bayesian networks in interactive narrative environments. In: Proceedings of the sixth AAAI conference on artificial intelligence and interactive digital entertainment. AAAI Press, pp 57–62
41. Stahlhut C, Navarro-Guerrero N, Weber C, Wermter S (2015) Interaction is more beneficial in complex reinforcement learning problems than in simple ones. In: Interdisziplinärer workshop kognitive systeme: mensch. Teams, Systeme und Automaten, pp 142–150
42. Sutton RS, Barto AG (1998) Reinforcement learning: an introduction. MIT press
43. Szegletes L, Forstner B (2013) Reusable framework for the development of adaptive games. In: 2013 IEEE 4th international conference on Cognitive infocommunications (CogInfoCom). IEEE, pp 601–606
44. Szegletes L, Köles M, Forstner B (2014) The design of a biofeedback framework for dynamic difficulty adjustment in games. In: 2014 5th IEEE conference on cognitive infocommunications (CogInfoCom). IEEE, pp 295–299
45. Tastan B, Sukthankar G (2011) Learning policies for first person shooter games using inverse reinforcement learning. In: Proceedings of the seventh AAAI conference on artificial intelligence and interactive digital entertainment. AAAI Press, AIIDE'11, pp 85–90. http://dl.acm.org/citation.cfm?id=3014589.3014604
46. Tesauro G (1990) Neurogammon: a neural-network backgammon program. In: IJCNN international joint conference on neural networks. IEEE, pp 33–39. https://doi.org/10.1109/IJCNN.1990.137821
47. Vernon D, Metta G, Sandini G (2007) A survey of artificial cognitive systems: Implications for the autonomous development of mental capabilities in computational agents. IEEE Trans Evol Comput 11(2):151–180. https://doi.org/10.1109/TEVC.2006.890274
48. Wendel VM (2015) Collaborative game-based learning—automatized adaptation mechanics for game-based collaborative learning using game mastering concepts. PhD thesis, Technische Universität Darmstadt

Chapter 20
A Study on a Protocol for Ad Hoc Network Based on Bluetooth Low Energy

Atsushi Ito and Hiroyuki Hatano

Abstract In this chapter, we propose a new protocol to construct ad hoc network using Bluetooth Low Energy (BLE). Especially, we are focusing on using advertising packet of BLE to construct ad hoc network, since the reach of advertising packet of BLE is longer than ordinary Bluetooth and WiFi. We would like to use ad hoc network to develop support tool for disaster victims, especially disabled people, group management tool to keep safe during school trip or beacon network management for navigation. In this application, we have to consider power consumption and longer reach. To solve this problem, we designed an ad hoc network protocol to construct ad hoc network efficiently using advertising packet of BLE. We implemented this protocol on iOS and performed experiments using 9 terminals. As the result of experiments, the latency of end-to-end was 2.2 s and that of one hop was less than 0.3 s. We think that the result shows the feasibility to construct ad hoc network for our purpose.

20.1 Introduction

In this chapter, we propose a new protocol to construct an ad hoc network using Bluetooth Low Energy (BLE). We are especially focusing on using an advertising packet of BLE to construct network and send information, since the reach of advertising packet is longer than WiFi or ordinary Bluetooth.

Figure 20.1 shows information delivery model through a network. This figure shows that the most appropriate media will be selected and used based on the environment in an unusual situation. There is an idea of a cognitive network to select the most appropriate network according to the environment [1, 2]. However if we are in an unusual situation such as a disaster, blackout is happened. We cannot access ICT based media even if there is a cognitive network. We have to select only non-ICT

A. Ito (✉) · H. Hatano
Information System Science Department, Graduate School of Engineering,
Utsunomiya University, 7-1-2 Yoto, Utsunomiya, Tochigi 321-8505, Japan
e-mail: at.ito@is.utsunomiya-u.ac.jp

© Springer International Publishing AG, part of Springer Nature 2019

R. Klempous et al. (eds.), *Cognitive Infocommunications, Theory and Applications*,
Topics in Intelligent Engineering and Informatics 13,
https://doi.org/10.1007/978-3-319-95996-2_20

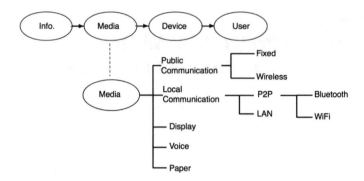

Fig. 20.1 Information delivery model through network

media such as blackboard, voice, and paper to deliver information [3]. We do not consider non-ICT media in the following part of this section. However, we always consider what is the most useful and appropriate ICT media to deliver information in an unusual situation.

According to the progress of ICT, the public mobile communication network such as 3G/LTE/5G, and near-field communication technology such as Wi-Fi and Bluetooth are widely used with applications around us such as in home, office, school, university, public space such as air port, train station, restaurant and shopping center. Also, our life has become more convenient according to the rapid progress of smartphone and cloud service. We could access many kinds of service and contents in a cloud by mobile devices and PCs.

However, if we become a victim of disaster or we are visiting deep rural area, we cannot access high speed network and separated from convenient services. If we are in the evacuation shelter after a larger earthquake and staying there, we would like to receive information, such as delivery of food, promptly. If we are in the rual area where is difficult to access network, we would like to receive minimum information to survive. If we are in the difficult situation, we usually turn on our smartphone and use applications to receive information as soon as possible. However, in such situation, we have to preserve battery. There was a research that people in an evacuation shelter turned off their smartphone to keep their battery level and prevent battery exhaustion and could no access necessary information [4]. We think that ad hoc network construction method by using BLE is one of the most appropriate ways to solve this problem.

Preserve battery of a smartphone is also important during sightseeing. For example, if we would like to receive sightseeing information in the historic area, many applications of navigation require GPS. However, GPS consumes battery quickly. Let's assume that we visit sightseeing point after breakfast (such as 9 a.m.) to return to the hotel (such as 8 p.m.). In this case, we cannot charge battery for almost half day and we should turn off GPS or a smartphone itself to preserve battery. As our research in Nikko, a world heritage in Japan, many foreign tourists did not use GPS [5]. They would like to use power efficient application to receive information for sightseeing. Otherwise they would like to turn off their smartphones. We think that ad hoc network construction method by using BLE is one of the most appropriate ways to solve this problem.

In the remaining part of this paper, we would like to discuss on ad hoc network construction protocol using advertising packet of BLE.

In Sect. 20.2, we introduce benefit and limitation of BLE technology and our previous works that use ad hoc network. Then we introduce the outline of Bluetooth and related works in Sect. 20.3. In Sect. 20.4, we explain problems of the limitation of packet size of BLE if we try to implement ordinary ad hoc network protocols on BLE advertisement. In Sect. 20.5, we describes the details of the design of our ad hoc network protocol. In Sect. 20.6, we explain and discuss the result of the experiment. At last, we mention conclusion.

20.2   Bluetooth Low Energy and Related Works on Ad Hoc Network

There are two types of Bluetooth technology, one is Bluetooth 2.1 (Classic Bluetooth) [6] and another is Bluetooth 4.0 (BLE) [7] and they are different technology as described in Table 20.1.

There is long history of studies on ad hoc network using Bluetooth 2.1 [8], however, they are few studies on ad hoc network using BLE [9] since this technology is very young. Recently, researches on ad hoc network on BLE is started. We would like to introduce some of them as follows.

- a protocol for WiFi/BLE co-existing environment [9]:

    - Proprietary protocol selectively uses WiFi and BLE
    - Designed for connection less communication

Table 20.1 Comparison of Bluetooth 2.1 and Bluetooth 4.0.

Characteristics	Bluetooth 2.1	Bluetooth 4.0
Purpose	Designed for hands-free phone or earphones	Designed to send small data from sensors such as data from thermos meter
Pairing	Requiring airing before starting communication	Not required pairing
Discovering	Only eight neighbors can be found at discovery phase before starting communication	No restriction
Power consumption	High	Low (1/10–1/100 of Bluetooth 2.1)

- ANT based ad hoc network [10]:

  – Routing algorithm based on ANT
  – This paper evaluates the performance of the protocol
  – Not implement their algorithm in a real system.

20.3   Ad Hoc Network Protocols

There are two types of ad hoc network protocol, one is reactive and another is proactive. The most popular protocol of the proactive type protocol is OLSR (Optimized Link State Routing protocol) [11] and the most popular reactive type protocol is AODV (Adhoc On-Demand Distance Vector) [12]. Characteristics of these two protocols are described in Table 20.2.

AODV uses four messages such as RREQ (Route Request), RREP (Route Reply), RERR (Route Error), RREP-ACK (Route Reply Acknowledgment). For example, RREQ uses 24 octets.

OLSR also uses several messages such as Hello and TC (Topology Control) message. Hello message is defined to use at least 16 octets and TC message is defined to use 12 octets.

Let's consider how to transmit messages used in each protocol using BLE. For example, RREQ of AODV uses 24 octets, and Hello message of OLSR uses at least 16 octets and TC message of OLSR used 12 octets. Figure 20.2 displays packet format of an advertisement packet of BLE. The maximum octets we can use is 8 octets (displayed as shaded box in Fig. 20.2), so that it is impossible to use BLE to send messages of AODV and OLSR without additional technique such as to separate message into small pieces. So that, it is valuable to develop a new protocol to construct ad hoc network.

20.4   Our Previous Work

In this section, we would like to introduce two of our projects using ad hoc network. One is developing sightseeing support system by using BLE beacon. If a traveler

Table 20.2 Characteristics of AODV and OLSR

Protocal	Routing	Delay	Typical messages
AODV	Before data packet is sent	Large	RREQ (Route request) RREQ (route reply) RER R (route error) RREP-ACK (route reply acknowledgement)
OLSR	Regular interval	Small	Hello TC (topology control)

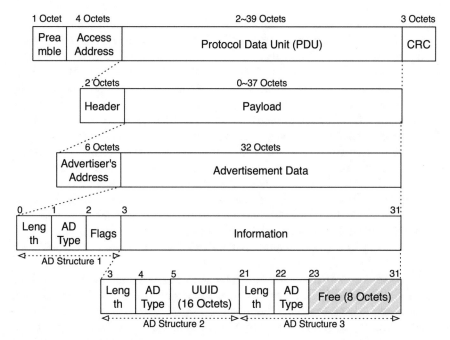

Fig. 20.2 Packet format of BLE advertisement

passes near a BLE beacon, they communicate each other and travel information is displayed on the screen of a smart phone. This system is introduced in Nikko, a world cultural heritage in Japan. Another is a system designed for deaf people to receive disaster information at the large disaster. This system, called Information Delivery System for Deaf People at a Major Disaster (IDDD), constructed with LED dot matrix displays connected by ad hoc network.

20.4.1   Nikko Beacon

Traveling is a good way to get away from daily life. Also, visiting unknown locations and looking incredible scenery is one of the great pleasures for us. When we visit a traditional area, we can discover fresh surprises such as a hidden history of a village, original and unique culture, and wild natures such as animals and flowers. Discover is constructed by two words such as dis = unveil + cover. It means that a travel brings to us new and surprising information. Traditionally, our passion to discover unknown parts of the world brings us to undiscovered places, cultures, food etc.

On the other hand, there are many web sites to provide travel information. However, there is a problem that the satisfaction of a travel is not proportional to the amount of information. When we started our project we considered what is the most attractive aspect of travel. And also, we would like to design a method to increase expectation and satisfaction of travel in Nikko. There are 17 world cultural heritages in Japan and Nikko is one of them. Nikko is famous of the Toshogu-shrine [14] that is a gorgeous grave of Ieyasu Tokugawa who is the first Shogun of Tokugawa Era. Unfortunately, Nikko is not so popular for foreigners right now. A research by a travel agency in 2015 shows the shocking result that Nikko was not listed in top 30 locations where foreigners would like to visit [15]. There are huge amount of information of Nikko on web sites, however, they are not working effectively.

We are interested in to rebuild a traditional area such an Nikko using ICT. Based on the above situation, we started our projects to develop a sightseeing support system in Nikko [13] and to rebuild a tradition of Nikko by ICT. The project started in 2014 and finished in 2016. We developed a sightseeing support system using BLE beacon in this project [5, 18, 19]. This study was selected as one of research themes of SCOPE (Strategic Information and Communications R&D Promotion Programme) [16] funded by Ministry of Internal Affairs and Communications of Japan (MIC) [17].

22 BLE beacons to navigate a trip from Tobu Nikko Station to Toshogu-shrine were installed. Also, 15 BLE beacons to introduce shops on the road were installed. The position if each beacon is described in Fig. 20.3. Beacons on the road were located on signboards as shown in Fig. 20.4. Beacons on the road use a button shape battery to make a beacon small to hide its presence. The diameter of the PCB bord of a beacon is 35 mm and that of the shell of a beacon is 50 mm (Fig. 20.5).

Beacons in shops were 15 as described in Fig. 20.6. These beacons use two 3 A battery since there are no needs to make it small.

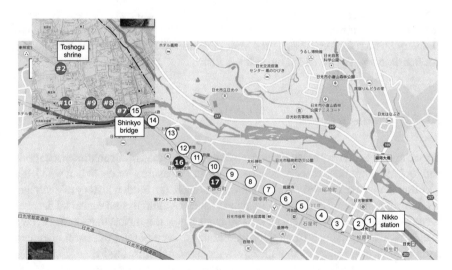

Fig. 20.3 Beacon MAP of outdoor beacons

Fig. 20.4 BLE Beacon on a sign board

Figure 20.7 explains the software components of this application. On the operating system (OS), BLE access function always scans advertising message. If the OS catches an advertising message, the information of the advertising message is forwarded to the application. For example, Core Location Framework of iOS (7 or later) provides three properties such as proximity UUID (Universal Unique Identifier), major and minor.

Android 5.0 or later also provides the similar function. If the information such as UUID, major and minor is received from a beacon, the application retrieves information that matches triples (UUID, major and minor). For example, if UUID = cb86bc31-05bd-40cc-903d-1c9bd13d966a, major = 1, minor = 1, the information

Left (board): R=32mm   Right (case): R=50mm

Fig. 20.5 Inside of a BLE beacon

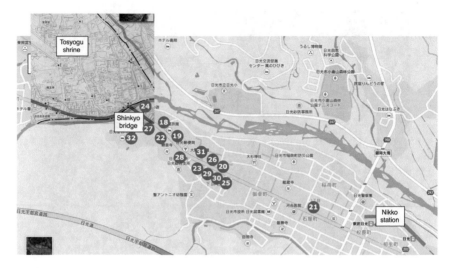

Fig. 20.6 Beacon MAP of shop beacons

relating to the beacon located in the Nikko Station is retrieved from DB and the information is displayed on the screen of the smart phone. Each beacon provided information related to its location. For example, when the application received a signal from beacon #13, it displays the following information "On the left, there is a slope. At the end of the slope, there is an old temple named Kannonji; 180 m to Shinkyo Bridge and 1,250 m to Nikko train station". The visitor could use this information to find a small, historic temple near the location.

One of a next application of our system is a school trip. In the case, the safety of students is the most important issue, so that, we think that we can provide a solution to realize this requirement. Our idea is as follows. Each students and teachers have a device with BLE such as smartphone. Among the devices, an ad hoc network is constructed and the number of students on the network can be counted and the result is informed to teachers.

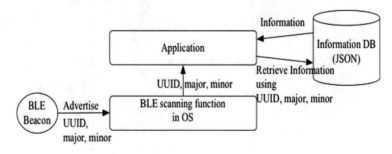

Fig. 20.7 Behavior of the sightseeing application using BLE beacon

Also, we are planning to use an ad hoc network for managing BLE beacons. To maintain beacons are one of the most important problems to realize this kind of service. If beacons can construct a network, it might be easy to find broken or unstable beacons in the network. Usually, we can use common ad hoc network protocol such as AODV and OLSR. However, it is difficult to use these protocols on BLE advertisement because of the limitation of packet size as described in Sect. 20.3.

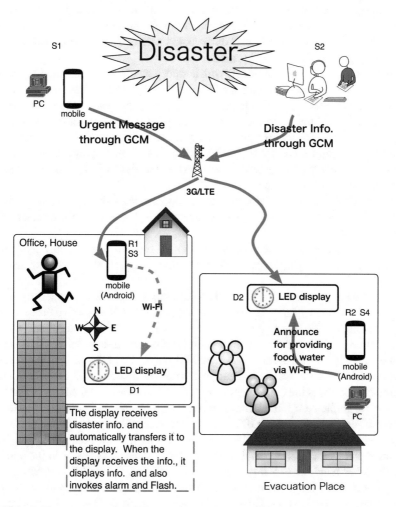

Fig. 20.8 Outline of behavior of IDDD

20.4.2 Information Delivery System for Deaf People at a Major Disaster (IDDD)

We have been developing Information Delivery System for Deaf People at a Major Disaster (IDDD) for 11 years [3] (Fig. 20.8).

5:46 a.m., January 17th, 1995 a large earthquake at a magnitude of 7.3 occurred near Kobe, west part of Japan, and caused a catastrophic disaster. Cities near the center of the earthquake were shaken with a seismic intensity of 7. 6,434 lives were lost and 43,792 were injured. Half of the 6,434 sacrifices at Kobe earthquake were the elderly person and disabled person.

We made a survey of individuals with auditory handicaps requiring support after the Great Hanshin-Awaji earthquake [20] and received 185 replies. We found many of the deaf were left without support during the disaster. Some of them were left in a house and could not go to shelter. Even in a shelter, life was not comfortable, the battery was dead, the equipment was broken and the surrounding was very noisy. They also required a battery for their hearing aids, large displays for information, radios and mobile phones. Lack of these devices meant that it was difficult for them to get important information.

58.1% of the deaf was informed of the disaster immediately: however, 49.1% knew of the disaster with a delay of half a day. 3.7% knew about the disaster after a week.

Based on these survey results, we developed Information Delivery System for Deaf People in a Larger Disaster (IDDD) using mobile phones and ad hoc network technology.

This system consists of several LED dot matrix displays connected by ad hoc network. Also some of the displays are connected to WAN to receive disaster information from outside. Through from WAN or local PC or local smartphone, disaster information, such as safety information, lunch delivery time, is delivered to victims of a disaster in an evacuation shelter. Stability of ad hoc network and the distance between LED dot matrix displays are the problems to be solved to realize IDDD. Figure 20.9 explains the block diagram of IDDD. We are using OLSR on WiFi ad hoc network to implement IDDD, however, this implementation is not suitable for our purpose. We have been testing IDDD at the deaf school in Sendai, Miyagi prefecture, Japan. The corridor of the school is about 50 m. If we use WiFi dongle on Raspberry Pi, the reach is about 20 m, so that, it is required some relay nodes in the corridor. On the other hand, as explained in Table 20.3, the reach of BLE advertisement is longer than WiFi dongle. It is reasonable to use BLE advertisement as the basis of developing ad hoc network. However, the packet size of BLE advertisement is very small, only 28 octets and free area is only 8 octets as described in Fig. 20.2. It is so that impossible to implement ordinary ad hoc network protocols such as OLSR and AODV. So that, we need to design a new ad hoc networking protocol that works on BLE Advertisement packet.

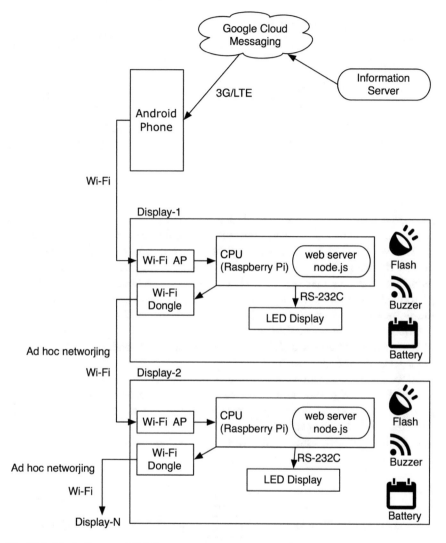

Fig. 20.9 Block diagram of IDDD

20.5   Designing Ad Hoc Network Protocol for BLE

In this section, we would like to explain how we design our ad hoc network construction protocol.

As mentioned in Sect. 20.2, one of the target of our new ad hoc network protocol is evacuation shelter. In the disaster, a gymnasium is typically used as an evacuation shelter. The size of a typical gymnasium is about 50 m × 50 m. We assume that this

Table 20.3 The reach of BLE advertisement

Location	Reach (m)
On the pavement between the Tobu Nikko station and Shinkyo-bridge	About 80
Around the Toshogu-shrine: on the main approach	About 80
Around the Toshogu-shrine: in the forest	About 40
Bus stops near the Tobu Nikko station	About 80
Shops near the Tobu Nikko station: in front of a shop	About 10
Forest of Senjyogahara (summer)	About 20
Forest of Senjyogahara (winter)	About 20

protocol can be used in the shelter of such size to receive daily information such as the time to start lunch or exchange information about friends of people in the shelter.

Another example is school trip. Teachers have to take care their students and always checking the number of students. Our protocol can be used to count the number of students automatically. Of course, at the school trip, their activity is performed in the wide area such as a large park. It is clear that this protocol works if the protocol can support multi-hop. In Table 20.3, the reach of an advertising message in different situation is described based on our experiments [5]. For example, the reach of advertising message of BLE is about 80 m in the good condition and about 20 m in the worst condition as described.

In the case of evacuation place, it is clear that the advertising packet can each end to end of a typical gymnasium. In the case of school trip, lets assume the size of a park as 1 km × 1 km. By 12 hops, a message from a teacher can reach a student in the different side of the park.

We think that advertising of BLE meets our requirements to deliver information.

Table 20.4 Three ideals to prevent infinite loop of a message.

Idea	Purpose	Behaviour
Re-send protect DB	To prevent an infinite loop	If a message is handled in a terminal, the data ID and destination are recorded in this DB
Short UUID	To use space efficiently	UUID of destination is shortened by using hash function since there is no space in this pocket
ID management DB	To provide hashed UUID	The hashed ID of local UUID is recorded in this DB and broadcasted to other terminals when a terminals joined existing ad hoc network

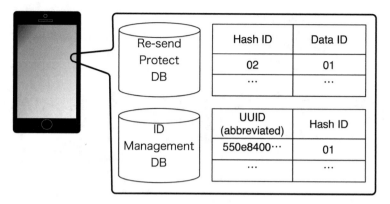

Fig. 20.10 Databases in each terminal

20.5.1 Outline of the Protocol

If we design an ad hoc network protocol, we have to prevent an infinite loop of messages. For that purpose, three ideas were
introduced to solve that problem as described in Table 20.4 and Fig. 20.10.

20.5.2 Messages

We defined four messages in Table 20.5 to define the behavior protocol. Two of them are used for getting new ID. Other two are a message and its ACK.

Table 20.5 The messages used in the protocol

Message	Description
ID_req	This message is used to check existing hash ID. A new terminal asks to existing terminals asks to existing terminals the collision oh hash ID
ID_rep	This message is used to send the result for the ID_req. If thew hash ID is already used, NG is returned and the terminal has to re-calculate hash ID
Msg	Message to the destination terminal
ACK Msg	ACK for the destination terminal to source terminal

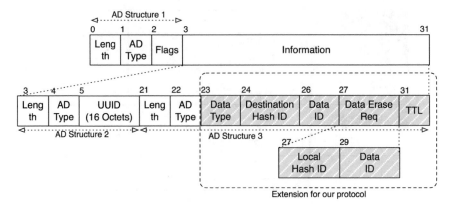

Fig. 20.11 Adding new features in BLE advertisement

Table 20.6 The information used in the protocol

Name of the information	Description	Details
Data type	Messgage type	0–9: Control
		0: ID_Req
		1: ID_Rep
		2: Msg
		3: ACK Msg
		4–9: reserved
		10–15: ID of message
Destination hash ID	Destination address	
Local hash ID	Local address	
Data ID	Sequence number of messgage	
Data erase request	Information to deleted data in re-sent protect DB	
TTL	Indicate messgae type	

20.5.3 Packet Format

Figure 20.11 describes the packet format of BLE advertising that is used in our protocol. According to the limitation of BLE advertising packet size, only 28 octets can be used. The protocol contains the information in an advertising packet as described in Table 20.6.

Fig. 20.12 Message
sequence to join the existing
network

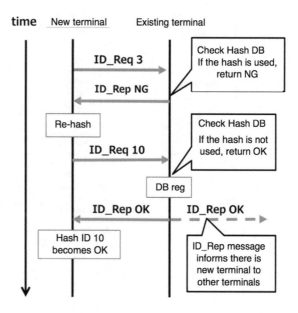

20.5.4 Join a Network

Firstly, each terminal should generate hash ID before joining the existing network
or constructing a new network. If the generated hash ID is already used, the terminal
has to re-generate different hash ID as described in Fig. 20.12.

Figure 20.12 shows how to join the existing network. The new terminal should
send ID_Req to the network to request to check the uniqueness of the generated
hash ID. The result of the request is returned to the terminal by using ID_Rep the
possibility to use the hash ID.

If there is no existing network, each terminal sends ID_Req message that contains
self UUID and hash ID. The receivers of the message ID_Rep check conflict of hash
ID. If there is an existing hash ID that is as same as an ID in ID Management DB, the
answer (ID_Rep) should request to change the hash ID. If there is no ID_Rep in a
specific waiting time, the terminal decides that there is no existing terminal around
it.

If there is an existing network around the terminal, a new terminal broadcasts
ID_Req to neighbor terminals in the network. The flowchart to explain the detailed
procedure is described in Fig. 20.13. The procedure is almost same, however, not
all terminals can receive ID_Rep message. If a terminal would like to send OK
by ID_Rep, not only sends ID_Rep to the new terminal but also sends the same
message to other terminals since not only all the terminal can receive ID_Req. Then,
all the terminals in the network can register UUID and hash ID of a newly registered
terminal in the local ID Management DB.

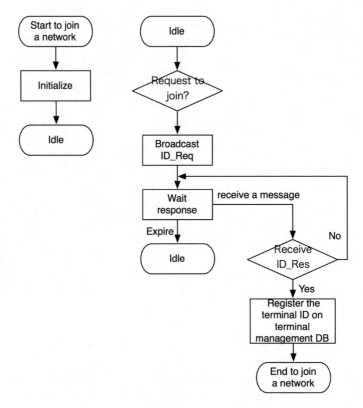

Fig. 20.13 Flowchart to join the existing network

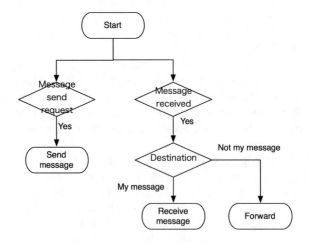

Fig. 20.14 Flowchart of top level behavior of the system

Fig. 20.15 Flowchart to
send a message

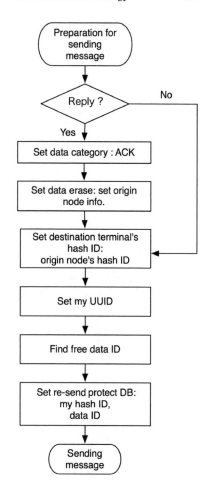

20.5.5 Message Delivery

Figure 20.14 displays the overall behavior of the protocol. The behavior is separated into three parts, sending a message, receiving message and forwarding message as described in Fig. 20.15, Fig. 20.16 and Fig. 20.17 respectively.

The message flow is described in Fig. 20.18. As mentioned before, the format of advertising packet that was used in our system is shown in Fig. 20.10. We used 28 octets in the advertising packet based on the specification of BLE.

Fig. 20.16 Flowchart to
receive a message

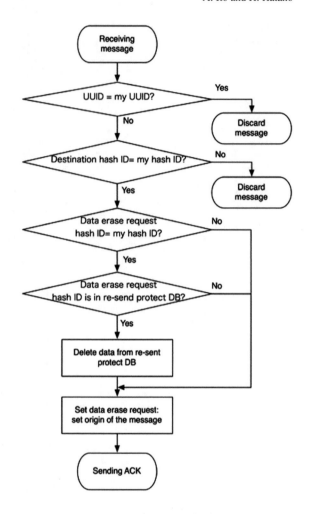

In this message sequence, the message to Terminal 3 is described as 'Msg.to 3'.
On the contrary, ACK to Terminal 1 is described as 'ACK Msg.to 1'.

Let's assume three terminals. Terminal 1 can communicate with terminal 2 and
terminal 2 can communicate with terminal 3, and there is no direct route between
terminal 1 and terminal 2.

- Terminal 1 sets there following data in the packet as described in Fig. 20.11 and
  broadcasts it.

  - Data Type: 2: Msg
  - Destination Hash ID: hash ID of terminal 3
  - Local Hash ID: my hash ID
  - Data ID: Sequence number of message
  - Data Erase Request: 0 since this message is not ACK

   – TTL: lifetime of packet

- Terminal 2 receives the message from terminal 1. Then decrements TTL and decides to forward or terminate. In this case, destination is the terminal 3, so that terminal 3 sets re-send protect DB (hash ID of origin and data ID) and broadcasts again.
- Terminal 3 receives message from terminal 2. Then decrements TTL and decides to forward or terminate. In this case, the destination is the terminal 3, so that terminal 3 sends ACK to the origin (terminal 1) and sets re-sent protect DB (hash ID of origin and data ID).

   – Data Type: 9: ACK
   – Destination Hash ID: hash ID of terminal 1
   – Local Hash ID: my hash ID
   – Data ID: Sequence number of message
   – Data Erase Request: hash ID of terminal 1 and data ID
   – TTL: lifetime of packet

- Terminal 2 receives ACK from terminal 3. According to the Data Erase Request in ACK, terminal 2 deletes information in ID Management DB and registers information of ACK (hash ID of terminal 3 and data ID).
- Terminal 1 receives ACK from terminal 2. According to the Data Erase Request in ACK, terminal 1 deletes information in ID Management DB.
- In the terminal of the origin of ACK and forwarding ACK, there is no procedure to erase information of ACK in re-sent protect DB. To erase such information, we use a timer to erase this information.

   A typical example is described in Fig. 20.18.

20.6 Experiment

The network of the experiment is described in Table 20.7. The experiment was performed in the Gymnasium of Utsunomiya University. The size of the gymnasium was 38 m × 34 m. The iPhone and iPod with the same OS (8.4.1) were used for the experiment.

The experiment used 9 terminals (iPhone and iPod) as described in Fig. 20.19. The typical behavior of the network of the experiment was as follows:

(Forward route)Terminal 1 sends a message to Terminal 9.

(Return route)Terminal 9 sends back an ACK message to Terminal 1 when it received the message from Terminal 1.

It is not required that the message forwarding route and the message returning route should be the same. Any available terminals were used for forwarding and returning.

Fig. 20.17 Flowchart to
forward a message

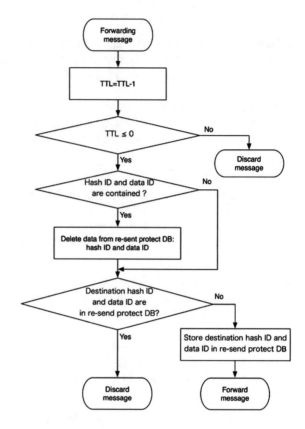

Table 20.7 The detailed condition of experiments.

Parameters	Details
Devices	iPhone 5 s × 6, iPod touch × 3
OS version	iOS 8.4.1
Bluetooth version	Bluetooth 4.0 (BLE)
Field	In the Gymnasium of Utsunomiya University (38 m × 34 m)
The number of hops	max 8 hops
The number of trials	20

According to the progress of the experiment, each terminal joined the network and the pair of UUID and hash ID of the existing terminals were stored in the ID Management DB as described in Fig. 20.20.

The result of the experiment is described in Fig. 20.21. The performance of this protocol in this experiment was 0.3 sec/hop including the overhead of the OS of a terminal.

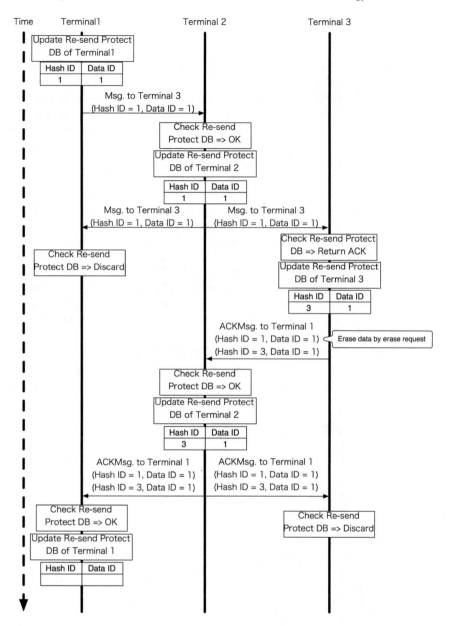

Fig. 20.18 An example of message delivery sequence

20.7 Conclusion

In this paper, we proposed an ad hoc network protocol using BLE for IoT devices. We designed this protocol for outdoor use and disaster situation to support disaster victims and teachers who manage a school trip. We designed this protocol to reduce the overhead. The following three functions were introduced in the ad hoc network protocol; (1) Re-sent protect DB: this DB stores send or forwarded message to prevent infinite loop, (2) Short UUID of destination: UUID is shorten by hash function, and (3) ID management DB: This DB provides unique short UUIDs.

We performed experiments of this protocol using iPhone and iPod. As the result of our experiment, the packet transmission time between two neighbor nodes was 0.3 s in average. It means that it is possible to transfer information for 33 nodes in 10 s. The performance of this protocol satisfies the requirements of IDDD.

There are five further study issues as follows. (1) to design efficient algorithm to erase over lapped data using re-sent protect DB, (2) to design efficient algorithm to generate and manage hashed UUID, (3) to design function to separate a long data into several packets and construct the original data them in receiver side, (4) to contract mathematical model of the protocol and estimate behavior of this protocol in the large number of terminals, (5) to implement useful cognitive infocommunication applications using this protocol.

Cognitive infocommunications (CogInfoCom) [21, 22] is a new idea to describe communications, especially combination of informatics and communications. Traditionally, communications have three aspects, such as media, informatics, and com-

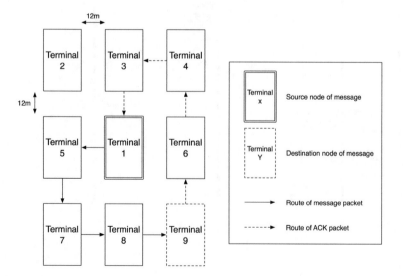

Fig. 20.19 Network of experiment

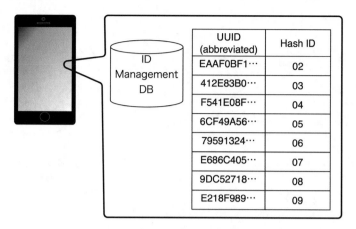

Fig. 20.20 Complete list of pair of UUID and hash ID

munications. However, the border of these aspects is not clear now because of a progress of technology of multimedia, mobile communication, and virtual reality.

CogInfoCom defines two phases of communications; intra-cognitive communication and inter-cognitive communication. From the viewpoint of communication technology, we can assume as follows. Intra-cognitive communication is a metaphor near-field communication and inter-cognitive communication is a metaphor of communication infrastructure such as 3G/LTE network.

The left side of Fig. 20.22 explains the outline of cognitive features [22]. The right side of Fig. 20.22 explains the position of two examples explained in Sect. 20.2.

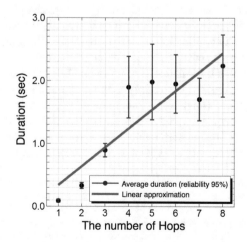

Fig. 20.21 Result of experiment

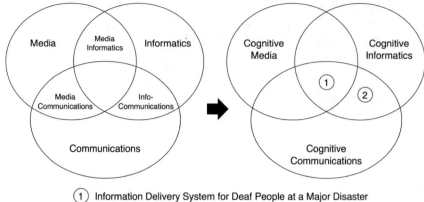

① Information Delivery System for Deaf People at a Major Disaster

② Technological Support for Tourism using BLE Beacon

Fig. 20.22 Cognitive features in our system

The Nikko Beacon has three faces, The first is an IoT based system (Cognitive Communications), the second is a navigation system (Cognitive Informatics) and the third is a multi-media service (Cognitive Media). On the other hand, IDDD has two faces. The first is an IoT based system as connected displays (Cognitive Communications) and the second is a disaster information system (Cognitive Informatics).

The idea of CogInfoCom is the good starting point to analyze and categorize information system.

There are several studies on CogInfoCom relating to multi-media service, ad hoc network, sensor network, and IoT. For example, [23] describes new HCI, cognitive informatics between human and machine, using gesture analysis for video conference. Garai et al. [24] describes inter-cognitive representation for bridging communication modality as an application of IoT for eHealth. Hideg et al. [25] discusses the relationship between sensor network and cognitive informatics. Heikkila et al. [26] describes cognitive informatics of robot to find fire detection.

At this moment, CogInfoCom mainly focuses on the border between informatics and communications (human-machine interaction), especially, human-machine blending. In the next step, we would like to enhance CogInfoCom from the viewpoint of the combination of media, communication protocol and human aspects such as mental state and brain activity to explain social aspects of communications.

Acknowledgements Authors would like to express special thanks to Mr. Funakoshi of Nikko Tourism Association and members of the "The committee of increasing travelers satisfaction in Nikko". Auhors thank to Mr. Takamura and Mr. Yoshida of Hatsuishi-kai that is an association of shopping street of Nikko, Mr. Nakagawa of Kounritsuin Temple, and Dr. Nagai who is a Professor Emeritus of Utsunomiya University. This research was performed as a project of SCOPE (Strategic Information and Communications R & D Promotion Programme) funded by Ministry of Internal Affairs and Communications in Japan. Authors also would like to express special thanks to Dr.

Tsunoda and Dr. Yabe of Tokyo Medical Center, who are the member of the research team of IDDD.

References

1. Clark D, Partridge C, Ramming C, Wroclawski J (2003) A knowledge plane for the internet. Proc ACM SIGCOMM 2003:3–10
2. Thomas R, Friend D, DaSilva L, MacKenzie A (2005) Cognitive networks. In: Proceedings of the first IEEE international symposium on new frontiers in dynamic spectrum access networks
3. Ito A, Murakami H, Watanabe Y, Fujii M, Yabe T, Hiramatsu Y (2013) Information delivery system for deaf people at a larger disaster. Biomed Res 1(2):17
4. Honjo S, Yuhashi H (2013) Information society that is strong in disaster. NTT Press (in Japanese)
5. Ito A, Hatano H, Fujii M, Sato M, Watanabe Y, Hiramatsu Y, Sato F, Sasaki A (2013) A Trial of navigation system using BLE beacon for sightseeing in traditional area of Nikko. In: IEEE International conference on vehicular electronics and safety (ICVES 2015)
6. https://www.inf.ethz.ch/personal/hvogt/proj/btmp3/Datasheets/Bluetooth_11_Specifications_Book.pdf, 24 Mar 2017
7. http://blog.bluetooth.com/bluetooth-sig-introduces-new-bluetooth-4-1-specification/,   24 Mar 2017
8. Frodigh M, Johansson P, Larsson P (2000) Wireless ad hoc networking the art of networking without a network. Ericsson Rev (4)
9. Turkes O, Scholten H, Havinga P (2015) BLESSED with opportunistic beacons: a lightweight data dissemination model for smart mobile ad-hoc networks. In: Proceedings of the 10th ACM MobiCom workshop on challenged networks CHANTS '15, pp 25—30
10. Miyashita S, Li Y (2015) ARMPP: an ant-based routing algorithm with multi-phase pheromone and power-saving in mobile ad hoc networks. In: Proceedings of 2015 third international symposium on computing and networking, pp 154–160
11. Optimized Link State Routing Protocol (OLSR), RFC-3626 https://www.ietf.org/rfc/rfc3626.txt,2003.10, 24 Mar 2017
12. Ad hoc On-Demand Distance Vector (AODV) Routing, RFC-3561 https://www.ietf.org/rfc/rfc3561.txt,2003.7, 24 Mar 2017
13. http://www.city.nikko.lg.jp.e.tj.hp.transer.com, 24 Mar 2017
14. http://www.toshogu.jp/english/index.html 24 Mar 2017
15. Trip advisor (2015) The most popular spot for visitors to Japan 2015. http://tg.tripadvisor.jp/news/ranking/inboundattraction_2015, 24 Mar 2017
16. http://www.soumu.go.jp/main_sosiki/joho_tsusin/scope/, 24 Mar 2017 (in Japanese)
17. http://www.soumu.go.jp/english/index.html, 24 Mar 2017
18. Ito A, Hiramatsu Y, Hatano H, Sato M, Fujii M, Watanabe Y, Sato F, Sasaki A (2016) Navigation system for sightseeing using BLE beacons in a historic area. In: IEEE 14th international symposium on applied machine intelligence and informatics (SAMI 2016)
19. Hiramatsu Y, Sato F, Ito A, Hatano H, Sato M, Watanabe Y, Sasaki A (2016) A service model using bluetooth low energy beacons—to provide tourism information of traditional cultural sites. Serv Comput
20. Yabe T, Haraguchi Y, Tomoyasu Y, Henmi H, Ito A (2016) Survey of individuals with auditory handicaps requiring support after the Western Tottori earthquake. Jpn J Dis Med 12(2)
21. Baranyi P, Csapo A (2012) Definition and synergies of cognitive infocommunications. Acta Polytech Hung 9(1):67–83
22. Baranyi P, Csapo A, Sallai G (2015) Cognitive Infocommunications (CogInfoCom). Springer International. ISBN 978-3-319-19607-7

23. Szczesny P, Nikodem J, Kluwak K (2016) Visual communication in expanding of human-computer interactions. 7th IEEE international conference on cognitive infocommunications (CogInfoCom 2016), pp 391–396

24. Garai A, Attila A, Pentek I (2016) Cognitive telemedicine IoT technology for dynamically adaptive ehealth content management reference framework embedded in cloud architecture. In: 7th IEEE international conference on cognitive infocommunications (CogInfoCom 2016), pp 187–192

25. Hideg A, Blazovics L, Csorba K, Gotzy M (2016) Data collection for widely distributed mass of sensors In: 7th IEEE international conference on cognitive infocommunications (CogInfoCom 2016), pp 193–198

26. Heikkila M, Pieska S, Jong S, Elsinga Ch (2015) Experimenting industrial internet with a mobile robot: expanding human cognitive functions In: IEEE 6th international conference on cognitive infocommunications (CogInfoCom 2015), pp 51–56

Index

A

Acoustic features, 118
Acoustic model, 154
   context-dependent, 165
   context-independent, 165
Actigraphy, 228, 229
Ad hoc network protocol, 436, 443
Adaptive e-Health content management, 311
Adaptive gameplay, 414, 418, 422, 423
   attention, 417
   deep reinforcement learning, 416
   flow, 414
   player model, 414
   psychophysiological feedback, 418
Affect, 112
Affective
   state, 114
Affective robotics, 222, 223
Affective scene, 110
   framework, 111
Air traffic control, 178
Air traffic management, 177
   security, 179
   security concept, 190
   voice communication, 182
Anger, 113
AODV, 436
Arduino
   IDE, 291
   Nano, 291
Artificial neural net
   deep rectifier neural net, 11
Automatic segmenter, 117

B

Bagging, 238, 241
Behaviour, 112
   affective, 112
   signals, 112
Behavioural
   pattern, 114
Behavioural analytics, 110
Big Data, 325
Bladder, 287
Bluetooth Low Energy (BLE), 433, 435
   advertisement, 436, 444, 446
   beacon, 436
Brain activity, 33
Brain-computer interface, 255, 262, 263

C

Call-center interactions, 110
Clinical interoperability, 313
Cloud architecture
   cloud-based telemedicine services, 319
   dynamic cloud architecture, 310
   hybrid cloud-base architecture, 325
CogInfoCom, 454
Cognitive architecture, 28, 29
Cognitive capabilities, 154
Cognitive design, 73
Cognitive infocommunication, 366, 426
Cognitive infocommunication-doctrines, 307
Cognitive process
   element, 330
   hiden, 330
   operation, 335

© Springer International Publishing AG, part of Springer Nature 2019
R. Klempous et al. (eds.), *Cognitive Infocommunications, Theory and Applications*,
Topics in Intelligent Engineering and Informatics 13,
https://doi.org/10.1007/978-3-319-95996-2

Cognitive tool, 51
Compassion, 202–223
Constraint
    programming, 393
    satisfaction problem, 395
Constructive solid geometry, 369, 380, 383
CORBA, 308
Corpus
    annotation, 113
Cultural conventions, 66

D
Da Vinci System, 278
DCOM, 308
Decision table, 235, 236, 250
Decision variables, 395
Deep reinforcement learning, 413, 420, 425
    asynchronous advantage actor-critic, 418,
        426
    deep Q-learning, 418
Desire not to act, 207, 211, 213
Desire to act, 207, 211, 213, 214, 222
Dissatisfaction, 113
Driver model, 28, 34, 41

E
ECG, 315
EEG, 255, 256, 258
Emotion, 112
    segment classifier, 122
    sequence labeller, 122
Emotional
    segment, 111
    state, 110, 112
Empathy, 112, 202–207, 210, 214–216, 221
Empatia, 205, 214–216, 219–223
Error repairs, 160
Executive functions
    neurobiological system, 80
    personality, 81
Expression of annoyance, 133
    annotation difficulties, 136
    annotation procedure, 134
    extended scale, 138
    perception difficulties, 133
Eye tracking
    analysis, 334
    data, 336
    examination, 330
    parameters, 333
    solutions, 332
    systems, 334

F
Feature selection, 119
Flow lines, 56
    streamlines, 59
fNIRS, 33, 39
Force sensor resistor, 230, 250
Forntal lobe, 256
Frustration, 113
Fused deposition modeling, 373

G
Game artificial intelligence, 419, 420
    asset adaptation, 428
    automated gameplay, 428
    deep learning, 418
    deep reinforcement learning, 420
    game testing, 428
    non-player character behaviour, 428
    real-time strategy, 421
    reinforcement learning, 420
    scenario control, 428
    tower defense, 421
    validation, 429
Gaze
    direction, 331
    length, 334
    motion, 334
General game playing, 417
    general video game playing, 417

H
Hesitation fillers, 154
Hierarchical classification, 20
HuComTech corpus, 3
Human-computer interaction, 426, 427, 429
    feedback, 427
Human-machine dialogue, 110
Hyperpipe, 235, 238, 241

I
IHTSDO, 309
Information delivery model, 434
Interactive deep reinforcement learning
    framework, 424
Interactive reinforcement learning, 413, 426
Interactive shaping, 426
Internet of Robotic Things (IoRT), 254, 259
Internet-of-Things, 306
Isolines, 56
    contour line, 59
    isotherms, 62
    pseudo-isolines, 60

Isopleth, 60
   choropleth, 60

K
Kappa measure, 113

L
Language model, 153
Lazy learners, 237
Learning curve, 279
Learning from human demonstration, 426
Learning system, 235, 237
Lexical features, 119
Logistic regression, 235, 236

M
Majority voting, 122
MakerBot, 385
Map, 51
   carte figuratives, 59
   cognitive, 69
   conventional signs, 53
   definition, 52
   diagrammatic, 52
   heatmap, 62
   orientation, 66
   thematic, 55
MedSol, 313
Methods
   GRID, 207–213, 221–222
   language corpus, 207, 209, 214, 221
   sorting, 203, 207, 209, 215, 219, 222
Mirror neurons, 92
Mirror system, 92
Mobile robot, 255, 259, 265, 270
Modal model, 113
Multiclass clssifier, 125
Multisensory effects
   illusory-flash, 65
   ventriloquism, 65
   vestibular sensation, 65
Multitasking, 36

N
NaÃve bayes, 235, 236, 241
N-back task, 32
Non-verbal, 111

O
OLSR, 436
OMG, 308
Open telemedicine interoperability hub-system, 306
OpenEHR, 309

OpenSCAD, 382
Oversampling, 120

P
Particle filter, 43
Pauses
   filled, 154, 160
   silent, 154, 160
Pediculus, 290
Pelvic floor muscles, 287
Performance measure, 122
Perplexity, 170
Personalized medical care, 305
Phrase modifiers, 160
Plexus santorini, 290
Politowanie, 214, 216, 218, 219, 221, 222
Prefrontal cortex, 256
Probabilistic sampling, 10
Problem
   direct, 405
   forward, 393, 400
   reverse, 400, 406
Production
   flow, 393
   orders, 392, 393, 399
   planning, 392
   routes, 395, 397, 404
Prolongations, 154, 160
Psycholinguistic features, 119

R
Radical prostatectomy, 287
Random forest, 238, 250, 251
Rectum, 287
Reference frame, 71
Reference model, 394
Reflexes, 159
Reinforcement learning, 413, 420
Relief, 119
Repeating words, 160
Repetitions, 154, 160
Representation, 51
   external, 51
   internal, 51
Rubosil SR-20, 289

S
Sampling rate, 113
SanTrain, 421
Satisfaction, 113
Segment classifiers, 118
Sentence restarts, 160
Serious games, 412, 421
Simulation

open-loop, 31, 34
SketchUP, 375
Sleep
    assessment, 228
    behavior, 228, 229, 239
    dataset, 228
    disorders, 228
    features, 228
    monitoring, 228, 250
    understanding, 229
SMOTE, 121
SNOMED-CT, 309
Social
    interactions, 110
SolidWorks, 378
Speaker verification, 180, 182, 183, 193, 197
Speakers' emotions, 110
Speech recognition, 153
Speech transcripts, 157
Spontaneous speech, 154
Statistical functionals, 118
STEAM, 369, 374
STL file format, 370, 377, 379
Stress
    detection, 180, 184, 194
    in the voice communication, 194
    levels, 186
Stuttering, 160
Supervised learners, 235, 240
Support Vector Machine (SVM), 11, 120
Surgical
    education, 280
    simulation, 281
    training, 280
Sympathy, 202–207, 214–223
Sympatia, 205, 215, 218–222

T
Tactical combat casualty care, 421
Task model, 34, 42
Telemedicine
    cloud-based adaptive services, 319
    open telemedicine interoperability, 306
    remote healthcare monitoring, 305
3D modeling software
    OpenSCAD, 382

SketchUP, 375
SolidWorks, 378
TinkerCAD, 378
Top 25, 375
3D printing, 371
    benefits, 372
    cognitive aspects, 368
    fused deposition modeling, 373
    higher education, 386
    in education, 385
    inspired events, 386
    milestones, 371
    support, 373
3D scanning, 374
TinkerCAD, 378
Topical unit, 2

U
Undersampling, 120
Unfinished words, 160
Unweighted average recall, 10

V
Valence, 204–207, 212, 214, 219–222
Verbal, 111
Visual categorization, 64
Visualization, 51
    diagrams, 56
    embedded and situated, 72
    graphs, 56
    human-computer interaction, 73
    interactive, 71
    mapping, 51
    medical imagery, 68
    social network, 69
    three dimensional, 66

W
Wearable devices, 228, 229, 232
Word error rate, 170
Workload, 40
    assessment, 27
    driver, 27
Workload assessment, 35
Współczucie, 205–207, 209, 212–223

Printed in the United States
By Bookmasters